BRITISH GEOLOGICAL SURVEY

British Regional Geology

South of Scotland

FOURTH EDITION

P Stone
A A McMillan
J D Floyd
R P Barnes
E R Phillips

Contributors:

M T Dean
G S Kimbell
J W Merritt
B É Ó Dochartaigh
A W A Rushton

BRITISH GEOLOGICAL SURVEY, NOTTINGHAM 2012

First published 1935
Second edition 1948
Third edition 1971

ISBN 978 085272 694 5

The grid, where it is used on the figures, is the National Grid taken from Ordnance Survey mapping. Maps and diagrams in this book use topography based on Ordnance Survey maps.

Definitions of stratigraphical units mentioned in the text may be found via the British Geological Survey's website in the Lexicon of Named Rock Units.

Printed by Butler Tanner & Dennis, Frome, Somerset.

Bibliographical reference

STONE, P, McMILLAN, A A, FLOYD, J D, BARNES, R P, and PHILLIPS, E R. 2012 *British Regional Geology: South of Scotland* (Fourth edition). (Keyworth, Nottingham: British Geological Survey.)

Authors
P Stone
A A McMillan
J D Floyd
R P Barnes
E R Phillips

Contributors
M T Dean
G S Kimbell
J W Merritt
B É Ó Dochartaigh
A W A Rushton

Front cover The sea cliffs at Pettico Wick, north-west of St Abb's Head, Berwickshire, showing folded and steeply inclined turbidite sandstone beds of the Kirkhope Formation (Ettrick Group, Silurian) (P000816).

Contents

ILLUSTRATIONS

Figures

Tables

Plates

Map

Frontispiece Dob's Linn, Moffatdale. The waterfall cascades down the slightly overturned base of a thick turbidite sandstone bed at the base of the Queensberry Formation (Gala Group, Silurian) into a gorge cut through the underlying mudstones of the Moffat Shale Group (Ordovician–Silurian) (P220190). The internationally agreed stratotype for the base of the Silurian lies within this section.

Foreword to the fourth edition

The south of Scotland is dominated by the bold, rolling hills of the Southern Uplands, a region underpinned by hard, intractable sandstone and mudstone of Ordovician and Silurian age, cut through in places by granitic plutons. Though these Lower Palaeozoic strata stretch from coast to coast, from the Firth of Clyde and Solway Firth in the west to the North Sea in the east, Upper Palaeozoic rocks fringe the southern and eastern parts of the Uplands, producing a softer landscape. To the north of the Uplands, and also in marked scenic contrast, lies the Upper Palaeozoic Midland Valley of central Scotland. To the south, though geology generally shows no respect for national boundaries, the rocks and landscape of northern England also stand in broad contrast to those of southern Scotland.

The Palaeozoic geological assemblage in the south of Scotland, together with some minor Mesozoic sedimentary and Cenozoic intrusive contributions, spans almost 500 million years of geological development — all topped and shaped by the deposits and erosive effects of the vast ice-sheets that finally retreated from the region only some 11 000 to 12 000 years ago. Perhaps most notably, the rocks of the Southern Uplands bear testament to the Early Palaeozoic growth and destruction of an Atlantic-scale ocean and the ensuing collision of the continents that once formed its shores. This tectonic saga has only been fully appreciated in relatively recent times, but the imposing landscape it produced clearly hinted at the drama within, inspiring the likes of Robert Burns, Sir Walter Scott, Robert Louis Stevenson and John Buchan. And quite apart from the literary associations, the more prosaic products of the Sanquhar and Canonbie coalfields, and the Leadhills–Wanlockhead and Galloway metalliferous mining fields have historically contributed much to Scotland's economic well-being. Scott, though probably not Burns, would have appreciated the use of gold from Leadhills to augment the Scottish crown jewels as early as 1540.

The first edition of this guide to the geology of the South of Scotland was written by J Pringle and published in 1935. Pringle then produced a second edition, with relatively few changes from the first, in 1948. A third, radically revised edition was compiled by D C Greig and published in 1971, but its preparation was mostly carried forward during the late 1960s, before the full effects of the plate tectonic revolution had permeated all aspects of geology. This fact alone would justify a new edition, but the huge volume of additional data generated over the last 30 years makes it long overdue. Prominent amongst those advances are the refinements in graptolite biostratigraphy and radiometric dating that have enabled the internationally agreed stratotype for the base of the Silurian System to be established in the heart of the Southern Uplands.

The fresh insights and interpretations presented here owe much to those who have contributed to the revolution in our understanding of the south of Scotland's geological development that has been carried forward since publication of the third edition of this guide: colleagues within the British Geological Survey, academic researchers, energy and mineral exploration geologists, civil engineers and, increasingly, scientists working with a broadly based environmental brief. Accordingly, this fourth edition is an eclectic work, developed from the perspective of dynamic geological development; tectonic processes, sedimentation and magmatism are integrated rather than being treated separately in the traditional fashion. The book gives a comprehensive account of the geology of the region that it is hoped will prove informative to a wide range of users, from students and those seeking a better understanding of their surroundings, to professional scientists, planners and engineers.

John N Ludden, PhD
Executive Director
British Geological Survey

Acknowledgements

This edition of British Regional Geology: South of Scotland has been compiled by P Stone and edited by him, S G Molyneux and J E Thomas.

Chapters have been written by the following authors:

One	**Introduction** *P Stone*	
Two	**Ordovician and Silurian of the Girvan–Ballantrae district** Ballantrae Complex *P Stone* Sedimentary succession *J D Floyd and P Stone*	
Three	**Ordovician and Silurian of the Southern Uplands** Ordovician *J D Floyd and P Stone* Silurian *R P Barnes and P Stone* With contributions on biostratigraphy by *A W A Rushton* and on the Ordovician, Bail Hill Volcanic Group by *E R Phillips*	
Four	**Caledonian structure and magmatism** *R P Barnes, P Stone and E R Phillips* With a contribution on geophysical modelling by *G S Kimbell*	
Five	**Devonian: the 'Old Red Sandstone' and associated volcanic rocks** *P Stone and J D Floyd*	
Six	**Carboniferous** *A A McMillan, P Stone and J D Floyd* With contributions on palaeontology and biostratigraphy by *M T Dean* and on magmatism by *E R Phillips*	
Seven	**Permian and Triassic** *A A McMillan* With a contribution on geophysical modelling by *G S Kimbell*	
Eight	**Jurassic to Palaeogene** *P Stone and E R Phillips*	
Nine	**Quaternary** *A A McMillan* With a contribution by *J W Merritt*	
Ten	**Geology and man** Fuel and energy *J D Floyd and P Stone* Bulk minerals *J D Floyd and P Stone* Building stones *A A McMillan and J D Floyd* Metalliferous and associated minerals *P Stone* Geological hazards *P Stone* Hydrogeology and water supply *B É Ó Dochartaigh and J D Floyd*	

The authors are grateful to the following (British Geological Survey unless otherwise indicated):

Mike Browne, Diarmad Campbell, Doug Fettes, Dick Merriman, Jon Merritt, Roger Musson and Adrian Rushton for their independent reviews of all or parts of the text.

Peter Brand advised on aspects of Carboniferous palaeontology; Robin Cocks (The Natural History Museum, London) advised on Silurian brachiopod taxonomy. Mark Woods and Sally Wild located specimens in the BGS palaeontology collections. Regional geochemical images were prepared by Andreas Scheib.

Figures were produced by Craig Woodward, Keith Henderson, Louise Wilson and Lesley Oliver.

Book design by Amanda Hill.

Photographs from the BGS archive were prepared by Brian McIntyre and Fergus MacTaggart.

Index compiled by Audrey Jackson.

one

Introduction

This account describes the geology of Scotland northward from the border with England up to a line running roughly south-west from Dunbar, on the North Sea coast, to Girvan, on the Firth of Clyde (Figure 1). It covers Scotland's two southernmost administrative regions: Dumfries & Galloway and Borders, thence extending northward to include parts of Strathclyde and Lothian. For much of its length the northern boundary line to the South of Scotland geological region coincides with the Southern Upland Fault, one of a series of major structures that partition mainland Scotland into five tectonic blocks known as terranes. The Southern Uplands form the southernmost of these but the north-west of the region described here extends beyond the Southern Upland Fault into the Girvan district, a part of the adjacent Midland Valley terrane. This allows the coeval Ordovician and Silurian successions of Girvan and the Southern Uplands to be integrated as part of the same geological exposition.

The south of Scotland region is underlain by a wide variety of rocks (Figure 2) with a geological history spanning almost 500 million years and discrete (though complementary) geological successions in each of its two terranes, separated by the Southern Upland Fault (Table 1). However, despite the overall variety, by far the largest area is underlain by Ordovician and Silurian, deep-marine sedimentary rocks, now much deformed, which create the rolling hills and rocky crags of the Southern Uplands. In their midst, at Dob's Linn (Frontispiece), is the internationally agreed global stratotype for the base of the Silurian Period. Several large granitic plutons were intruded into the western part of the Southern Uplands sedimentary sequence in Early Devonian times, and these give rise to a rugged, mountainous terrain that includes Merrick, which at 843 m above sea level is the highest peak in the region (Plate 1). It is formed from thermally metamorphosed sandstone adjacent to the Loch Doon granitic pluton. Elsewhere, the Southern Uplands' rolling hills form an undulating plateau, rising to 840 m at Broad Law but more generally about 500 to 600 m above sea level, dissected by deep, glaciated valleys. Where these rocks reach the coast they form dramatic sea cliffs

Plate 1 *Merrick (843 m), the highest point in the Southern Uplands, as seen looking north from Lamachan Hill, Galloway (P774191).*

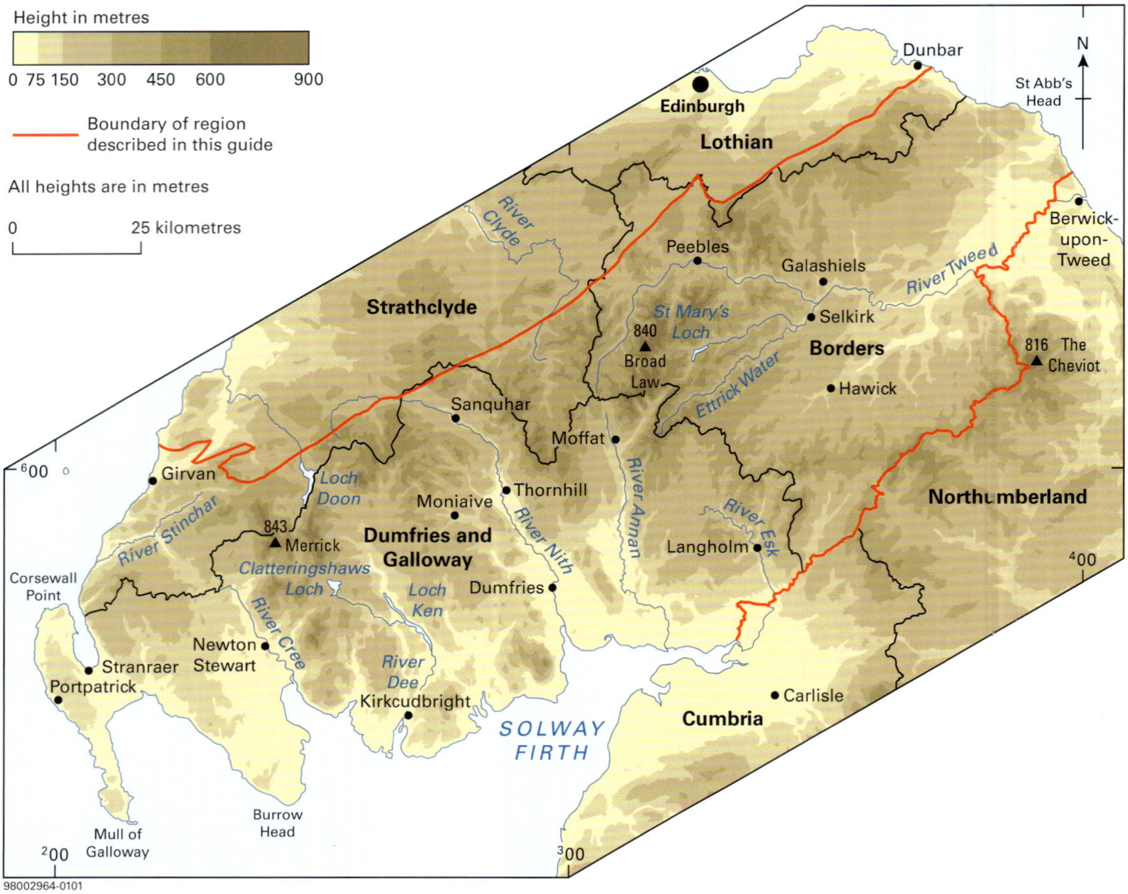

Figure 1 *Topography of southern Scotland and adjacent areas.*

(Cover photograph and Plate 2). Glacial erosion has sculpted and smoothed the higher ground, but the depositional legacy of the ice age can be seen in many lower areas. Most spectacular are the spreads of drumlins — elongate mounds of glacial detritus — that demonstrate the passage of ice from two major dispersal centres: south-westward across Galloway (Plate 3) from the mountainous area round Merrick, and eastward from the Moffat Hills across the Tweed Basin.

To the north of the Southern Upland Fault, the Ordovician and Silurian strata around Girvan are rather more varied and less deformed than their equivalents in the Southern Uplands. They form a more subdued topography of hills and valleys than is seen to the south, but unconformably overlie an assemblage of unusual, mostly igneous rocks (the Ballantrae Complex) that locally gives rise to prominent hills and sections of rocky coastline.

Eastwards, in both the Girvan district and the Southern Uplands, and also southwards in the latter, the topography gradually becomes more muted as the Ordovician and Silurian rocks are buried beneath an increasing thickness of Upper Palaeozoic strata that overstep with marked unconformity. Upper Palaeozoic beds are also preserved in a series of isolated structural outliers that originated as ancient valley systems cutting across the Southern Uplands. These prominent fault-controlled valleys, orientated approximately north-north-west and orthogonal to the regional strike of the Ordovician and Silurian strata, are a long-standing feature of the terrane, dating back more than 300 million years. They are filled with Devonian to Permian (possibly even Triassic) shallow-marine to terrestrial strata (Figure 2). The Upper

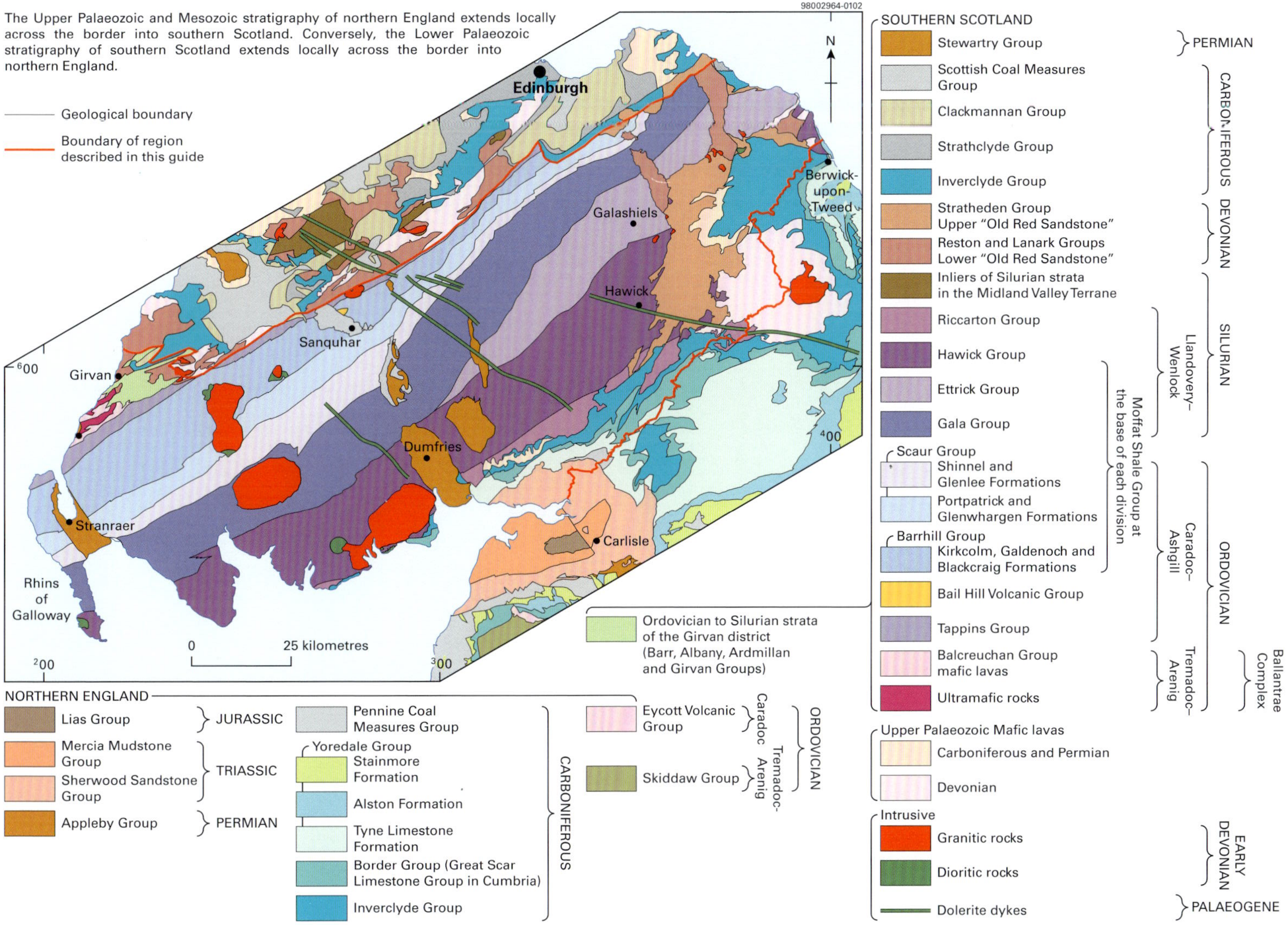

Figure 2 Outline geology of southern Scotland and adjacent areas.

Palaeozoic sedimentary rocks give rise to undulating lowland (Plate 4) but contemporaneous igneous rocks, mostly Devonian lavas or Carboniferous intrusions, form prominent landscape features locally (Plate 5).

Throughout the south of Scotland, the underlying geology has had a strong influence on the scenery and pattern of land use. That influence has been locally moderated by the erosive and depositional effects of the Quaternary ice age. Some upland areas were scoured by ice and are now typically covered by a veneer of frost-shattered rock, whilst some lower-lying districts were plastered with a thick layer of glacially transported debris — hence the drumlin swarms — in the form of till, moraine or outwash sand and gravel deposited by meltwater. And of course the development of today's landscape has not escaped anthropogenic effects. Farming, forestry, quarrying and the use of local building materials are widespread influences, whilst coal and metal mining have been of great historical importance in several areas. Underground coal working in the Canonbie and Sanquhar coalfields has a long history but expanded substantially through the 19th century before operations ceased in the 20th century: in the 1920s in the Canonbie coalfield and in the early 1970s at Sanquhar. Opencast working still continues in the Sanquhar coalfield at the time of writing, though individual pits have a relatively short operational life and are then filled and landscaped. Gold and base metal mining in the Leadhills–Wanlockhead district can be traced back to the 16th century but came to an effective end in 1929; base metal mining in Galloway mostly took place during the 19th century.

The geological features that shape and define the character of southern Scotland have arisen over hundreds of millions of years through the interplay of diverse geological processes; some with local cause and effect, others reflecting the local response to large-scale crustal events. This dynamic geological evolution is explored in the following account

Table 1a *Geological succession of the rocks and deposits in the South of Scotland region: Midland Valley terrane.*

a

Ma					Main episodes of magmatism: ☆
	QUATERNARY	Holocene Pleistocene		Alluvium, peat, coastal and estuarine sand and mud Glacial till and moraine, glaciofluvial sand and gravel, glaciolacustrine silt and clay	
2.6					
	Unconformable cover of Upper Palaeozoic strata				North-directed thrusting in late Silurian
423					
	SILURIAN	Wenlock Llandovery	GIRVAN GROUP	Sandstone, siltstone, conglomerate	Extension in forearc region
444					
	ORDOVICIAN	Ashgill Caradoc	ARDMILLAN GROUP	Sandstone, siltstone, mudstone, conglomerate, rare limestone	"Midland Valley arc"
		Llanvirn	BARR GROUP	Conglomerate, sandstone, limestone	☆
		Arenig Tremadoc	BALCREUCHAN GROUP as part of the BALLANTRAE COMPLEX	Basaltic lava and breccia, rare chert and mudstone, volcaniclastic sandstone Ultramafic rock and gabbro	☆ Obduction of Ballantrae Complex
488					
	CAMBRIAN		?		

980029-T0101a

EARLY PALAEOZOIC IAPETUS OCEAN

Scotland's southern geographical border coincides, more or less, with one of the most fundamental geological boundaries in Britain. This is the Iapetus Suture, the trace of a long-vanished, Early Palaeozoic ocean obliterated by the convergence and ultimate collision of the ancient continents that it once separated. The Iapetus Ocean (as a forerunner of the Atlantic

Table 1b Geological succession of the rocks and deposits in the South of Scotland region: Southern Uplands terrane.

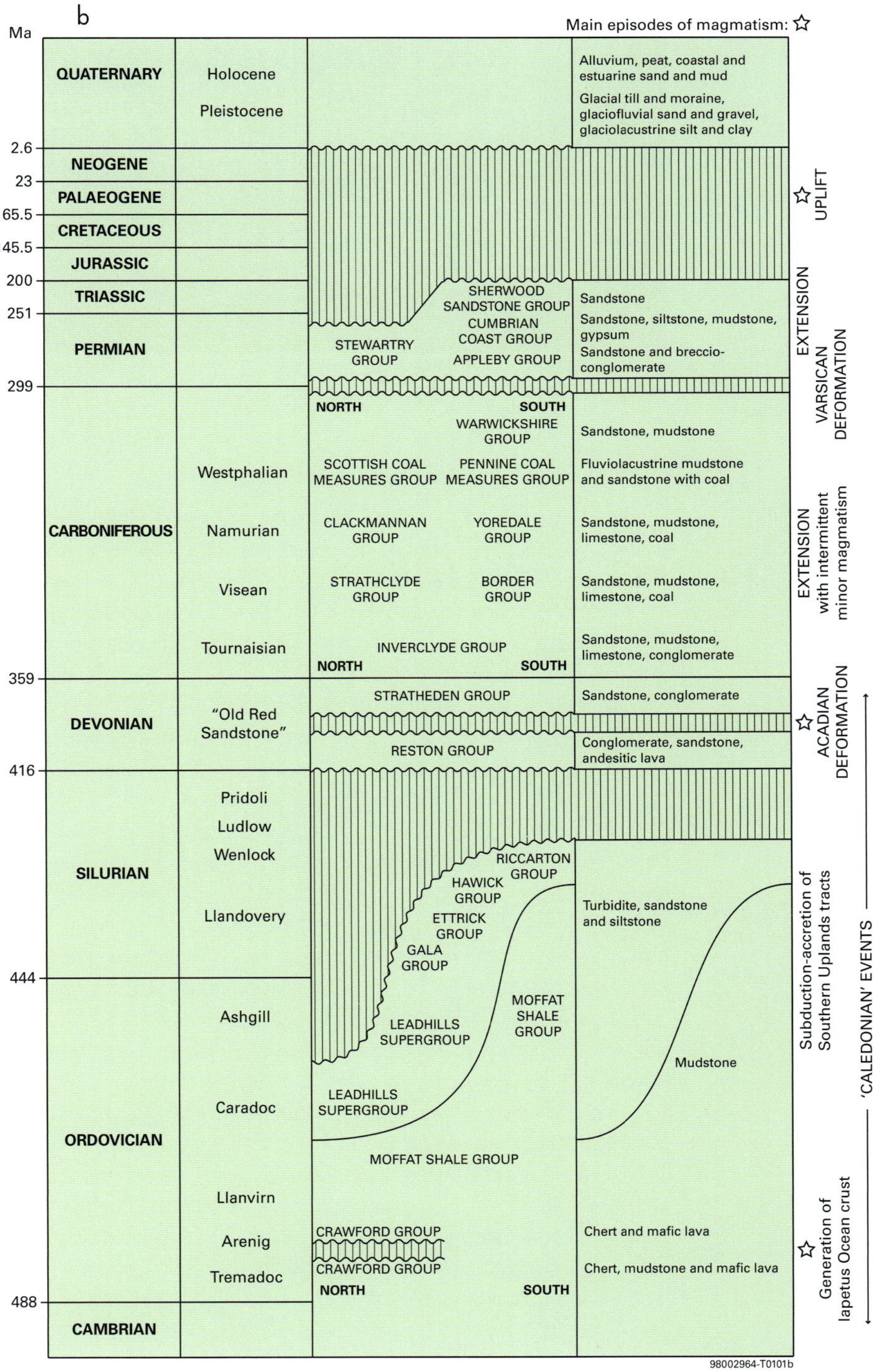

Plate 2 *Sea cliffs of Silurian sandstone on the west coast of the Rhins of Galloway south of Port Logan (P008484).*

Ocean it was named after the father of the eponymous Atlas) was initiated during late Neoproterozoic times and grew to its maximum width by the beginning of the Ordovician Period (Figure 3a). Thereafter, subduction at its margins wrought its eventual destruction and drove the series of collisional events that built up the Caledonian Orogen, a major tectonic zone that can be traced from Scandinavia, through Britain and Ireland, and on into Greenland and maritime North America. There are particularly clear geological links from southern Scotland, through Ireland, into Newfoundland, Canada.

Along the northern margin of the Iapetus Ocean at the beginning of Ordovician times, the continent of Laurentia lay in subtropical latitudes (Figure 3a). The Archaean and Proterozoic crystalline basement rocks of Scotland formed a part of this continent, and subduction of Iapetus oceanic crust beneath its margin led to the sequential accretion of oceanic rock complexes, both volcanic and sedimentary. These now make up much of southern Scotland. Forming the southern margin of the Iapetus Ocean during Ordovician times, and in a latitude of about 60° south, lay the shores of the Gondwanan continent, from which a fragment had broken away early in Palaeozoic times. This continental fragment, known as Avalonia, drifted north, towards Laurentia, as the intervening Iapetus Ocean closed (Figure 3b). The Lower Palaeozoic inliers of the English Lake District and Cross Fell reveal parts of the northern margin of Avalonia and show how it developed in response to the changing geotectonic regime, as described in the companion volume for Northern England.

Some other parts of the Avalonian continental margin provide evidence from volcanic rocks that southward subduction of the Iapetus Ocean commenced during late Cambrian times, but in the Lake District and Cross Fell inliers the earliest subduction-related volcanic activity occurred late in the Ordovician with eruption of the Borrowdale and Eycott Volcanic groups. The relative brevity but great intensity of the Borrowdale–Eycott volcanic episode may have been the result of the subduction of the Iapetus Ocean spreading ridge at its Avalonian margin. The structural weakness of the ridge zone might then have facilitated detatchment of the 'Avalonian' oceanic crust to create a break in the descending slab and so allow an upwelling of hot mantle material. This process would also have effectively transferred Avalonia

Plate 3 *Drumlin topography near Wigtown, Galloway (P741207).*

onto the oceanic plate that was being subducted northwards beneath the Laurentian margin of the ocean, creating an asymmetric oceanic plate configuration that would, ultimately, influence the tectonic style of the collision between the Laurentian and Avalonian continents. To consider that phenomenon here is to get a little ahead of the geological story, but the principal stages in the destruction of the ocean are illustrated in the series of cross-sectional sketches shown in Figure 4.

Biodiversity

One important 'trans-Iapetus' feature is the variation through time of the fossil assemblages found in the rocks. The Laurentian and Avalonian shelly fossils, remains of animals such as trilobites and brachiopods that lived in shallow, coastal marine environments, are quite different in the Ordovician, so that distinct 'faunal provinces' can be identified, but became progressively more cosmopolitan through the Silurian. This neatly illustrates the narrowing of the Iapetus Ocean that, by the late Silurian had changed in character and all but disappeared (Figure 3c).

Whilst the shelly faunas from the continental margins demonstrate provinciality, the biostratigraphy of the deeper-water, oceanic successions, depends on graptolites. This long-extinct group of colonial animals formed a major part of the oceanic macro-zooplankton during Ordovician and Silurian times, and their fossils are of importance in establishing age and order in the Ballantrae Complex and the Girvan succession (Chapter 2), and crucially so in the Southern Uplands (Chapter 3). Graptolites are commonly slender and delicate and a few centimeters in length, though the length may range from almost microscopic to a metre or so in extreme cases. During the Early Palaeozoic they evolved an extraordinary variety of shapes, and it is this variety that is key to the biostratigraphical zonation now based on their remains. The zonal scheme was established towards the end of the 19th century by Charles Lapworth, who was the first to recognise that the graptolite assemblages changed systematically through time. Lapworth's most important work on graptolite biostratigraphy was carried out in the Moffatdale area of the Southern Uplands and around Girvan, though his zonal scheme, only slightly modified by subsequent research, has international application. A selection of Ordovician and Silurian graptolites from the south of Scotland is illustrated in Figures 5 and 6.

Island arcs and obduction

At its northern, Laurentian margin, the first stage in the closure of the Iapetus Ocean was the Early Ordovician development of oceanic subduction zones and associated volcanic island

Plate 4 *Looking across the Lochmaben Permian basin towards the distant hills of the Southern Uplands (P681907).*

arcs (Figure 4a). A modern analogue of this situation is provided by the active volcanic arcs of the south-west Pacific Ocean. One of the Early Ordovician, Iapetus Ocean arcs was to become the Ballantrae Complex of south-west Scotland. This Tremadoc to Arenig assemblage of oceanic, mostly igneous rocks collided with and was tectonically emplaced (obducted) onto the Laurentian continental margin at about 470 Ma (Figure 4b), to form what is known as an ophiolite complex. Its obduction played a peripheral part in the large-scale collision of a volcanic-arc complex (now forming the buried core of the Midland Valley terrane) with Laurentia that provoked the Grampian event of the polyphase Caledonian Orogeny.

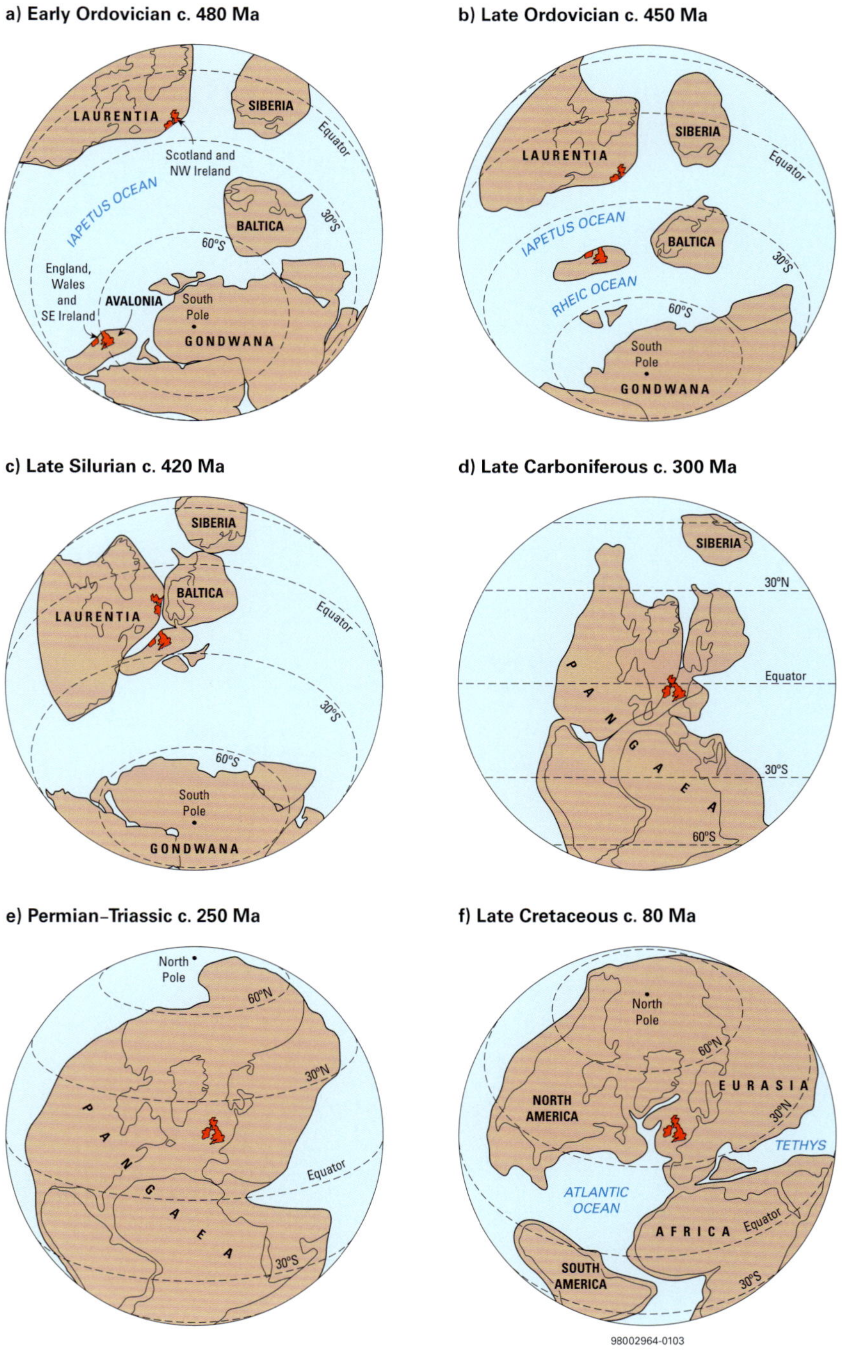

Figure 3 *A series of palaeogeographical reconstructions showing continental movements from the Ordovician to the Cretaceous (after Woodcock and Strachan, 2000).*

Within the Ballantrae Complex there is a bewildering array of rock types: ultramafic rock of mantle origin, volcanic lavas erupted in contrasting island-arc and within-plate settings, intrusive gabbros and pelagic sedimentary strata. All were tectonically juxtaposed and obducted as the Iapetus Ocean began to close. By early Llanvirn times the complex was in

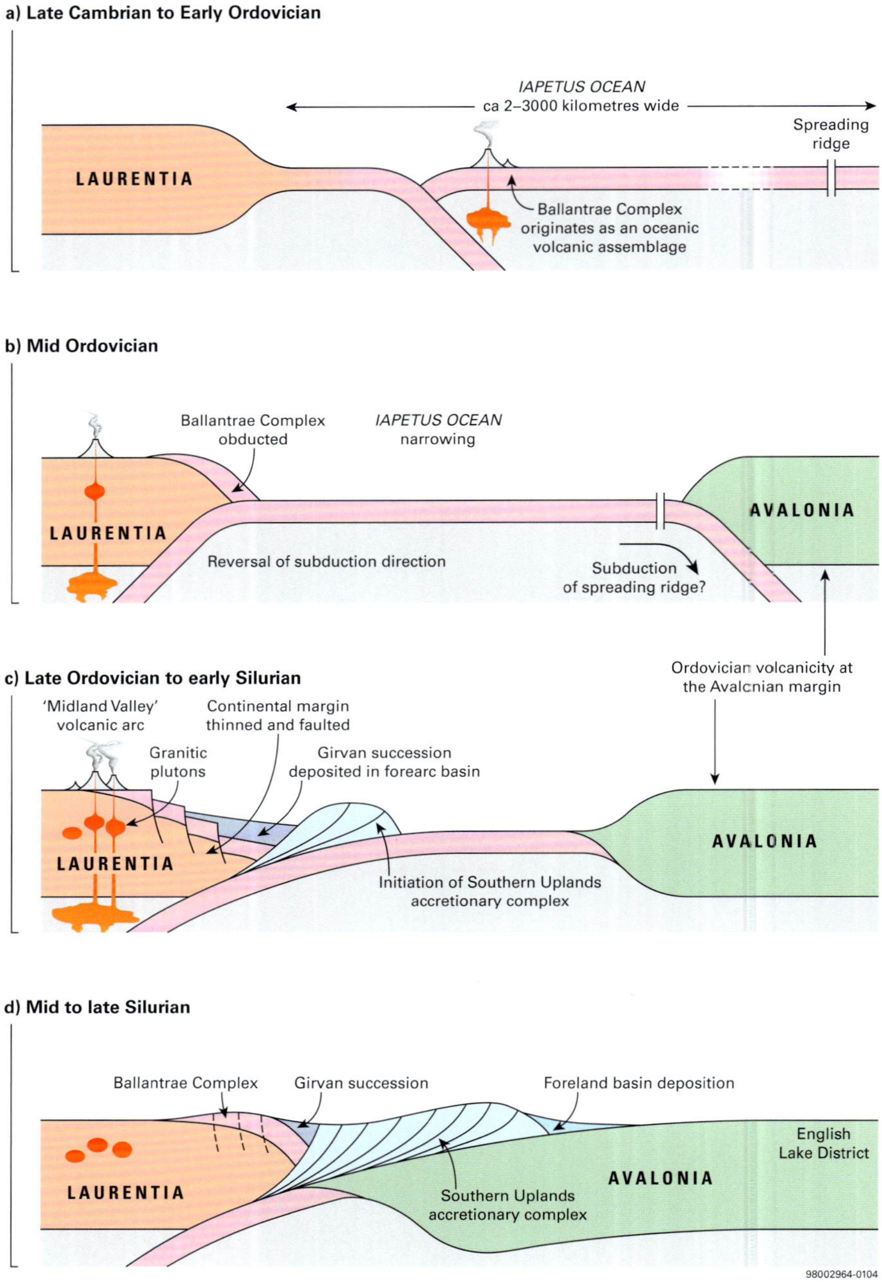

Figure 4 *Stages in the closure of the Iapetus Ocean focusing principally on events at the Laurentian margin; the complex evolution of the Avalonian margin is not illustrated.*
a and b *Late Cambrian to Mid Ordovician — the formation and obduction of the Ballantrae Complex.*
c *Late Ordovician to early Silurian — the depositional environment of the Girvan succession and growth of the Southern Uplands accretionary complex.*
d *Mid to late Silurian — the migration of the accretionary complex onto the Avalonian continental margin.*

Note that in the Mid to Late Ordovician (b/c) the Iapetus Ocean spreading ridge may have been subducted beneath Avalonia to instigate an intense volcanic episode (Borrowdale and Eycott Volcanic groups) and create an asymmetric plate pattern.

Plate 5 *Eildon Hills near Melrose, formed by a Carboniferous igneous intrusion (a composite laccolith) that cuts Devonian and Silurian strata (P219630).*

place and had been deeply eroded. Obduction had been accompanied by a switch in the polarity of subduction and as oceanic crust began to be consumed beneath the continental margin a volcanic arc was generated on what is now the basement to the Midland Valley of Scotland. Relative uplift caused by this 'Midland Valley' magmatism was accompanied by extension and relative subsidence of the continental margin to the south (Figure 4c). There, the Ballantrae Complex was progressively buried by a sedimentary cover sequence of shallow to deep marine strata that systematically overstepped northwards from the late Llanvirn to the early Wenlock (c. 460–428 Ma). This now forms the Girvan succession. Its northwards transgression was controlled by major faults, with downthrow to the south, stepping back sequentially into the Midland Valley arc zone. The eventual, probably late Silurian deformation of the Girvan succession involved the reactivation of those originally normal faults as northward directed thrust planes.

Subduction and accretion

Whilst the Ballantrae Complex was being buried beneath the thick Ordovician to Silurian sedimentary succession now seen around Girvan, a very different process was operating farther south. As the Iapetus oceanic crust was subducted beneath the margin of Laurentia, sections of the oceanic sequence and its sedimentary cover were intermittently stripped from the subducting plate and thrust beneath the stack of similar stripped-off slices to initiate an accretionary complex (Figure 4c). The Southern Uplands terrane represents the deeply eroded remains of this accretionary complex, which developed along the northern fringe of the Iapetus Ocean sequentially from late Llanvirn to mid Wenlock times.

The sedimentary units incorporated into the accretionary complex originated as sand and mud carried by turbidity currents from the continental shelf, via submarine canyons, and built up into huge depositional fans (now the Southern Uplands sandstone groups, Table 1b). The clastic turbidite deposits filled the supra-subduction-zone trench and encroached onto the oceanic plate, where they covered the sequence of hemipelagic mud (Moffat Shale Group), radiolarian chert and pillow lava (Crawford Group). As the submarine fans built out, they overstepped progressively younger oceanic sequences that were continually approaching the continental margin as the oceanic plate was subducted. Then, during the subduction process, discrete sections of the oceanic sequence and its cover of turbidite sandstone were sequentially stripped from the subducting oceanic plate and thrust beneath the stack of similar stripped-off slices that made up the growing accretionary complex. These slices, structurally rotated towards the vertical (and in places even beyond it), now give rise to the characteristic pattern of elongated and north-east-trending, fault-bounded tracts that define the Southern Uplands lithostratigraphical outcrop pattern (Figure 2). Polyphase folding was imposed on the strata as the accretionary complex built up. Early folds were formed in association with thrusting during the subduction process. Subsequent folds developed as accommodation structures when the early-formed part of the accretionary complex adjusted to continued subduction at its leading edge and responded to intervals of strike-slip movement rather than orthogonal compression.

This model of the Southern Uplands as a forearc accretionary complex is now generally accepted, following much discussion of possible alternative models that arose, in part, from the provenance contrasts evident between different sandstone tracts. In particular, the apparent introduction of volcanic detritus from the south, i.e. from the oceanic plate rather than the continental margin, has been cited in support of a backarc origin for all or some of the terrane. There has also been some discussion of the possible extension of the Girvan depositional setting, an extending and subsiding continental margin, into the northern part of the Southern Uplands terrane. Analyses of basin thermal history and radiometric dating of detrital minerals have now ruled out the backarc possibility, and whilst some uncertainties remain, the development of the Southern Uplands terrane as a supra-subduction-zone accretionary complex is now the consensus view.

There is however one important difference between the early and late stages in the development of the accretionary complex. The older tracts were accreted from subducting oceanic crust, but by mid Wenlock times the Iapetus Ocean had effectively closed and the complex, at the leading edge of Laurentia, overrode the margin of Avalonia (Figure 4d). The meeting of the two continental masses did not produce a climactic deformational event and instead the accretionary complex continued to advance through the foreland basin that formed ahead of it, above Avalonian continental crust depressed by the encroaching mass of Laurentia. In somewhat pedantic terms, the accretionary complex had become a foreland fold and thrust

***Figure 5** (opposite) Examples of Ordovician graptolites from the south of Scotland. All are shown about twice natural size.*

a Dicellograptus anceps *(Nicholson 1867)* — anceps *Biozone.*
b Orthograptus abbreviatus *Elles and Wood 1907 — chiefly* anceps *Biozone.*
c Climacograptus supernus *Elles and Wood 1906* — anceps *Biozone.*
d Leptograptus flaccidus *(Hall 1865)* — clingani *and* linearis *biozones.*
e Neurograptus margaritatus *(Lapworth 1876)* — clingani *Biozone.*
f Corynoides calicularis *Nicholson 1867* — bicornis *and* clingani *biozones.*
g Dicranograptus ramosus *(Hall 1847) — chiefly* clingani *Biozone.*
h Orthograptus quadrimucronatus *(Hall 1865)* — clingani *and* linearis *biozones.*
i Climacograptus caudatus *Lapworth 1876* — clingani *Biozone.*
j Climacograptus bicornis *(Hall 1847)* — bicornis *Biozone.*
k Normalograptus euglyphus *(Lapworth 1880)* — gracilis *and* bicornis *biozones.*
l Diplograptus leptotheca *Bulman 1946* — bicornis Biozone.
m Dicellograptus sextans exilis *Elles and Wood 1904* — gracilis *and* bicornis *biozones.*
n Nemagraptus gracilis *(Hall1847) — chiefly* gracilis *Biozone.*
o Dicranograptus ziczac *Lapworth 1876* — bicornis *Biozone.*
p Expansograptus? superstes *(Lapworth 1876) — chiefly* gracilis *Biozone.*
q Tetragraptus approximatus *Nicholson 1873* — approximatus *Biozone.*

belt. Thereafter, convergence of the two continental plates probably ceased in Ludlow times, to be replaced by intermittent lateral movement between them.

Turbidity currents

Before leaving the Iapetus Ocean, it is worth looking more closely at turbidity currents, which were an important mechanism in the deposition of both the Girvan and the Southern Uplands

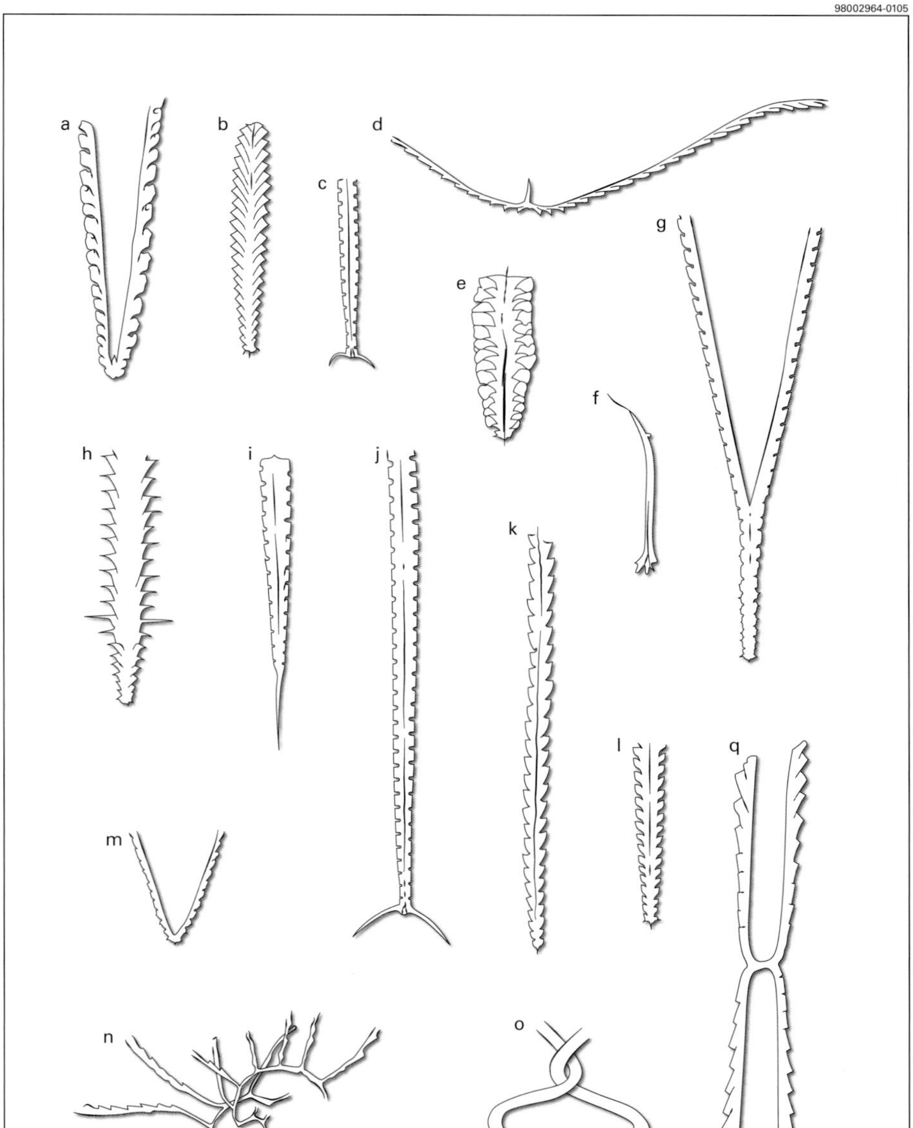

successions. Each current was created by the collapse and mobilisation of an unstable pile of sediment, built up on the continental margin, and its consequent flow down into deeper water, via submarine canyons, as a huge mass of turbulent, water-suspended sediment. As each flow spread out across the sea floor it decelerated and began to drop its load of suspended sediment, starting with the largest grains. A single sandstone bed would be deposited from each turbidite flow and may display a complete or partial array of distinctive features. The coarsest sand grade present is seen at the base of the bed and there is an upward grading ('graded bedding') to progressively finer sand that, towards to top of the bed, may be parallel- and/or cross-laminated (Plate 6). If substantial turbidity flows closely follow each other in time, a thick sequence may build up in which graded sandstone beds become amalgamated, with only subtle variations in grain size determining the margins of each bed. In more distal parts of the fan, or as a result of small, low-density flows, thin and fine-grained sandstone beds may be pervasively laminated and separated by thicknesses of silty mudstone, the finest-grained part of the submarine fan succession deposited in areas that were spatially or temporally removed from most of the depositional activity. Whilst the deposition of a single turbidite bed was, in geological terms, an instantaneous event, a considerable time interval (perhaps 100–1000 years) might pass between successive flows.

Where the base of a turbidite sandstone bed overlies the fine-grained top of its predecessor, it is usually sharp and clearly defined. Sole marks, common on the base of many sandstone beds, are casts of features formed on the substrate upon which the sand was deposited. They formed when solid objects were carried across the sediment surface, forming linear grooves, or by the brief erosive scouring of the sediment surface by intense vortices carried along in the flow of currents. These erosive hollows were then filled by the subsequent sand deposit to form positive features, protruding from the sandstone bed base, and are known respectively as groove and flute casts. Both types range from a few millimetres to several tens of centimetres in width, and though they may occur individually they are more commonly seen in swarms (Plate 7). The orientation of the palaeocurrent indicators can be used to determine the flow direction of the eroding current. In the Southern Uplands context this is important, since there the

***Figure 6** (opposite) Examples of Silurian (Llandovery) graptolites from the south of Scotland. All are shown about twice natural size.*

a Stimulograptus sedgwickii *(Portlock 1843) —* sedgwickii *Biozone.*
b Pristiograptus regularis *(Törnquist 1899) — chiefly* sedgwickii *Biozone.*
c Paradiversograptus runcinatus *(Lapworth 1876) —* turriculatus *Biozone.*
d Spirograptus turriculatus *(Barrande 1850) —* turriculatus *and* crispus *biozones.*
e Monograptus discus *Törnquist 1883 —* crispus *to* crenulata *biozones.*
f Monograptus crispus *Lapworth 1876 —* crispus *Biozone.*
g Streptograptus exiguus *(Nicholson 1876) —* turriculatus *to* griestoniensis *biozones.*
h Monograptus convolutus *(Hisinger 1837) —* convolutus *Biozone.*
i Dimorphograptus swanstoni *Lapworth 1876 —* acinaces *Biozone.*
j Coronograptus gregarius *(Lapworth 1876) —* revolutus *to* convolutus *biozones.*
k Monograptus limatulus *Törnquist 1892 — chiefly* convolutus *Biozone.*
l Monograptus triangulatus fimbriatus *(Nicholson 1868) —* magnus *Biozone.*
m Cephalograptus cometa *(Geinitz 1868) —* convolutus *Biozone.*
n Monograptus priodon *(Bronn 1834) —* griestoniensis *Biozone to lower Wenlock.*
o Parakidograptus acuminatus *(Nicholson 1867) —* acuminatus *Biozone.*
p Climacograptus trifilis *Manck 1923 —* acuminatus *Biozone.*
q Rhaphidograptus toernquisti *(Elles and Wood 1906) —* atavus *to* sedgwickii *biozones.*
r Atavograptus atavus *(Jones 1909) —* atavus *to* revolutus *biozones.*
s Rastrites longispinus *Perner 1897 —* triangulatus *to* convolutus *biozones.*
t Monograptus triangulatus triangulatus *(Harkness 1851) —* triangulatus *Biozone.*

marked compositional contrast between different suites of sandstones is most readily explained by their constituent sand having been derived from different source regions. Of even greater importance in the Southern Uplands, where successions of steeply inclined or vertical beds are commonplace, is the proof provided by graded bedding, sole marks and cross-lamination of the stratigraphical top and bottom of a sequence at any one locality; crucial data in any interpretation of the larger-scale structure.

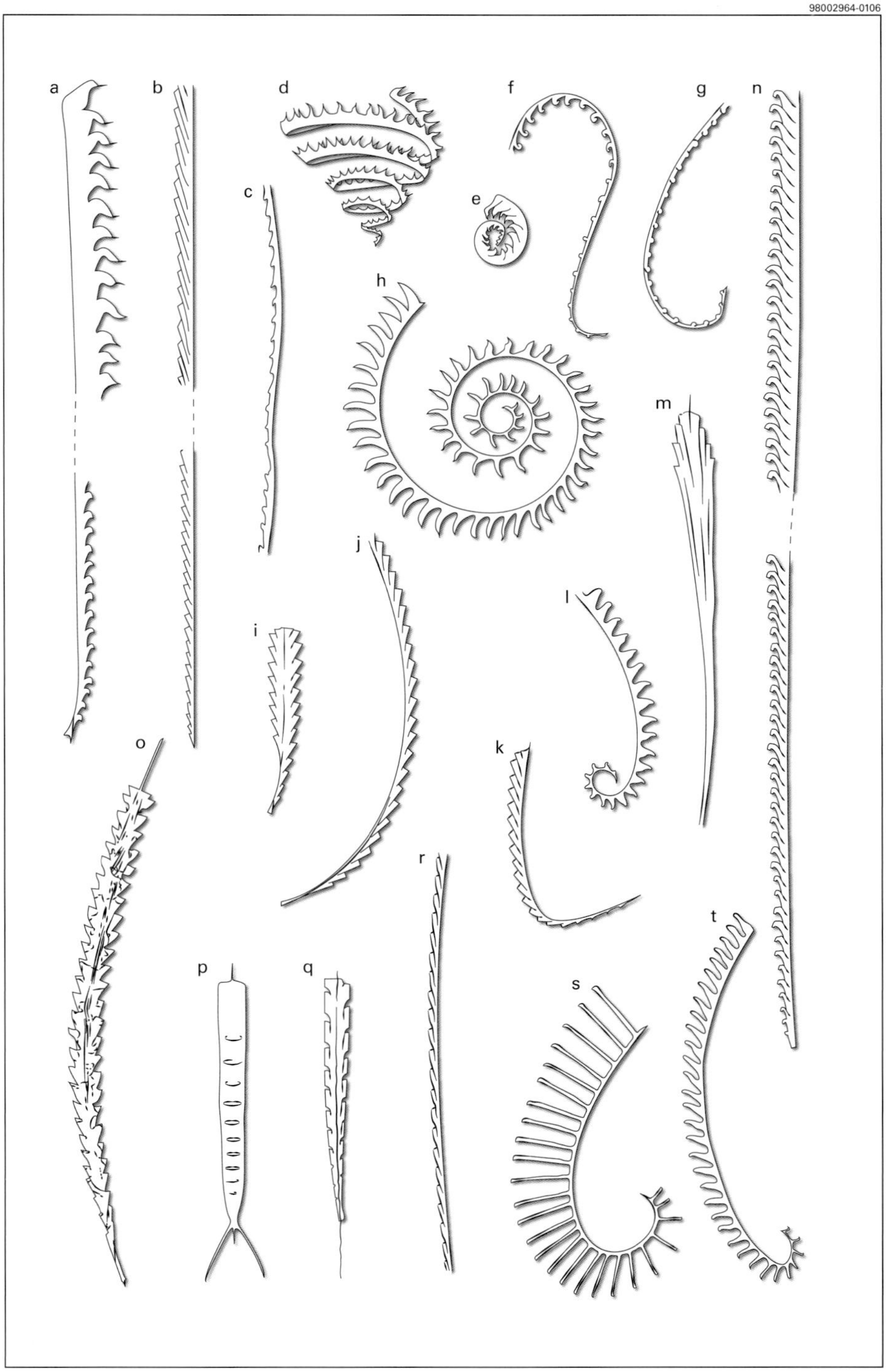

Plate 6 *Typical sequence of turbidite beds. This example of Silurian (Llandovery) strata is exposed on the coast near Port Logan [NX 092 402], Rhins of Galloway (P008471).*

IAPETUS SUTURE

By late Silurian times the Iapetus Ocean had all but closed (Figure 3c), though the ultimate continental collision was something of a tectonic anticlimax. It was not a mountain-building event of orogenic proportions and the effects are hard to find in the tectonic record preserved on either the Laurentian or Avalonian margins. There was instead something of a tectonic continuum, as the Southern Uplands accretionary thrust terrane overrode Avalonia and continued southwards as a foreland fold and thrust belt (Figure 4d). Initially, a load-induced, flexural foreland basin advanced ahead of the thrust front and was an influential control on sedimentation during the accumulation of the mid to late Silurian parts of the Windermere Supergroup in the south of the English Lake District. This situation did not last, and by the end of Ludlow times convergence between Laurentia and Avalonia had ceased,

Plate 7 *Flute casts, two examples from the Ordovician Kirkcolm Formation showing different styles and current directions: a) Finnarts Bay [NX 053 722] (P008425); b) Portobello [NW 960 665] (P008463). The shape of the linear flute casts from Finarts Bay indicates current flow from top right to bottom left; the linguiform flute casts from Portobello indicate current flow from top left to bottom right. In both cases the steeply inclined beds are slightly overturned and are viewed looking north.*

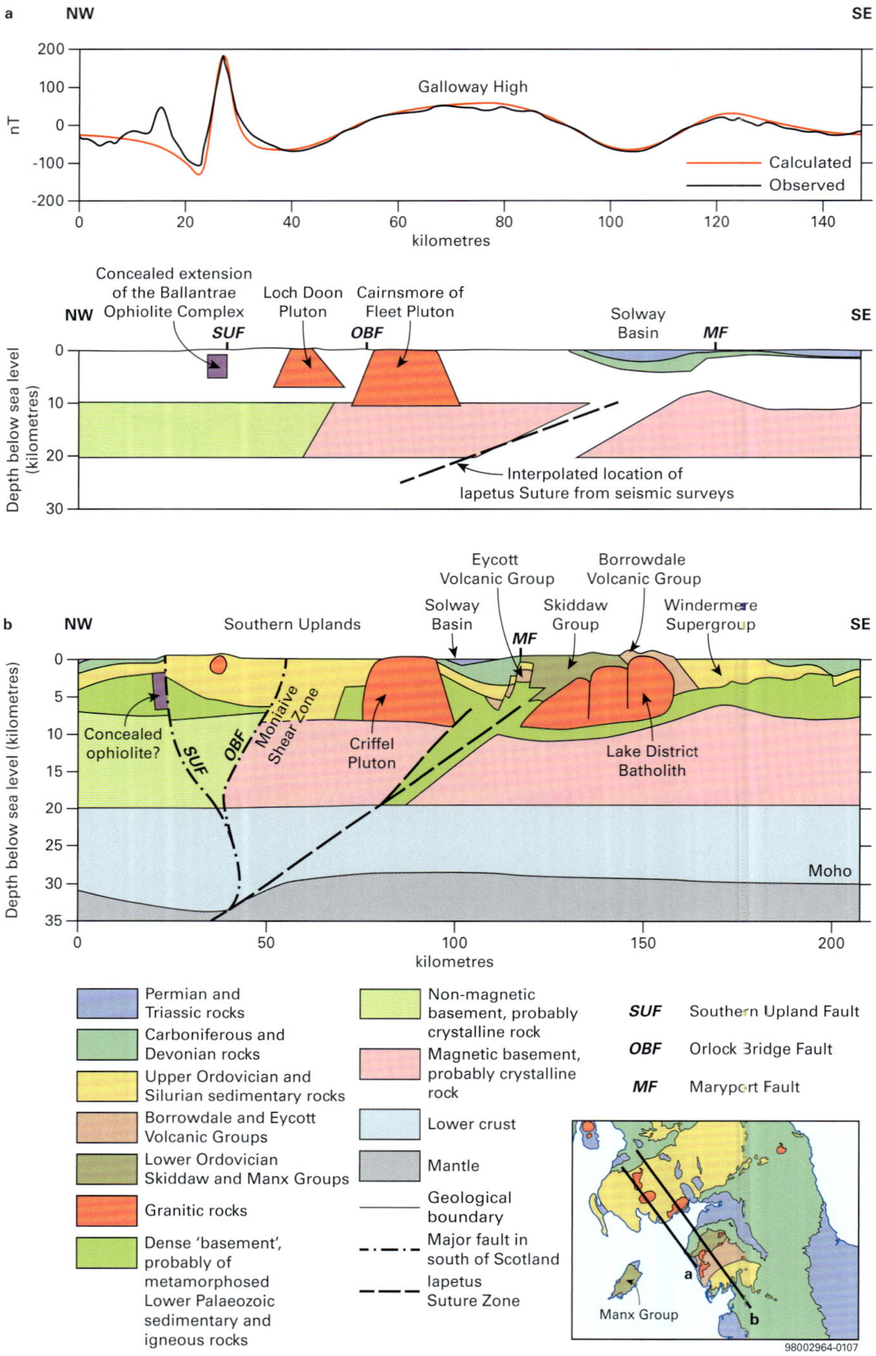

Figure 7 *Deep crustal sections across the Iapetus Suture Zone based on geophysical evidence.*

a Simplified model illustrating the long-wavelength magnetic anomaly known as the 'Galloway High'.

b A composite model based on a combination of gravity and magnetic data and linked to geological outcrop.

the foreland basin failed to migrate farther southwards, and isostatic adjustments reversed the earlier effects of loading.

The tectonic effects seen within the exposed rock sequence, though created by collision-related processes, give little indication of the deeper structure of the suture zone. This is more usefully modelled from geophysical data. A number of seismic lines have traversed the Iapetus Suture Zone and have generally been interpreted in terms of a north-west-dipping, reflective zone projecting to the surface close to the northern coast of the Isle of Man and thence striking north-east beneath northern England. When the seismic results are integrated with regional interpretations of gravity and magnetic data a rather more complicated picture emerges in which Avalonian-type crust is caught up in a compound suture zone that extends well to the north beneath the Southern Uplands terrane (Figure 7).

It is something of a geological paradox that the Lower Palaeozoic rocks of the Laurentian margin did not experience substantial deformation as a result of the Laurentia–Avalonia collision. One possible effect is seen in the Girvan district, where the normal faults that had controlled deposition of the Ordovician to Silurian forearc basin succession were reactivated as north-directed thrusts late in the Silurian. Another possible tectonic outcome of the collision is implicit in Figure 7, wherein Laurentian-type crystalline basement extends from the Midland Valley terrane beneath the northern part of the Southern Uplands. Perhaps large-scale northward thrusting of the accretionary complex onto the Laurentian margin accompanied the demonstrable north-directed thrusting of the Girvan succession. The considerable horizontal shortening of the accretionary complex that would have been likely in such circumstances could have been accommodated by the widespread rotation of bedding towards the vertical, an attitude widely seen throughout the Southern Uplands and one hard to attain throughout the terrane only by accretionary activity.

Despite the uncertainties, it is clear that the continental collision between Laurentia and Avalonia was not an orthogonal event. A wealth of evidence shows that a sinistral stress regime was important during the later stages of convergence, and indeed may have been the dominant final effect. With a sinistral shear sense applied to the major north-east-trending strike faults, a conjugate pattern of smaller, cross-cutting, late Caledonian strike-slip faults was established across both the Southern Uplands terrane and the Lower Palaeozoic outcrop in northern England. The conjugate fault system comprises strike-slip faults trending either generally north-west with a dextral sense of displacement, or generally east-north-east with sinistral displacement. Though individually minor, these faults were to have a profound structural influence during subsequent episodes of extensional tectonism when their reactivation controlled Late Palaeozoic basin development and geometry. More immediately, in the transtensional tectonic regime pertaining during latest Silurian to Early Devonian times, strike-slip basins opened across the region and were filled with the clastic, terrestrial sediments of the Old Red Sandstone lithofacies. The transtensional regime may also have been an important factor in the intrusion of the Early Devonian granite plutons.

AFTER IAPETUS

In the aftermath of the final closure of the Iapetus Ocean a range of igneous rocks were intruded into the Ordovician and Silurian strata of the accretionary complex: a regional swarm of late Caledonian (Silurian to Devonian) calc-alkaline felsic and lamprophyre dykes, several Early Devonian granitic plutons, and a number of smaller diorite-granodiorite-granite intrusions. Radiometric dating has confirmed that the first to be intruded were microdiorite and lamprophyre dykes, with ages ranging from 418 ± 10 Ma to 400 ± 9 Ma. The age of the larger, granitic intrusions varies across the Southern Uplands with the northern plutons (around 410 Ma, i.e. late Silurian) proving to be older than the southern plutons (around 397 Ma, i.e. Early Devonian) that were intruded closer to the Iapetus Suture.

Devonian

By the Early Devonian, terrestrial conditions were established with a high-relief terrain undergoing rapid erosion in a relatively arid environment. Conglomerate wedges built up against fault scarps and alluvial fans carried the finer-grained material out into transtensional strike-slip basins; large rivers traversed the region and reworked the terrestrial deposits into fluvial sequences. This assemblage is now preserved as the red sandstone and conglomerate of the Reston Group, which crops out mostly in the east of the region. Also in the eastern Southern Uplands, extensive flows of andesitic lava were extruded onto, and interfingered with, the fluvioterrestrial Reston Group strata. Some of these lavas were erupted in association with the intrusion of another granitic pluton at about 396 Ma. It is now exposed just south of the Anglo-Scottish border, where it forms the highest ground of the Cheviot Hills.

Broadly coincident with the granitic magmatism in southern Scotland and northern England at about 400 Ma, but focused away to the south, was a major deformation event caused by the collision of another Gondwanan continental microplate at the southern margin of Avalonia. This is commonly described as the Acadian event of the polyphase Caledonian Orogeny, but since it has no connection with closure of the Iapetus Ocean it has also been thought of as a separate orogeny in its own right. In northern Britain the strongest Acadian effects seen are the folding and cleavage developed in the Lower Palaeozoic rocks of the English Lake District. In southern Scotland, uplift in the Mid Devonian was a more distant and less intense Acadian effect of the collision farther south. Lower Devonian strata were tilted and disturbed so that when alluvial basins were re-established, the Upper Devonian strata deposited therein were laid down unconformably on their predecessors, both Lower Palaeozoic and Lower Devonian. Most of the Upper Devonian deposits were fluvial in origin though some also show an aeolian influence; they consist of red sandstone and siltstone with some conglomerate that are now assigned to the Stratheden Group. Across much of southern Scotland there is a conformable transition from these Upper Devonian strata into similar lithologies of early Carboniferous age.

Carboniferous

Extensional and transtensional tectonics were active for much of Carboniferous time, with the pre-existing Caledonian and Acadian structures exerting strong control on the orientation of the subsiding basins that resulted. Localised volcanic activity also arose from the extensional regime and produced both extrusive lavas and a range of minor intrusions. By this time the region formed part of the southern margin of Laurussia, an aggregation of Avalonia, Laurentia and Baltica (and ultimately Siberia) which formed the northern part of the Pangcea 'supercontinent' (Fig 3d). The British sector drifted slowly north through equatorial latitudes. An initially arid climate became more hot, humid and wet during that northward drift, and then reverted to arid conditions towards the end of the period. Dextral strike slip became progressively more important within the broadly extensional tectonic regime.

Throughout the Carboniferous, the Southern Uplands formed a relatively stable structural block, with subsidence occurring largely at its margins and within an internal series of small rifted grabens that developed by reactivation of Caledonian faults. The most complete sequence, preserved along the southern margin of the Southern Uplands block, is part of the Solway Basin succession, most of which crops out to the south of the border in north-west England. Similarly, the Carboniferous strata that surround the Cheviot block and extend across the Tweed Basin are the extremities of the more extensive succession seen in the Northumberland Basin of north-east England. These sequences contrast stratigraphically with those of the Carboniferous outliers within the Southern Uplands, for example at Sanquhar, Thornhill and Oldhamstocks, which originated as the infill of the internal grabens and have more in common with those seen to the north in the Midland Valley of Scotland.

The Carboniferous sedimentary record is the result of a complex interplay between several factors: subsidence rates, changes in sea level, limestone reef formation and the progradation of major sandstone deltas into the subsiding basins. Early Carboniferous sedimentation was in fluvial and lacustrine to paralic environments, building up the dominantly clastic Inverclyde Group. With continuing subsidence, a greater marine influence is seen in the succeeding strata in the south of the region, where the largely deltaic to shallow marine Border Group built up, whilst the rather more heterolithic but mainly clastic Strathclyde Group accumulated farther north, albeit in a broadly similar depositional environment. Through the middle part of the Carboniferous succession a major delta built out into the Northumberland–Solway basin system, depositing the Yoredale Group, with a similar succession accumulating independently farther north and now forming the Clackmannan Group. At that time cyclic sedimentation was a particular feature, with the sandy, alluvial and deltaic flats subject to intermittent marine incursions that allowed the development of limestone. The colonisation of the delta tops by lush, peat-forming swamp vegetation, and its subsequent burial and conversion to coal, is most evident in the upper Carboniferous successions of the Pennine and Scottish Coal Measures groups, exploited respectively in the Canonbie and Sanquhar coalfields. Towards the end of the Carboniferous Period, in the Canonbie Coalfield, deposition of Warwickshire Group strata occurred across a fluvial to deltaic plain under increasingly arid, oxidising climatic conditions; coal is largely absent and the strata are generally reddened.

By late Carboniferous times, the extensional and thermal subsidence that had largely controlled sedimentation patterns earlier in the period began to wane. Instead, the region experienced the peripheral effects of a major orogenic collision as, far to the south, Laurussia and the huge Gondwana 'supercontinent' came together to unite the Earth's land areas into the single vast expanse of Pangaea (Figure 3d). The compressive deformation associated with this event, the Variscan Orogeny, was most intense across mainland Europe and the southern parts of England, Wales and Ireland, south of the so-called 'Variscan Front'. Related deformation farther north was relatively weak, but Carboniferous basins were compressed and their strata folded and faulted, with dextral strike-slip movement imposed in places. The orientation of the resulting structures was much influenced by their interaction with the rigid structural blocks of the Southern Uplands and The Cheviot. Variscan deformation spanned an interval of approximately 15 million years from the late Carboniferous to the early Permian, and was accompanied by the large scale intrusion of basaltic magma to form sill and dyke swarms to both the north and south of southern Scotland. However, within the latter region, magmatism was limited to lava eruption in some of the Southern Uplands extensional grabens, with a more widespread scattering of minor intrusions.

Permian and Triassic

During Permo-Triassic times, global sea level was relatively low and Scotland was located within the interior of Pangaea and a little to the north of the Equator (Figure 3e); about 10°N at the beginning of the Permian, drifting to about 30°N by the end of the Triassic. Thus the preserved sequences are largely the result of terrestrial sedimentation in an arid environment. There was rapid erosion of the Carboniferous strata that had been uplifted by the Variscan compression, with several hundred metres likely to have been removed in some places. A renewal of tectonic extension early in Permian times then reactivated the broadly north-trending margins of the Carboniferous outliers in the Southern Uplands, and other pre-existing Caledonian structures, to form the margins of depositional graben and half-graben basins such as those seen at Thornhill, Lochmaben and Dumfries. North-east–south-west faults were also reactivated but mostly with strike-slip displacement. The early Permian extension was accompanied by some local volcanicity.

The lowest Permian strata seen in southern Scotland are fluvial and aeolian sandstones with conglomerates derived locally from the sides of the fault-defined depositional basins. Strata in the Scottish outliers within the Southern Uplands massif (and also to the north in the Midland Valley) are assigned to the Stewartry Group, but to the south, the succession bordering the Solway Firth is a continuation from north-west England of the Appleby Group. Along the Solway coast, the Appleby Group strata form the marginal deposits of the Solway Basin, wherein sedimentation continued from the Permian through into the Jurassic. In southern Scotland, representatives of this succession form the Cumbrian Coast Group (Permian) and the Sherwood Sandstone Group (Permian to Triassic). The former includes evaporite, formed on coastal mudflats and sabkhas, interbedded with fine-grained clastic lithologies deposited by a combination of wind and muddy sheet-flow. The Sherwood Sandstone Group strata were laid down by ephemeral rivers on braided alluvial plains and playa mudflats. The outcrops of the younger Triassic (and Jurassic) components of the Solway Basin succession do not extend into southern Scotland.

Jurassic to Palaeogene

It was the break-up of Pangaea during Late Triassic times that brought an end to the prolonged period of mainly terrestrial conditions across southern Scotland. Marine transgressions extended across an ever-widening area until, by Early Jurassic times, global sea levels were relatively high and much, possibly all of the region was submerged. Calcareous mudstone from this period, part of the Lias Group, is preserved around Carlisle in the centre of the Solway Basin.

A period of uplift and erosion in Mid Jurassic times is recorded by a widespread unconformity, with the maximum effect seen in North Sea basinal sequences. The commensurate fall in relative sea level continued through the Early Cretaceous and an extensive unconformity developed across the surrounding land areas, until rising sea levels brought renewed marine transgression through the later part of the Cretaceous. Continental drift had carried Britain to the latitude of about 35°N by the end of the Triassic and slow northwards progress continued during the Jurassic and Cretaceous periods (Figure 3f). The climate was strongly seasonal with warm, relatively dry summers and cool wet winters.

The later Mesozoic geological history is obscure, with no sedimentary rocks from that interval, or the succeeding Cenozoic, preserved in southern Scotland. There is some evidence that normal and possibly strike-slip faulting continued, and in the Solway Basin sedimentation probably continued into Early Cretaceous times. Thereafter, it is likely that continued subsidence allowed deposition of the Upper Cretaceous Chalk Group across much of the region, with the maximum post-Variscan burial probably achieved towards the end of the Cretaceous Period.

At the end of the Cretaceous and into the early Palaeogene Period, regional uplift began as a precursor to the major magmatism associated with the initial opening of the Atlantic Ocean. This was driven by development of the proto-Icelandic mantle plume, which had its maximum impact in what is now the Hebridean province of western Scotland and Northern Ireland and in Greenland, areas that were then adjacent (Figure 3f). There, from about 60 Ma to 55 Ma, immense volumes of basaltic magma were erupted with the commensurate intrusion of plutons, sill-swarms and swarms of dykes. Some of the latter, emanating from a volcanic centre on Mull, run south-eastward across southern Scotland and extend well into northern England, more than 400 km from their source; examples include the well-known, Cleveland and Acklington dykes.

Additional impetus was given to the regional uplift of northern Britain in Miocene times as a distant effect of the Alpine Orogeny. This resulted from collision between the European and African plates, but its main structural effects are not seen much beyond southern England and Wales. Around the south of Scotland, its influence was likely restricted to uplift of the Solway Basin sequences. Overall, the Palaeogene to Neogene uplift episodes created erosive conditions across southern Scotland that have continued to the present day. A considerable thickness of strata may have been removed in this time, perhaps as much as 2000 m in places, including all of the relatively soft, Jurassic and Cretaceous successions that most probably had once extended across much of the region. Further erosion of the Devonian and Carboniferous rocks followed, revealing the Caledonian, Lower Palaeozoic basement of the Southern Uplands structural block.

GLACIATION

Erosion was locally much more vigorous from about 2.6 million years ago as glacial conditions were established across northern Britain during the Quaternary 'ice age'. In fact, this was a period of alternating cold and more temperate interludes, with the most recent of the latter commencing only about 12 000 years ago (12 ka BP — where BP stands for Before

Present) and continuing to the present day. Ice-sheets repeatedly built up on the higher ground and fed glaciers that eroded deep valleys in places, whilst depositing a thick blanket of glacigenic sediment in others. The multiphase nature of glacial advance and retreat led to complex variations in ice flow direction through time, resulting in a complicated pattern of glacial landforms. This is particularly apparent in coastal areas of south-west Scotland and in the Solway lowlands, where there was interaction of different ice-sheets emanating from the Southern Uplands, the high ground of northern England and the Scottish Highlands, the latter skirting the Rhins of Galloway to enter the Irish Sea area.

As the ice-sheets waxed and waned so relative sea level changed. Short-term, eustatic changes were directly related to the growth of the ice-sheets, but longer term, isostatic changes arose from the depression of the Earth's crust by the enormous mass of ice, and its slow recovery once the ice had melted. These effects have led to a range of submerged and elevated coastal features, the most prominent of which are raised beaches and clifflines now abandoned several metres above present sea level.

Later glacial events tend to destroy, redistribute and incorporate the deposits of previous glaciations unless the latter are preserved in very special circumstances. The net result is that most of the till and morainic deposits now seen in the south of Scotland were deposited in the early part of the Late Devensian, during the Dimlington Stadial (about 29–14.7 ka BP), when extensive glaciers and ice-sheets grew in the region. A climatic amelioration followed, the Windermere Interstadial (about 15–13 ka BP), before ice returned to the highest parts of the region during the Loch Lomond Stadial (about 12.9–11.7 ka BP). But even when the higher ground was largely free of ice, extended periods of periglacial conditions led to the widespread development of a deep layer of frost-shattered rock, particularly over the more elevated Lower Palaeozoic strata. This material has subsequently been redistributed by solifluction, landslipping, and the development of screes. As deglaciation gathered pace, huge amounts of meltwater were released and deposited extensive sheets of sand and gravel, mostly across the lower ground. Much of this was then reworked into alluvial flood plains as the modern drainage system developed.

Relatively wet climatic conditions were prevalent in the immediately postglacial interval (Holocene, from about 11.7 ka BP) and encouraged the build-up of extensive peat deposits, both in upland areas of poor drainage and across lower but flatter terrain. The relatively lower rainfall of more recent times, coupled with the anthropogenic effects of overgrazing and drainage, has led to the drying-out and extensive erosion of many areas of upland peat.

HUMAN INTERVENTION

Since the last retreat of the ice from southern Scotland, humans have become a significant geological agent, modifying landscapes and sedimentary patterns through deforestation, reforestation, agriculture, mining and urban development, whilst driving many other species to extinction. Now global warming is once again leading to a rise in sea level, but this time accentuated by an anthropogenic contribution to the cause. Indeed, so profound has been Man's recent influence on global geological processes that it is becoming common practice to refer to the geological interlude that has followed the Industrial Revolution as the Anthropocene. How these changes will affect the future geological record remains to be seen. In the meantime, across southern Scotland, we continue to rely on geological sources of many raw materials, particularly for construction, road building and water supply. We can also enjoy a landscape of high scenic value created by the interaction of the underlying rocks with eons of geological change, and urban environments visually enriched by the use of local building stone in unique, vernacular architecture.

two

Ordovician and Silurian of the Girvan–Ballantrae district

In the south-west of the Scottish Midland Valley, between and inland from Girvan and Ballantrae, a structurally dismembered assemblage, mostly comprising Early Ordovician mafic and ultramafic igneous rocks, the Ballantrae Complex, is unconformably overlain by an Ordovician to Silurian sedimentary succession of clastic rock with some limestone (Figure 8). These Ordovician and Silurian rocks provide a geological link between the Southern Uplands and Midland Valley terranes. Whilst there are several other inliers of Silurian strata in the Midland Valley, none of them have anything like the same degree of overlap in age range and close

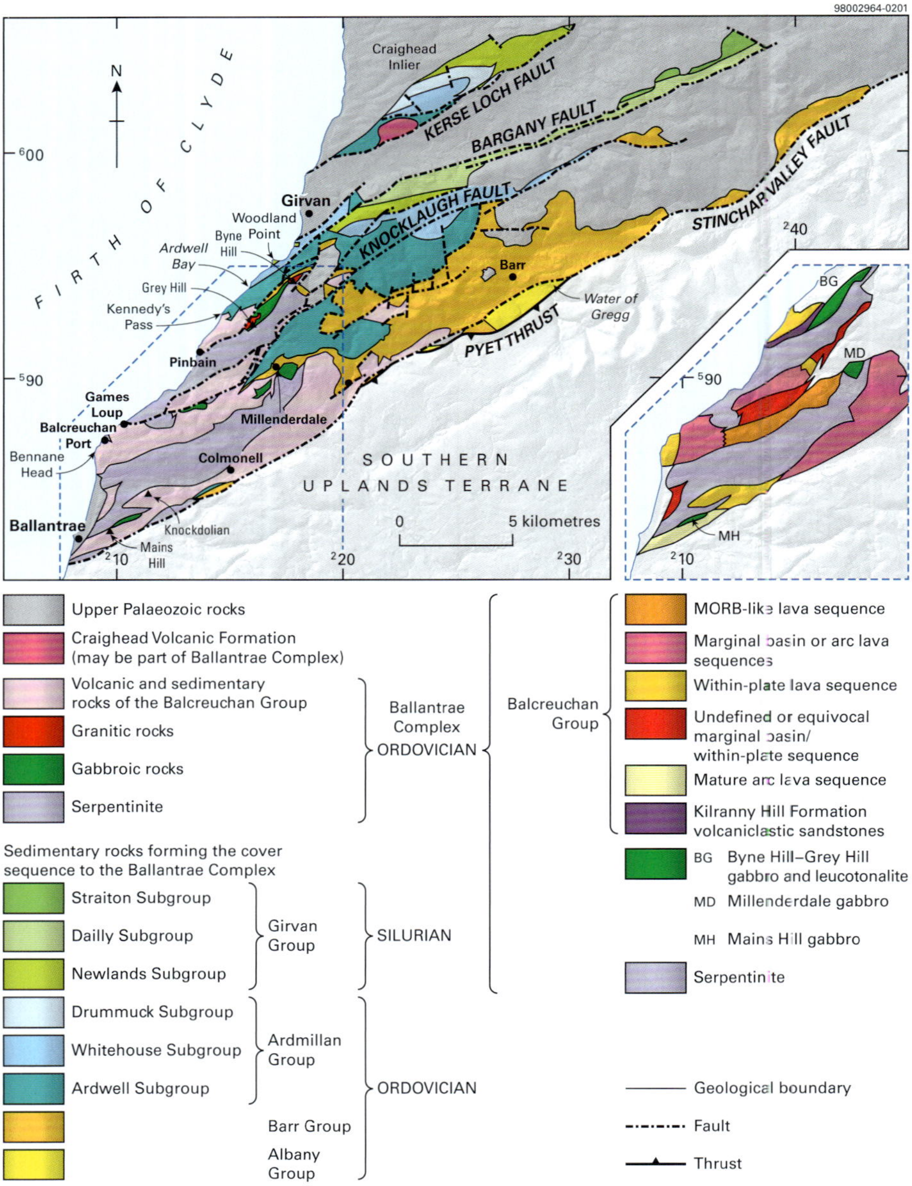

Figure 8 *Outline geology and location map for the Girvan–Ballantrae district.*

Figure 9 *Idealised ophiolite succession compared to the seismic structure of ocean crust and mantle (after Coleman, R G. 1977. Ophiolites: Ancient Oceanic Lithosphere? Springer-Verlag). As an indication of scale, beneath the oceans the depth to the geophysical moho is between 5 and 10 km.*

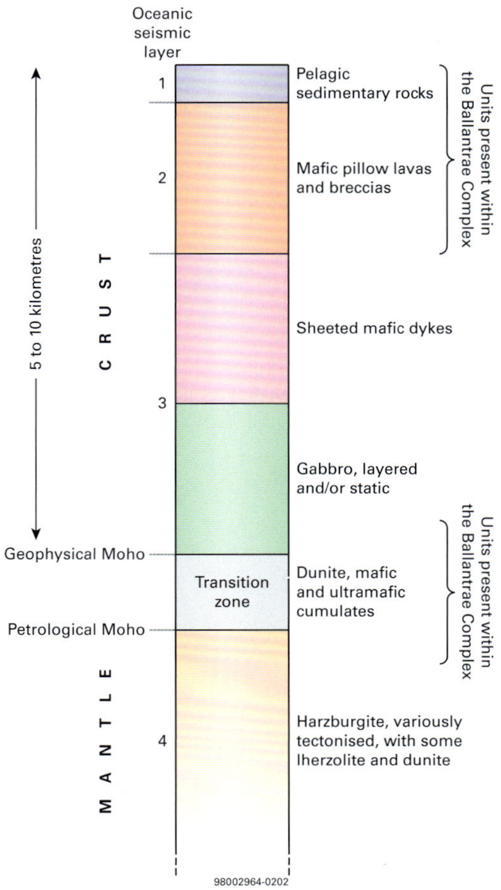

juxtaposition with the Southern Uplands as the succession seen in the Girvan–Ballantrae district. It is for this pragmatic reason that the Girvan–Ballantrae rock assemblage is included in this South of Scotland volume of the British Regional Geology series, along with the Southern Uplands, rather than in the complementary volume for the Midland Valley, wherein are described the other inliers.

ORDOVICIAN BALLANTRAE COMPLEX

The Ballantrae Complex is a tectonised assemblage of mainly igneous rocks that originated as the various components of the oceanic crust and mantle. When tectonically removed from their oceanic origins and accreted at a continental margin, such associations are commonly referred to as *ophiolite complexes*. In an ideal state they preserve the quasi-stratigraphical succession of the ocean crust and mantle (Figure 9): a basal ultramafic component passing upwards through gabbros and sheeted dykes into a volcanic and volcanosedimentary succession dominated by submarine pillow lavas, all capped by marine sedimentary rock. In common with most ophiolites, only disaggregated parts of this succession are present in the Ballantrae Complex, which crops out over about 75 km^2 between Girvan and Ballantrae, immediately to the north of the Stinchar Valley Fault, the local expression of the Southern Upland Fault and the southern boundary of the complex (Figure 8). Geophysical evidence suggests an eastwards extension beneath the unconformable cover of younger sedimentary rocks.

Within its relatively small outcrop area, the ophiolitic Ballantrae Complex demonstrates a bewildering variety of rock types. Different lithologies that are now in structural contact originated at contrasting levels in the oceanic crust and mantle, spanning a considerable depth range (Figure 9). Moreover, several entirely different oceanic regimes are represented, spanning island-arc and within-plate volcanoes, and it is clear that only tiny fragments derived from each have been preserved and juxtaposed. The environment of volcanism can be established by the trace element content of the basaltic lavas produced. Elements such as titanium, zirconium and yttrium are particularly useful in this respect and their ratios can be used to indicate island-arc, ocean-island (also known as within-plate or Hawaiian-type) or mid-ocean-ridge (MORB — where the B stands for basalt) eruption. A series of analytical studies has established that the Ballantrae Complex lavas are polygenetic, with all three of the main environment types represented (although MORB-type are relatively rare) and the island-arc varieties showing clear subdivisions; an indication of the distribution of the different lava types is shown in the inset map in Figure 8.

Compressive, subduction-related movement might produce the assemblage of rock types present in the Ballantrae Complex, but it could probably be achieved more readily with additional components of backarc extension and strike-slip tectonics. Figure 10 gives one idea of the possible original relationships between the different elements now present, but there are other, equally valid possibilities. Although the area has been the focus of considerable research, and an extensive geological literature is available, there is still much to be discovered about the Ballantrae Complex and it remains one of the more enigmatic geological features of southern Scotland.

The ophiolitic lithologies have structurally complicated relationships within the complex but two main elements dominate, interleaved by faulting: serpentinised ultramafic rocks derived from the oceanic mantle, and volcanic sequences representing the remains of island-arc and oceanic crust (Figure 9). The principal structures are north-east to south-west faults that split the complex into discrete lithological zones such that northern and southern serpentinite belts separate three areas of mainly volcanic rock (Figure 8). A late Cambrian to Early Ordovician age has been established for eruption of the island-arc volcanoes from radiometric (Sm-Nd) dating of their basalts at 501 ± 12 and 476 ± 14 Ma. The age of the within-plate components also fall into this range since the interbedded sedimentary strata contain early Arenig (ca 478 Ma) graptolite faunas (Figure 5q). Minor components of the complex include gabbro

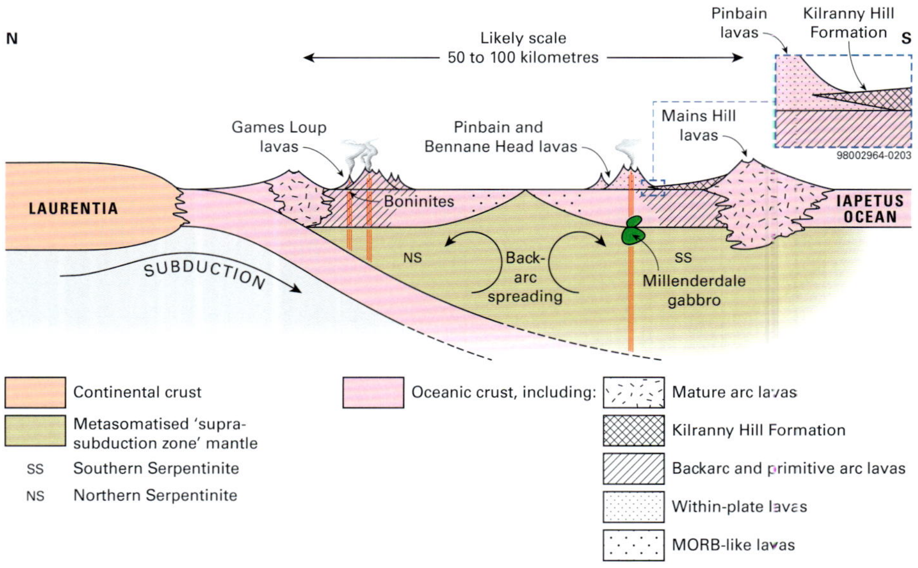

Figure 10 *Possible model for the development of the various components of the Ballantrae Complex within an oceanic volcanic arc and backarc basin, marginal to the Laurentian continent.*

and leucotonalite (traditionally described as trondhjemite or plagiogranite) intrusions, a fragment of a sheeted dyke assemblage, and sheared sedimentary mélange deposits. Zircons from the leucotonalite have been dated (U-Pb) at 483 ± 4 Ma. These minor components take on a disproportionate significance in the interpretation of the complex as an ophiolite.

The Ballantrae Complex ophiolite was finally emplaced (obducted) onto the Midland Valley continental basement, at that time the leading edge of Laurentia, during the mid to late Arenig. High-grade metamorphic rocks were formed at this stage as hot mantle material was thrust up and across the basalt lavas of the oceanic crust. These are now seen as a metamorphic 'sole' at the base (the south-east margin) of the northern serpentinite belt and have been radiometrically dated (K-Ar) at 478 ± 8 Ma, whilst upper Arenig sedimentary rocks that were probably deposited during the obduction process are now structurally included within the complex. The oldest strata within the unconformably overlying cover sequence (comprising the Barr Group) are of Llanvirn age. In a regional context, the obduction of the Ballantrae Complex would seem to be part of the large-scale, arc–continent collision at about 470 Ma that drove the Grampian phase of the Caledonian Orogeny.

Ultramafic rocks

Within the Ballantrae Complex, the ultramafic rocks have been pervasively altered, with serpentine replacing the original olivine to produce a dark green to black, generally fine-grained rock, locally containing yellowish green, bastite pseudomorphs of altered orthopyroxene. Complex veining by quartz and carbonate is a general feature of the faulted margins to the ultramafic bodies. Olivine was the dominant original mineral but since it has been largely altered to serpentine the resulting ultramafic rock is referred to generally as serpentinite. This lithology is relatively soft and readily eroded, so that exposure of the serpentinite is very limited. A small proportion of the ultramafic rock is composed mainly of varieties of pyroxene, so is appropriately referred to as pyroxenite.

Some of the olivine-rich ultramafic rocks originally crystallised within oceanic mantle at depths of up to 60 km. Their composition indicates that they are residues after the extraction of a basaltic melt. However, the detailed chemistry of rocks from both ultramafic belts, and also of the chrome spinel grains that they contain, is anomalous for normal ocean lithosphere and suggests that they formed in metasomatised mantle above a subduction zone. Subtle but significant differences in composition between the two main outcrops, the northern and southern serpentinite belts, show that they do not share the same history of formation. The northern serpentinite is the more metasomatised of the two and most probably originated in a sub-arc environment. The southern serpentinite, though still metasomatised, has more of the characteristics of mid-ocean mantle. To accentuate these differences, the following account will utilise rock names reflecting the original composition of the rocks, prior to serpentinisation.

The northern serpentinite belt (Figure 8) is composed mainly of harzburgite (olivine with accessory orthopyroxene) with some lherzolite (olivine with both accessory orthopyroxene and clinopyroxene) and a variety of pyroxenites; concentrations of finely disseminated chromite are present locally. There is a widespread tectonic fabric that is strongly developed in places. In contrast, the southern serpentinite belt has either no tectonic fabric or one that is only weak and localised, and also has a slightly different lithological assemblage to that seen in the north. In the southern belt, although harzburgite is again dominant, it is generally coarser grained than in the north and is accompanied by dunite (>90 per cent olivine), wherlite (olivine with accessory clinopyroxene) and troctolite (olivine-rich gabbro). At one locality, Poundland Burn [NX 167 882], a local concentration of chrome spinel grains forms a distinctive nodular lithology. The overall implication is that, quite apart from their different geotectonic environments, the ultramafic protolith of the southern serpentinite belt formed at a substantially shallower depth then its northern counterpart.

Plate 8 *Arenig (Ordovician) basaltic pillow lavas at Bennane Head [NX 094 875], north of Ballantrae on the coast of the Firth of Clyde (P005992).*

Volcanic rocks

The volcanogenic sequences within the complex are dominated by pillow lavas produced during submarine eruptions (Plate 8) and the breccias derived from them. A number of formations have been defined locally, but all of the volcanic rocks are included within the **Balcreuchan Group**, named after Balcreuchan Port [NX 098 876], the putative home in the early 17th century of the legendry Sawney Bean and his cannibal tribe. Some parts of the group comprise uniform accumulations of dark greenish grey, mainly aphyric, basalt pillows, whereas other sections show alternations of aphyric and coarsely feldsparphyric lavas (Plate 9), reddened in places, interbedded with sedimentary layers of chert and shale and merging vertically and laterally with thick breccia units. The lavas are tough and resistant to erosion so that they form most of the high ground within the complex and create some spectacularly rugged coastal sections.

The tectonic slicing of the complex into numerous fault-defined structural blocks prevents the establishment of a comprehensive stratigraphy, but similarities are apparent between some of the blocks: for example, the mixed volcanic–sedimentary sequences at Pinbain, Bennane Head and Knockdolian (Figure 8). These mixed lava (aphyric and feldsparphyric) and sedimentary rock associations prove to be of exclusively ocean island (within-plate) origin. At Bennane Head, graptolites from the shale and chert layers that are interbedded with the lavas establish an early Arenig age (Figure 5q) and imply fairly deep-water conditions. Conversely, some sedimentary features of these ocean island sequences (such as rounded clasts in breccias, reddened tops to some lava flows, and rare volcanic lapilli tuff beds) indicate relatively shallow water conditions or even sporadic and temporary emergence of the lava pile. A variable and unstable depositional environment seems certain, whilst one feature of the Pinbain assemblage suggests that different depositional regimes were in relatively close proximity. At the base of the Pinbain succession of within-plate lavas is a unit of volcaniclastic sandstone (the Kilranny Hill Formation) with the geochemical characteristics of a mature island arc, characteristics that are shared with the Mains Hill lavas farther to the south (Figure 8). The Kilranny Hill Formation also contains early Arenig graptolites.

The pillow lava accumulations of island-arc origin contain very little interbedded sedimentary rock. Most of the lavas are aphyric but some contain pyroxene phenocrysts, in particular those

Plate 9 Porphyritic basalt lava with large, partially aligned feldspar phenocrysts, exposed on the coast of the Firth of Clyde at Slockenray [NX 139 919], south of Girvan (P220326).

at Mains Hill and to a lesser extent those at Games Loup [NX 104 880]. The radiometric (Sm-Nd) dates of 501 ± 12 and 476 ± 14 Ma that have been obtained, respectively, from these two localities span the late Cambrian and Early Ordovician. Of particular interest in the Games Loup section is the presence of lavas with an unusual composition: relatively high silica content accompanied by high levels of MgO, Cr and Ni. Such lavas, known as boninites, are characteristic of eruption in an oceanic, but supra-subduction setting and modern examples are typically found in the forearc region of volcanic island arcs or as the earliest products of backarc spreading. If the Games Loup boninites are taken to indicate crustal extension in a backarc basin, the likely age difference between these relatively primitive arc rocks and the more evolved, mature arc basalts (e.g. those at Mains Hill and Bargain Hill), permits the latter to be interpreted as part of an early-formed volcanic arc that was split by supra-subduction zone extension. Subsequent, within-plate eruptions and gabbroic intrusion then took place in the ensuing backarc basin. This is the tectonic model illustrated in Figure 10.

Lavas with the geochemical characteristics of eruption in an extensional environment at a mid-ocean ridge (MORB-type) are relatively uncommon in the Balcreuchan Group. The closest association is shown by basalts in the central part of the Ballantrae Complex that lie in a zone along the northern margin of the southern serpentinite belt (Figure 8). Though the composition of these basalts differs in several subtle respects from the normal MORB type, it compares closely with that of basalts generated at some backarc spreading centres.

Intrusive gabbro and granite

Coarse-grained intrusive rocks are a widespread, minor component of the Ballantrae Complex and form substantial, composite bodies at three localities: between Byne Hill and Grey Hill, around Millenderdale and north-east of Mains Hill (Figure 8). At Byne Hill

[NX 179 947], an intrusive body ranges in composition from a leucotonalite core, through dioritic lithologies, to gabbro at the margins that is chilled against the host northern serpentinite belt. Zircons from the leucotonalite have provided a U-Pb cooling age of about 483 Ma, but despite the similarity in age, the relationship of the Byne Hill–Grey Hill intrusion to the rest of the Ballantrae Complex is uncertain. The geochemistry of the gabbro and leucotonalite suggests an origin in association with magmatism in an extensional, mid-ocean-ridge setting but with an additional supra-subduction influence, whilst the field relationships are more indicative of post-obduction intrusion. A smaller gabbro intrusion into the northern serpentinite belt, known as Bonney's Dyke [NX 135 911] is markedly pegmatitic. Near Millenderdale [NX 172 906], intrusive gabbro bodies are banded and foliated and are cut by several generations of doleritic dykes, the earlier of which are themselves foliated. Most of the dykes and parts of the gabbro bodies have been subjected to high-temperature metamorphic recrystallisation and are now granular-textured metamorphic rocks. The relationship of the Millenderdale rocks to the rest of the Ballantrae Complex remains conjectural, although their geochemistry suggests an origin in association with ocean island (within-plate) magmatism. Within the southern serpentinite belt, foliated and metamorphosed doleritic rock forms a large number of small tectonic inclusions.

Several small gabbro bodies are intruded into the southern serpentinite belt in the vicinity of Mains Hill [NX 093 829], and are chilled against the host ultramafic rock, but the largest gabbroic mass in this part of the complex appears to have entirely faulted margins. These may be post-obduction in age, as are numerous small intrusions of dolerite and gabbro in the central part of the Ballantrae Complex, between Knockdaw and Balsalloch. There, both serpentinite and Balcreuchan Group lavas have been intruded and on the north side of Carleton Hill [NX 127 894] intrusive dolerite cuts across the metamorphic zone at the base of the northern serpentinite belt. None of the post-obduction intrusions cut the Llanvirn and younger sedimentary succession that overlies the Ballantrae Complex.

Origins and emplacement of the Complex

There is general agreement that the Ballantrae Complex is an assemblage of polygenetic, oceanic rocks, most of which originated in one or more island arcs and adjacent, actively spreading backarc zones. There is less of a consensus as to how and when the various components were brought together and what tectonic processes were responsible. It is possible to derive an idealised, island-arc and backarc basin model in which most of the components can be accommodated (Figure 10). In this interpretation, horizontal compression is largely driven by subduction, but tectonic juxtaposition of the various different parts is then eased if significant strike-slip movement is invoked during closure of the marginal basin and collision of the arc with the Laurentian continental margin. Even so, major problems of timing remain. Not least of these is the requirement to uplift mantle ultramafic rocks so that they might cool prior to the intrusion of the gabbro bodies at Byne Hill and Mains Hill.

The thrusting of hot, mantle ultramafic rock through and across the crustal sequences is recorded by the dynamothermal metamorphic aureole seen at the southern margin of the northern serpentinite belt. Radiometric dates from these metamorphic rocks, 505 ± 11 Ma (Sm-Nd) and 478 ± 8 Ma (K-Ar), suggest that this process was either protracted or polyphase. It was also contemporaneous with the continuing igneous development of the complex since ultramafic rock has been intruded by gabbro and leucotonalite, at Byne Hill for example. There, the igneous bodies have chilled margins, demonstrating that the host serpentinite had cooled by the time of intrusion, 483 ± 4 Ma (U-Pb) in the case of the leucotonalite. Elsewhere, at Carleton Hill, the dynamothermal metamorphic aureole is cut by dolerite dykes that do not penetrate the Llanvirn and younger sedimentary cover, so constraining magmatism to the late Arenig.

Note the overlap between the production of the metamorphic aureole by the emplacement of hot ultramafic rock at 478 ± 8 Ma but the intrusion of leucotonalite and gabbro into cool ultramafic rock at 483 ± 4 Ma. The two sets of data come from adjacent areas of the

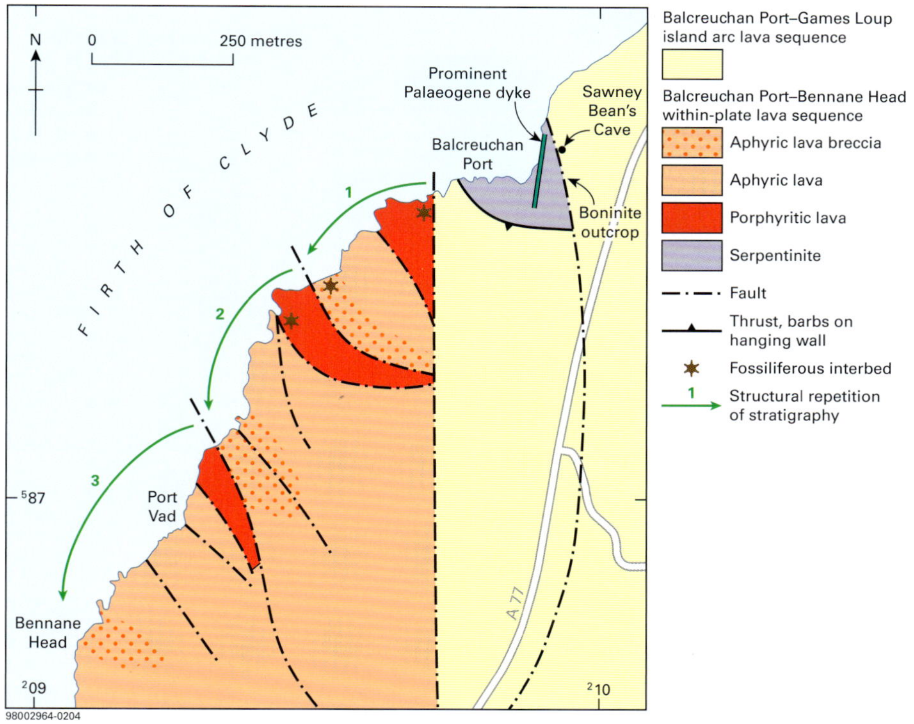

Figure 11 *Geological sketch map of the Balcreuchan Port to Bennane Head area showing the structural repetition of the volcanosedimentary succession.*

northern serpentinite belt and suggest that the later stages of the complex's development were of short duration. That message is reinforced by the appearance in sedimentary mélange deposits, themselves intimately associated with the Balcreuchan Group, of amphibolite schist clasts probably derived from the dynamothermal aureole. The mélange deposits also contain a range of volcanic and intrusive rock types, probably sourced locally during the final stages of tectonic assembly of the complex, as well as more exotic lithologies including blue-amphibole schists. The latter were formed at considerable depth and were most likely derived from a pre-existing subduction complex.

The youngest strata known to be associated with the Ballantrae Complex form a clastic sequence at North Ballaird [NX 121 878] that contains a late Arenig graptolite fauna. All of its margins are faulted against parts of the Balcreuchan Group. Granules of altered serpentinite in one of the upper Arenig beds have algal rims, demonstrating that mantle-derived, ultramafic rock was available for erosion into shallow water by the late Arenig. This presumably marks the final stage of obduction and tectonic assembly, processes that were essentially complete by the early Llanvirn when the lowest beds of the unconformable sedimentary cover sequence were deposited. The detail of the obduction-related structure is difficult to decipher and has most probably been further complicated by subsequent tectonic events. Some indication of that complexity is shown by the tectonic repetition, probably a syn-obduction effect, of a graptolitic mudstone–chert–lava succession between Bennane Head and Balcreuchan Port (Figure 11).

SEDIMENTARY SUCCESSION: ORDOVICIAN

The Arenig rocks of the Ballantrae Complex, which form the irregular 'basement' to the younger, Ordovician and Silurian sedimentary sequence in the Girvan area, had been obducted onto the southern margin of the Midland Valley terrane by Llanvirn times. These 'basement' rocks were then deeply eroded prior to deposition of a mostly shallow- to moderate-depth

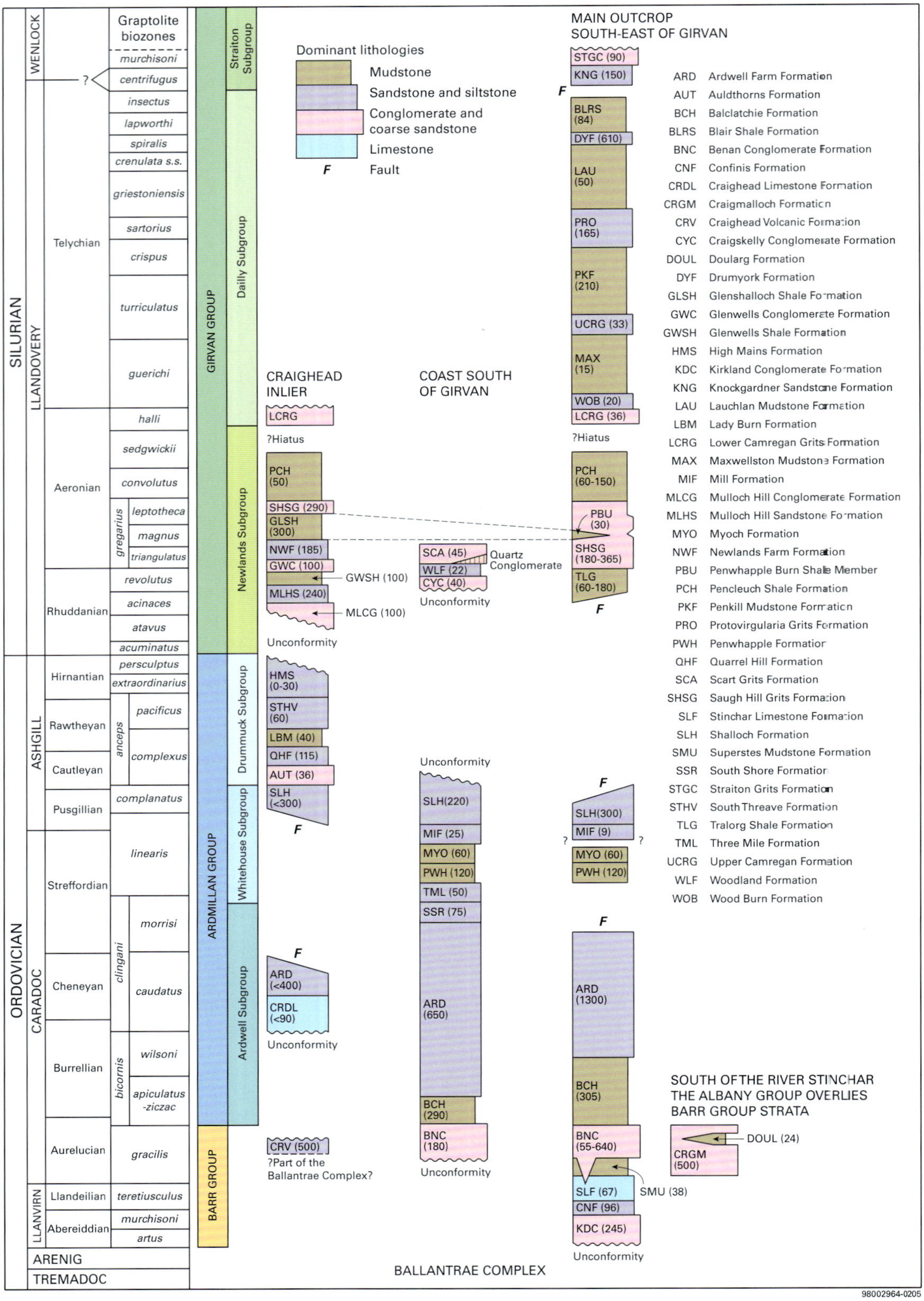

Figure 12 *Stratigraphical columns for the Ordovician (Llanvirn) to Silurian (Wenlock) sedimentary successions in the Girvan area. Formation and member thicknesses are shown in metres.*

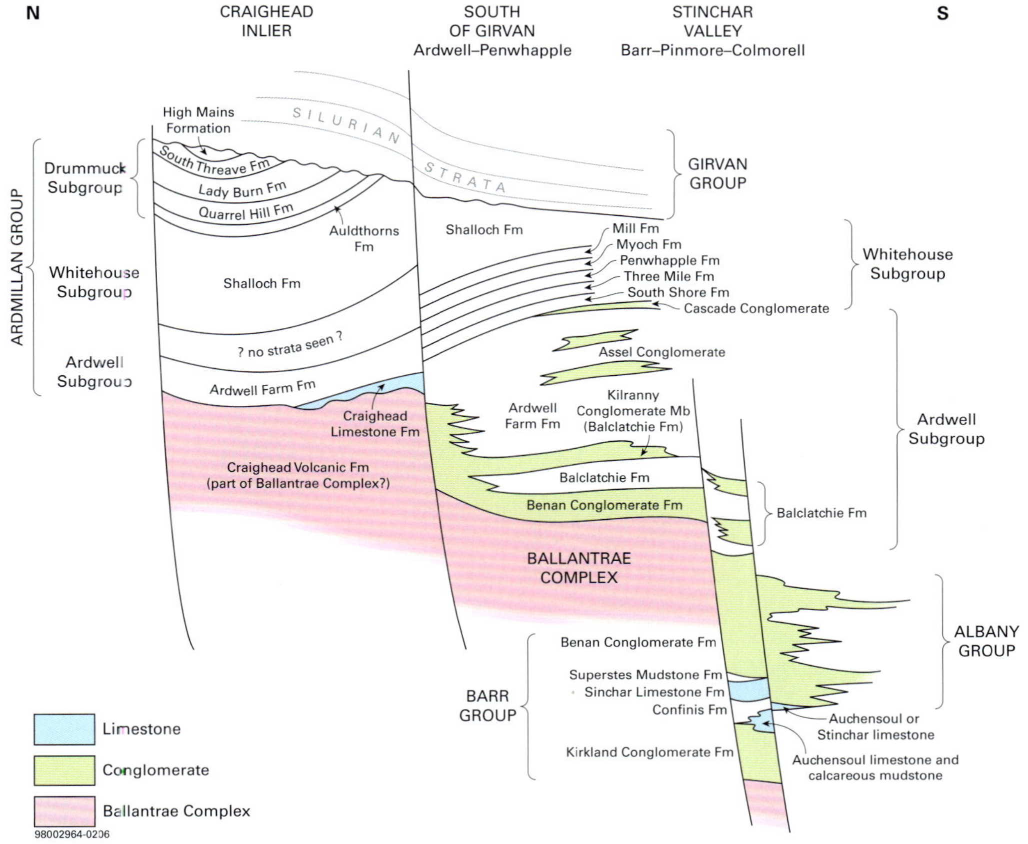

Figure 13 *Syndepositional fault control on Ordovician stratigraphy in the Girvan area. After Williams (1962).*

marine 'cover sequence' of conglomerates, sandstones, siltstones, mudstones and minor reefal limestones. The 'cover sequence' shows distinctive variations across its principal areas of outcrop (Figure 12): a Main Outcrop inland from Girvan and along the coast to the south of the town, and the Craighead Inlier to the north-east of Girvan (Figure 8). There is evidence of tectonic instability and significant fault control of sedimentation, with major changes in thickness of (particularly) Ordovician conglomerate units across contemporaneous, broadly east–west faults. The faults developed sequentially northwards, with downthrow to the south, and so produced a northward transgressive onlap of the sedimentary sequence (Figure 13). The counterpart of this is a southward transition to more distal, deeper-water lithofacies at any given stratigraphical level. An example of this effect is preserved on the south side of the River Stinchar, south-westward from its confluence with the Water of Gregg.

Water of Gregg

In general, the southern margin of the Ordovician–Silurian 'cover sequence' is formed by the Stinchar Valley Fault, but there is one important exception. On the south side of the River Stinchar, and cropping out intermittently for about 10 km south-west from the river's confluence with the Water of Gregg, Ordovician sedimentary rocks appear on the south side of the Stinchar Valley Fault, in the footwall of the Pyet Thrust (Figure 8). This thrust dips to the south and has transported Tappins Group strata northwards; as described in Chapter 3, the Tappins Group forms the northernmost division of the Southern Uplands terrane. Along strike to both the north-east and the south-west, the Pyet Thrust is cut out by the Stinchar Valley Fault.

Part of the succession between the Stinchar Valley Fault and the Pyet Thrust can be assigned to the Barr Group, which is more fully developed to the north and is accordingly described below from the Main Outcrop. However, to the south of the River Stinchar, the relatively shallow-water deposits of the Barr Group appear to be overlain by more than 500 m of deep-water deposits: wacke sandstone, mudstone and conglomerate. These form the **Albany Group** (Figures 8 and 12), much of which comprises the **Craigmalloch Formation**, a clastic unit of conglomerate, sandstone and siltstone. Towards the top of the succession, the mudstones of the **Doularg Formation**, about 25 m thick, interfinger with the Craigmalloch Formation conglomerates, and contain a significant deep-water fauna, mainly of trilobites. This shelly fauna, together with graptolites from the Craigmalloch Formation, establishes the *gracilis* Biozone and allows biostratigraphical correlation with the upper part of the Barr Group in the Main Outcrop (Figures 12 and 13).

Main Outcrop and coastal section

Barr Group

The oldest part of the succession in the Main Outcrop, the **Barr Group**, crops out in the south along and on the north flank of the Stinchar Valley, where the **Kirkland Conglomerate Formation** (thicknesses for this and all of the other units subsequently described are given in Figure 12) forms the core of an anticline plunging gently north-eastwards. Though the base is nowhere seen, the conglomerate is assumed to rest on Ballantrae Complex rocks and contains a large proportion of ophiolitic and igneous detritus, both within the matrix and as clasts. This rudite is overlain by the **Confinis Formation**, a flaggy sandstone unit with minor calcareous members (e.g. the Auchensoul algal limestone and calcareous mudstone). The calcareous aspect increases significantly upwards into the succeeding **Stinchar Limestone Formation**, a massive reef unit containing both shelly and conodont faunas. The conodonts define the local base of the Llandeilian Stage within the reef limestone sequence. A return to fine-grained clastic sedimentation is seen in the overlying graptolitic **Superstes Mudstone Formation**, which is in turn overlain, and channelled into, by the thick **Benan Conglomerate Formation**. An important feature of this conglomerate is its dramatic change in thickness (640 m to 55 m) when traced north-west across strike, a feature thought to have been caused by its deposition against steep submarine fault scarps (Figure 13), the faults having downthrow to the south-east and most probably forming a listric set stepping sequentially northward. As one result of this transgression, the Benan Conglomerate is the only part of the Barr Group seen in the Girvan coastal section, where it is presumed to rest unconformably on Ballantrae Complex lavas (Figure 12), the exposed contacts being faulted.

Although many of the clasts in the Benan Conglomerate can be matched with the subjacent Barr Group sedimentary rocks and Ballantrae Complex lithologies, there are many boulders of igneous rock (e.g red granite) for which no obvious source is known at outcrop. Radiometric age-dating of some of these granitic boulders has demonstrated a relatively short time span between their intrusion (cooling dates cluster around 560 Ma and 470 Ma) and their subsequent erosion into the conglomerate (depositional age ca 460 Ma). The most likely source of the granitic boulders is a magmatic arc to the north, within the Laurentian, Midland Valley terrane, that was initiated by the establishment (or rejuvenation) of northward subduction following the arc–continent collision that emplaced the Ballantrae Complex. This most probably involved a reversal of subduction direction (cf. Figure 10) to establish oceanic subduction northward beneath the Midland Valley terrane. This would have initiated suprasubduction magmatism there and led, ultimately, to the development of the Southern Uplands accretionary complex as described in Chapter 3.

Ardmillan Group

The **Ardmillan Group** succeeds the Barr Group and although it contains numerous conglomerate lenses, it is generally a more sandstone- and mudstone-dominated succession than its predecessor; turbidity currents dominated the depositional regime. The **Ardwell, Whitehouse and Drummuck subgroups** are recognised, though the latter is restricted to the Craighead Inlier in the north (Figures 8 and 12).

Plate 10 *Folding of thinly bedded turbiditic strata from the Ardwell Farm Formation, exposed on the coast of the Firth of Clyde to the south-west of Ardwell Bay [NX 152 933], south of Girvan (P005823). The rock face is about 4 m high.*

At the base of the Ardmillan Group (and the Ardwell Subgroup), the **Balclatchie Formation** is a thick mudstone unit with important faunas, both shelly and graptolitic, at several levels; the graptolitic strata include the palaeontologically renowned Laggan Member, which has yielded remarkable 3D graptolites dissolved out from calcareous nodules. The presence of both shelly and graptolitic faunas has been of particular importance for biostratigraphical correlations. Some substantial conglomerate units also occur within the Balclatchie Formation, including the Kilranny Conglomerate Member, which has formerly been much confused with the older Benan Conglomerate. The succeeding **Ardwell Farm Formation** is well exposed on the coast northward from Kennedy's Pass [NX 150 933] to Ardwell Bay and consists mainly of fissile turbidite sandstone beds, some micaceous, some calcareous, with sporadic lenticular conglomerate members interspersed; the conglomerates are mainly confined to the inland part of the outcrop. In its coastal outcrop the formation's strata are folded into a spectacular series of angular 'box' folds (Plate 10) for which a soft-sediment slumping origin has been proposed, though not universally accepted.

The shore section between Ardwell Bay and Girvan displays an excellent strike section through the middle part of the Ardmillan Group, the Whitehouse Subgroup and its constituent formations. At the base of the subgroup (Figure 12), the **South Shore Formation** is a calcareous turbidite succession of wacke-type sandstone and mudstone, with coarse shelly debris in places. The overlying **Three Mile Formation** comprises a fine-grained and laminated, sandstone–siltstone–mudstone sequence characteristic of deposition from low-density turbidite flows. Black graptolitic mudstone characterises the succeeding **Penwhapple Formation** with a *linearis* Biozone fauna indicating equivalence to the upper part of the Lower Hartfell Shales in the Southern Uplands terrane (see Chapter 3). This lithology is succeeded by the distinctive red and green silty mudstones of the **Myoch Formation**, a deep water deposit with an extensive shelly fauna that includes unusual trilobite species which were either blind or had very large eyes. Above this, the **Mill Formation** is lithologically diverse with mottled and dark mudstones, fissile, calcareous sandstone, and mud-clast conglomerate; it is formally subdivided into several members. One of the calcareous sandstone members at the top of the formation (Stacks Member) has yielded a diverse and abundant brachiopod fauna that establishes a Pusgillian age. The **Shalloch Formation**, at the top of the Whitehouse Subgroup, is a fairly uniform succession of thinly bedded turbidite sandstone with a few interbedded green mudstone beds and some turbidite units composed largely of limestone detritus. The important zonal graptolite *Dicellograptus anceps* occurs within this formation, indicating correlation with the Upper Hartfell Shales of the Southern Uplands (see Chapter 3). These are the youngest Ordovician strata seen in the Main Outcrop at Girvan, being cut off by the unconformity beneath the lowermost Silurian rocks, as seen at Woodland Point [NX 170 953]. The Shalloch Formation provides a correlative overlap with the youngest part of the Girvan Ordovician succession, which is only preserved in the Craighead Inlier (Figure 12).

Craighead Inlier

Located some 7 km north of the Main Outcrop, the Craighead Inlier contains a sequence that spans the Ordovician–Silurian boundary. The local 'basement' is a volcanic assemblage of basaltic pillow lavas and associated (perhaps infolded) bedded chert, the **Craighead Volcanic Formation**, which has commonly been assumed to form the northern outpost of the Arenig Ballantrae Complex. However, the cherts have yielded conodonts of Llanvirn (late Abereiddian or Llandeilian) age and, depending on the exact relationship of the cherts with the lavas, it is possible that the latter are younger than those of the Ballantrae Complex. In the disused Craighead Quarry [NS 235 014], the **Craighead Limestone Formation** (here the basal unit of the Ardwell Subgroup) rests with marked unconformity on an irregular surface of lava. It includes algal and pelmatozoan limestones and limestone breccias with an interesting, mid Caradoc shelly and ostracod fauna of North American affinity. Some interfingering mudstone members (e.g. the Kiln and Sericoidea members) are also richly fossiliferous and provide useful correlation between the graptolitic (basal *clingani Biozone*) and shelly biozonal schemes. The overlying **Ardwell Farm Formation** is locally in faulted contact with the Craighead Limestone and, as in the more extensive outcrops farther south, consists of fissile

Plate 11 *(opposite) Ordovician and Silurian fossils — trilobites, brachiopods and a starfish — from the Girvan district and a detrital Ordovician coral from the Southern Uplands that was derived from the Midland Valley Ordovician succession:*

Brachiopods

a Leptaena (Leptaena) valentia *Cocks 1968. (P773359) GSE 853. Silurian (Llandovery). Newlands Subgroup. Kirk (Mulloch) Hill, Girvan [NS 268 043].*

b Clorinda undata *(J de C Sowerby 1839). (P773356) GSE 929. Silurian (Llandovery). Newlands Subgroup. Newlands, Girvan [NS 275 045].*

c Eostropheodonta mullochensis *(Reed 1917). (P773355) GSE 875. Silurian (Llandovery). Newlands Subgroup (Woodland Formation). Woodland Point, Girvan [NX 170 953].*

Trilobites

d Acernaspis woodburnensis *Clarkson, Eldredge and Henry 1977. (P521140) GSE 5780. Silurian (Llandovery). Dailly Subgroup. Bargany Burn, Barr [NX 250 990].*

e Stenopareia shallochensis *(Reed 1904). (P521133) GSE 482. Ordovician (Ashgill). Drummuck Subgroup. Lady Burn, Girvan [NS 235 032].*

Starfish

f Tetraster *sp. (P740610) GSE 12671. Ordovician (Ashgill). Drummuck Subgroup (South Threave Formation). Lady Burn, Girvan [NS 245 037].*

Coral

g Kilbuchophyllia clarksoni *Scrutton 1994. (P521144) GSE 9935. Ordovician (Caradoc). Allochthonous in the Kirkcolm Formation. Wandel Burn, Abington [NS 968 262].*

a
2 cm
b
2 cm
c
2 cm
d
2 cm
e
2 cm
f
5 mm
g
5 mm

Plate 12 *Lower Silurian, Craigskelly Conglomerate Formation, as exposed on the coast of the Firth of Clyde at Horse Rock, south of Girvan. Ailsa Craig on the horizon (PC05981).*

calcareous sandstones and siltstones (the latter dominating the Plantinhead Member) with a graptolite fauna indicative of the *clingani* Biozone.

The component formations of the Ardwell Subgroup, described above, crop out in the southern part of the Craighead inlier and are in faulted contact with a succession of younger Ordovician and Silurian strata that form the northern part of the inlier (Figure 8). The outcrop pattern of this younger, northern succession is controlled by its distribution around a large anticlinal structure plunging gently towards the north-east. The lowest stratigraphical unit at outcrop, the **Shalloch Formation**, is the only visible representative of the Whitehouse Subgroup and consists of a cyclic sequence of fissile, wacke-type sandstone and siltstone similar to that occurring to the south in the Main Outcrop. The succeeding Drummuck Subgroup, known only from the Craighead Inlier, includes five conformable formations extending up to the local base of the Silurian (Figure 12).

The oldest division of the Drummuck Subgroup, the **Auldthorns Formation**, comprises coarse green sandstone, conglomerate and minor mudstone. Rich faunas have been obtained from thin shelly horizons throughout the formation. The succeeding green and purple mudstone with conglomerates and mass-flow deposits constitute the **Quarrel Hill Formation**, with well-preserved brachiopod faunas at various horizons throughout. The grey, silty mudstone of the overlying **Lady Burn Formation** has yielded only a few brachiopods and trilobites. However, the **South Threave Formation**, comprising mudstone, sandstone and siltstone, has several highly fossiliferous members including, at its base, the Farden Member with the well-known Starfish Beds. The latter are thought to be mass-flow deposits which have transported and redeposited a rich shelf faunal assemblage (Plate 11) into deeper water. The youngest unit seen in the Ordovician succession of the Craighead Inlier is the **High Mains Formation**, a fine- to medium-grained grey-green sandstone body cropping out only in the core of the north-east-plunging anticline. It has a scattered shelly fauna of Hirnantian age that includes trilobites and brachiopods. The Ordovician succession is here truncated by the gently unconformable base of the Silurian, with, in chronostratigraphical terms, relatively little Ordovician strata missing (Figure 12). Although the principal structure in the northern part of the Craighead Inlier is anticlinal, the sub-unconformity distribution of the Ordovician strata is broadly synclinal (cf. Figure 13); hence the restriction of the High Mains Formation to the anticlinal hinge zone.

SEDIMENTARY SUCCESSION: SILURIAN

At Girvan, the base of the Silurian succession rests with gentle unconformity on Ordovician strata in the coastal section to the south of the town, and in the Craighead Inlier to the north-east. At Woodland Point, on the coast, the angle of unconformity is about 8° and a significant part of the Ordovician Ashgill succession (the entire Drummuck Subgroup) is missing. In the Craighead Inlier the stratigraphical break is smaller but the unconformity is a little more abrupt. All of the Silurian strata are encompassed by the **Girvan Group**. The succession arose from two major shallow- to deeper-water transgressive cycles during the Llandovery, followed by a marine regression during the early Wenlock. This depositional pattern is reflected by the arrangement of the Girvan Group's three subgroups: Newlands, Dailly and Straiton (Figure 12). The lowest of these, the Newlands Subgroup, contains all of the units in the first of the major transgressive cycle and crops out in three distinct areas: in the coastal section, in the inland part of the Main Outcrop, and in the Craighead Inlier; there are subtle differences in lithofacies between the three areas. Strata deposited during the second of the major transgressive cycles make up the Dailly Subgroup which is fully developed in the inland part of the Main Outcrop, but elsewhere has only its basal division preserved in the Craighead Inlier. At the top of the Girvan Group, the Straiton Subgroup occurs only inland within the Main Outcrop and includes the strata deposited during marine regression and the transition to terrestrial conditions. A rich shelly fauna is preserved at a number of stratigraphical levels throughout the succession, with brachiopods and trilobites well represented (Plate 11).

Girvan coastal section

All of the Silurian strata in the coastal section lie within the Newlands Subgroup. The **Craigskelly Conglomerate Formation** is the oldest unit seen and rests on strata of the Ordovician Shalloch Formation with slight unconformity. It is a coarse conglomerate, variably clast- to matrix-supported, and contains interbedded sandstone lenses (Plate 12). The range of clast compositions is broadly similar to that seen in the Ordovician conglomerates occurring lower in the sequence. Radiometric (U-Pb) dating of detrital zircon crystals from the interbedded sandstone shows a largely bimodal population: Early Ordovician and Mesoproterozoic. The Craigskelly Conglomerate is succeeded by the **Woodland Formation**. This division features a basal member of massive, carbonate-rich sandstone beds that contain a shelly fauna of brachiopods and trilobites, a middle, more thinly bedded member, and an upper member of siltstone and graptolitic mudstone; the graptolites establish the *revolutus* (known until recently as *cyphus*) Biozone.

There is then a return to a coarse clastic facies in the overlying **Scart Grits Formation**, the youngest Silurian unit seen on the coastal section. It consists of massive, coarse-grained turbidite sandstone beds, with a distinctive quartz-pebble conglomerate at the base (the Quartz or Cow Rock Conglomerate) that has disturbed and channelled into the underlying strata of the Woodland Formation. The quartz-rich character of the conglomerate stands in marked contrast to the compositions of the conglomerates lower in the sequence.

Craighead Inlier

Newlands Subgroup

In the Craighead Inlier, the basal unit is the **Mulloch Hill Conglomerate Formation**, a grey-buff polymict rudite containing well-rounded pebbles but with a mid section dominated by sandstone showing hummocky cross-stratification. This feature suggests deposition in a relatively shallow environment within reach of storm wave base, a conclusion reinforced by the presence of a sparse, low-diversity shallow-water shelly fauna of crinoids and brachiopods. The conglomerate grades up into the overlying **Mulloch Hill Sandstone Formation**, a sequence of grey-green sandstone, with siltstone and mudstone interbeds, which again features hummocky cross-stratification in its lower part. It contains a diverse shelly fauna that is indicative of a shallow-water environment for much of the formation, but suggests a change to deeper water conditions towards the top. The Mulloch Hill Sandstone is succeeded in turn by the interlaminated siltstone and mudstone of the **Glenwells Shale Formation**, which may be turbiditic in origin. The lower part of the formation mostly comprises calcareous siltstone, whilst the upper part contains graptolitic mudstone proving the *revolutus* Biozone.

Another conglomeratic unit, the **Glenwells Conglomerate Formation**, overlies the Glenwells Shale. The Glenwells Conglomerate is very coarse-grained and poorly sorted with clasts over 8 cm in diameter, and contains sporadic interbeds of coarse sandstone. It probably originated as a channel-fill deposit in a deep-water, submarine fan setting. The conglomerate is succeeded by the **Newlands Farm Formation**, an ochreous-weathering unit of blue-grey calcareous siltstone and thinly bedded sandstone containing a rich and well-documented shelly fauna interpreted as a deep shelf community. Graptolitic, blue-grey mudstone follows and comprises the **Glenshalloch Shale Formation**, the graptolites indicating the *magnus* and possibly the *leptotheca* biozones, before a return to coarse sandstone comprising the **Saugh Hill Grits Formation**. In this unit, the constituent strata are greenish grey turbidites with a few thin mudstone interbeds and sporadic layers of pebble conglomerate. Overlying the Saugh Hill Grits, and at the top of the Newlands Subgroup, is the graptolitic mudstone of the **Pencleuch Shale Formation**, the graptolites proving the *convolutus* Biozone and possibly the *sedgwickii* Biozone.

Daily Subgroup

In the Craighead Inlier, only the lowest unit of the Dailly Subgroup is preserved at outcrop. It comprises the red-stained sandstone of the **Lower Camregan Grits Formation** which contains a sparse brachiopod fauna. Exposure of this unit is very limited at Craighead and its character is best described from the more extensive development in the Main Outcrop.

Main Outcrop

Newlands Subgroup

The Newlands Subgroup succession in the Main Outcrop is slightly less varied than in the Craighead Inlier, so that only three formations are recognised, roughly correlating to six or seven at Craighead (Figure 12). The lowest part of the subgroup to be seen at outcrop, the **Tralorg Shale Formation**, comprises a range of mudstones, some brown, some grey-green, some black and graptolitic with a fauna indicative of the *revolutus* Biozone. Brachiopods contained in concretions from the lower part of the sequence are of deep-water character. The Tralorg Shale Formation is faulted against Ordovician strata so that the stratigraphical base is not seen; it is succeeded abruptly by the **Saugh Hill Grits Formation**, which has a stratigraphically more extensive development than at Craighead (Figure 12). As there, the formation is made up of sandstone turbidites with thin interbeds of grey-green mudstone and sporadic conglomeratic layers. A conspicuous mudstone towards the middle of the formation has been separately defined as the Penwhapple Burn Shale Member, and can be tentatively correlated with the Glenshalloch Shale Formation which underlies the Saugh Hill Grits at Craighead. In the Main Outcrop, the proportion of mudstone increases towards the top of the Saugh Hill Grits Formation and graptolites recovered there prove the *triangulatus* or *magnus* biozones. Above the Saugh Hill Formation, and at the top of the Newlands Subgroup, lies the **Pencleuch Shale Formation**, a sequence of grey and black mudstone which is pyritic in its upper part and contains an abundant and diverse graptolite fauna. The graptolites are from the *gregarius, convolutus* and *sedgwickii* biozones, but their distribution suggests that this part of the succession has been tectonically shuffled.

Dailly Subgroup

Within the Main Outcrop, the Dailly Subgroup is made up of alternating formations dominated by either sandstone or mudstone. At the base, as in the Craighead Inlier, lies the **Lower Camregan Grits Formation**. It comprises medium- to thickly bedded turbidite sandstone with sporadic development of hummocky cross-stratification showing that deposition took place within storm wave base. A shelly fauna of shallow-shelf character is present at several levels, and though there is some evidence for this having been transported, such redistribution is likely to have been relatively local. The succeeding pale grey sandstone and siltstone of the **Wood Burn Formation** also features some examples of hummocky cross-stratification and contains shelly faunas, but in this case there is a stratigraphical variation in the fossils that suggests deepening shelf conditions. The Wood Burn Formation is abruptly succeeded by the distinctive purple mudstone of the **Maxwellston Mudstone Formation**. This unit was possibly deposited from distal, low density turbidite flows and contains graptolites indicative of the *guerichi* Biozone. There follows the **Upper Camregan Grits Formation**, comprising generally thickly bedded turbidite sandstone, and then the **Penkill Mudstone Formation**, a variable mudstone sequence with thin sandstone interbeds in the upper part. Graptolites are abundant in parts of the Penkill Formation, and establish a stratigraphical level spanning the *turriculatus* and *crispus* biozones (Figure 12).

Continuing into the upper part of the Dailly Subgroup, the **Protovirgularia Grits Formation** conformably overlies the Penkill Mudstone; it is named after a distinctive trace fossil found therein. Thickly bedded turbidite sandstone dominates, with a few mudstone interbeds. One of the latter, near the top of the formation, carries graptolites that define the *crispus–griestoniensis* biozonal boundary. The succeeding **Lauchlan Mudstone Formation** is a sequence of red and purple, fissile sandstone and mudstone, with a few dark and pyritic mudstone interbeds. A sparse, deep-water brachiopod fauna has been recorded near the top of the formation, as have rare graptolites of the *spiralis* Biozone. The establishment of this biostratigraphical level is important since the overlying unit of fissile sandstone–mudstone turbidites, the **Drumyork Flags Formation**, has a similar graptolite fauna, and also an acritarch flora, which confirm that the top of the Drumyork Formation also lies within the *spiralis* Biozone. Thus, more than 600 m of turbiditic strata were deposited in a time interval equivalent to less than a single biozone (Figure 12), perhaps only half a million years or even less. In the succeeding **Blair Shale Formation**, a unit of thinly bedded grey to brown sandstone and mudstone, graptolites from the lower part of the formation confirm the *spiralis* Biozone. There is no faunal evidence for the age of the upper part of the Blair Shale, and the top is not seen since the formation is faulted against younger strata of the Straiton Subgroup.

Straiton Subgroup

The Straiton Subgroup occurs only in the north-east part of the Main Outcrop and includes two units recording marine regression and the onset of terrestrial deposition. The lower of the two units, the **Knockgardner Sandstone Formation**, is a sequence of thinly bedded, grey-green sandstone, laminated mudstone and siltstone, and sporadic thick sandstone beds. Some of the thick sandstones carry hummocky cross-stratification, indicating deposition within reach of storm wave base, and the relatively shallow depositional environment is confirmed by the presence of a brachiopod fauna of shallow shelf aspect. An early Wenlock age is suggested by the brachiopods and supported by a sparse acritarch flora. The Knockgardner Formation is succeeded by greenish, coarse-grained sandstone forming the lower part of the **Straiton Grits Formation**. This sandstone contains a sparse fauna of ostracods and bivalves, along with a low diversity acritarch flora, that suggests deposition in a very shallow marine or lagoonal to fresh-water setting. The upper part of the formation consists of unfossiliferous, coarse red sandstone and conglomerate with a clear terrestrial character. They are the youngest Silurian strata to be seen in the various Girvan outcrops and are unconformably overlain by Carboniferous beds of the Inverclyde Group.

three

Ordovician and Silurian of the Southern Uplands

A broad expanse of Ordovician and Silurian strata underlies the rolling hills of the Southern Uplands, forming a distinctive geological terrane that extends westward across Northern Ireland and into the Irish Republic. The terrane formed as an accretionary thrust complex at the Laurentian continental margin during Late Ordovician to mid Silurian subduction of the Iapetus Ocean. It was built up by a series of southward-propagating, imbricate thrust faults that structurally repeated an oceanic sequence. In the north of the terrane, the older fault-defined tracts consist of a thin basal assemblage of black, graptolitic mudstone (**Moffat Shale Group**) — in a few places accompanied by chert and basaltic lava (**Crawford Group**) — overlain by a very much thicker accumulation of turbiditic sandstone (Frontispiece) deposited in a series of huge submarine fans. The turbidite deposits dominate the succession such that thousands of metres of sandstone beds may overlie only a few metres of Moffat Shale Group mudstone. In the south of the terrane, the younger tracts do not contain a basal mudstone unit. Either the floor thrust of the accretionary complex climbed to a higher stratigraphical level, or oceanic mudstone was not deposited in the foreland basin setting that was the likely depositional environment for the younger turbidite successions (Figure 4).

As the submarine fans built out from the Laurentian continental shelf they encroached onto progressively younger levels of the oceanic sequence that was continually approaching the continental margin as the oceanic plate was subducted. As this happened, sections of the sedimentary package (and occasionally vestiges of the subjacent volcanic rocks) were intermittently stripped from the subducting oceanic plate and thrust beneath the stack of similar stripped-off slices that made up the growing accretionary complex (Figure 4). Occasionally, oceanic volcanoes were caught up in this process so that in a few places masses of lava are interleaved with the turbidites. Though the accretionary thrusts were originally at a relatively low angle, they were subsequently steepened, partly by the growth of the accretionary complex and partly by subsequent tectonic events, and now appear as near-vertical, major strike faults running from north-east to south-west and separating tracts of steeply inclined and north-east-striking beds (Figure 14).

Each tract has an internal sense of younging towards the north whereas the minimum age of each tract decreases southwards. Further, the time interval represented by the Moffat Shale Group increases southward through successive tracts showing that the onset of turbiditic sedimentation occurred progressively later southward (Figure 15). Although the sandstones comprise the vast majority of the succession, those in any one tract are either of the same biostratigraphical age as, or only slightly younger than, the youngest part of the underlying Moffat Shale Group; very close biostratigraphical control is provided by locally abundant graptolite faunas (Figures 5 and 6). The Crawford Group may range down to the Arenig whilst the overlying Moffat Shale Group is restricted to the Caradoc in the northern tracts but spans the Caradoc to Llandovery interval in the southern tracts. The age of the turbidite successions follows the top of the Moffat Shale Group in becoming younger southwards: Caradoc and Ashgill in the north of the terrane in the **Tappins, Barrhill** and **Scaur groups**, Llandovery (and early Wenlock locally) in the **Gala, Ettrick** and **Hawick groups**, and entirely Wenlock in the **Riccarton Group** which comprises the southernmost tracts.

For over 100 years, the region has been divided, for convenience, into three parts: a **Northern Belt**, consisting of the Ordovician sandstones (the Leadhills Supergroup — though in its original usage the 'Northern Belt' did not include the Tappins Group) with inliers of Moffat Shale Group and Crawford Group; a **Central Belt** of Llandovery sandstones, again with inliers of Moffat Shale and (rarely) Crawford Group rocks; and a **Southern Belt** of Wenlock sandstones without inliers of Moffat Shale or Crawford groups. The Northern and Central belts remain useful general concepts, with the boundary recognised as the Orlock Bridge

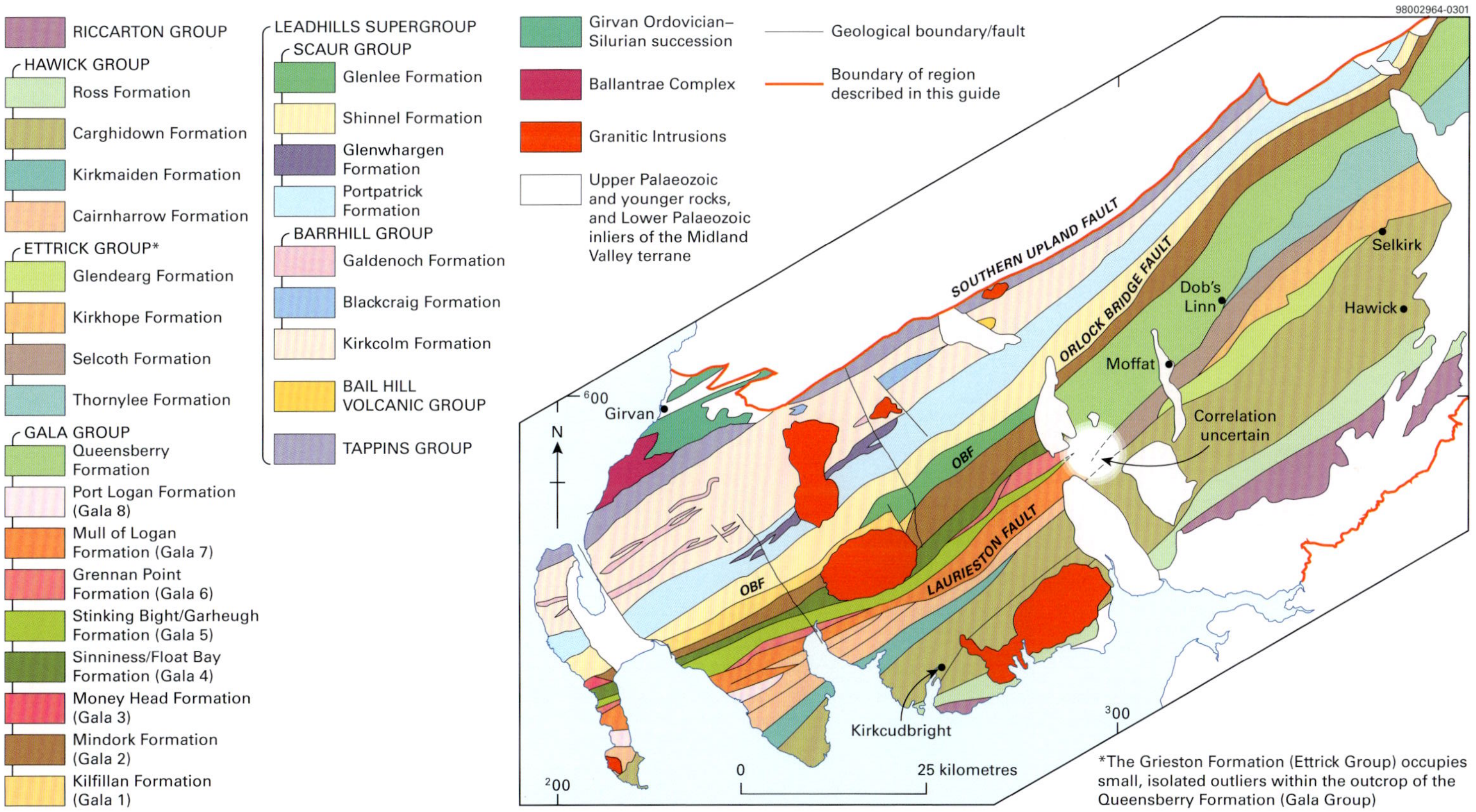

Figure 14 *Generalised representation of the principal Ordovician–Silurian structural tracts of the Southern Uplands terrane, which form the basis of the regional lithostratigraphy. For chronostratigraphical correlation see Figure 15.*

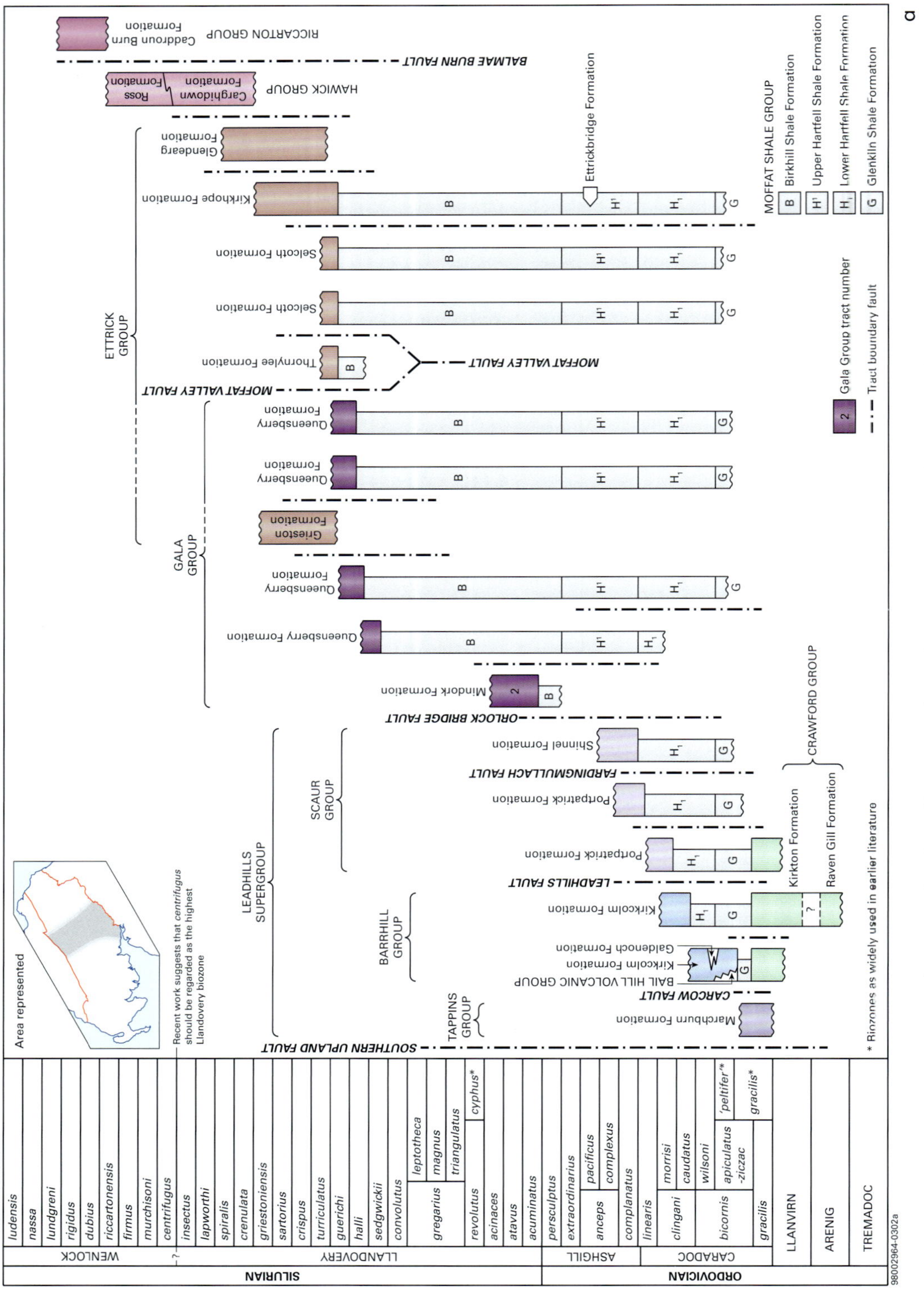

Figure 15 Representative structural-stratigraphical profiles across the Ordovician–Silurian Southern Uplands terrane: a) north-east Southern Uplands, b) south-west Southern Uplands. Additional details, particularly for the Queensberry, Selcoth, Ross and Raeberry Castle formations, are discussed and illustrated later in the chapter.

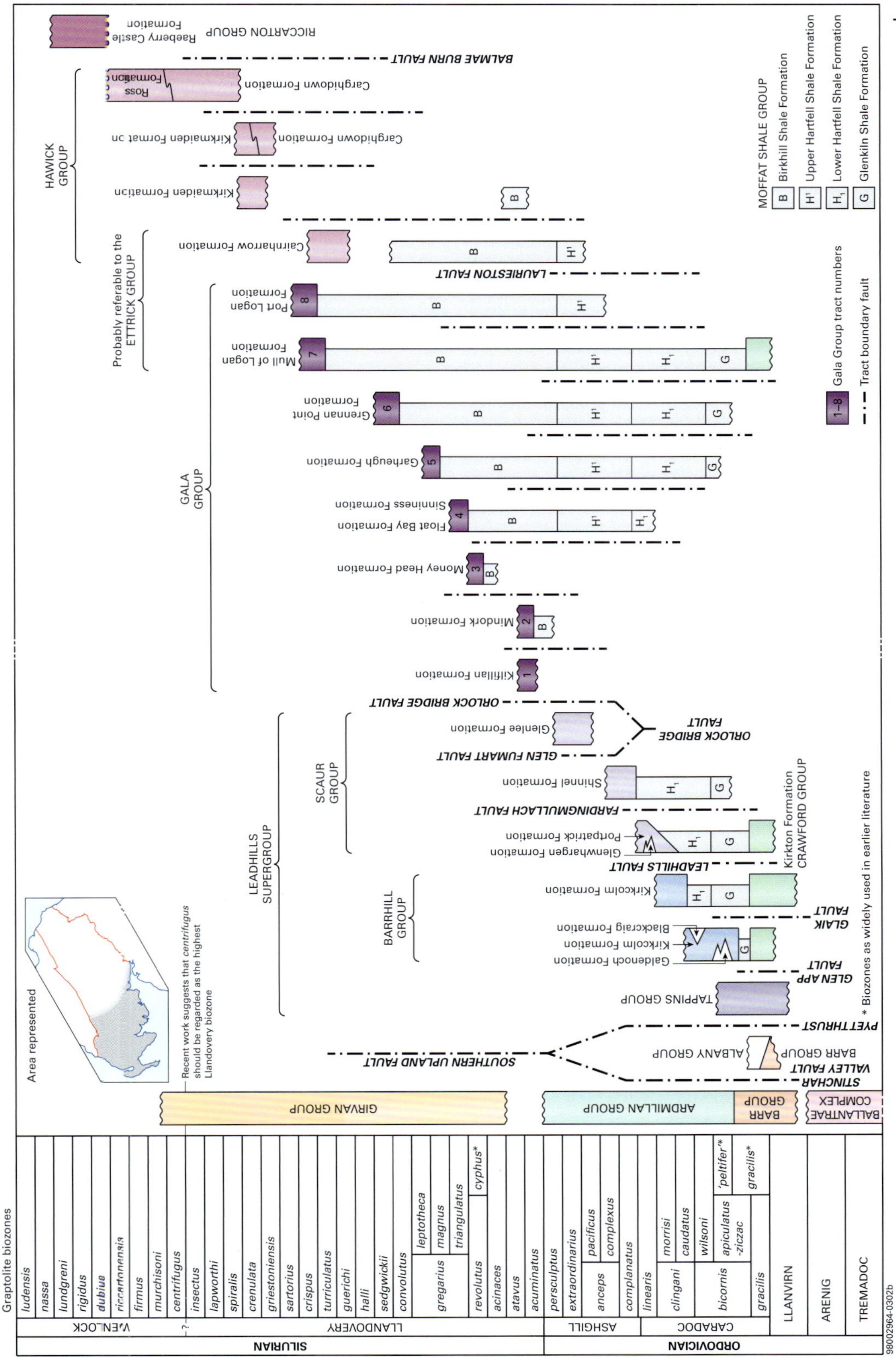

Figure 15 continued.

Fault, but with the recognition that Central Belt tracts extend up into the Wenlock, the distinction between it and the Southern Belt has diminished.

DEPOSITIONAL ENVIRONMENT

The thin successions of oceanic, pelagic mudstone and chert that form the Crawford Group and Moffat Shale Group accumulated very slowly across the broad expanse of the Iapetus Ocean, laid down on the volcanic rocks of the oceanic crust. Vastly more extensive volumetrically are the thick sandstone successions deposited in submarine fan systems adjacent to the Laurentian continental margin, with turbidity currents as the principal depositional agents as described in Chapter 1. There are three dominant lithofacies: channel-fill sequences composed of massive sandstone, coarse grained, thickly bedded and locally associated with intraclast (sandstone) breccia and conglomerate; lobe or unconfined sheet-flow deposits composed of well-bedded, fine- to medium-grained sandstone with subordinate mudstone; and interchannel or interlobe deposits composed of siltstone and silty mudstone. These lithofacies are interbedded in vertical alternations that range from a few tens of metres to several hundred metres in thickness, but they also pass laterally from one to another on a scale of hundreds of metres to several kilometres. Most of the sandstones contain a high proportion of fine-grained matrix enclosing ill-sorted grains and so can be classified as wackes.

Individual beds may display a complete or partial array of the ideal turbidite features. If substantial turbidity flows closely follow each other, a thick sequence may build up in which graded sandstone beds become amalgamated, with only subtle variations in grain size determining the margins of each bed. In more distal parts of the fan, or as a result of small, low-density flows, thin and fine-grained sandstone beds may be pervasively laminated and separated by thicknesses of silty mudstone, the finest-grained part of the submarine fan succession deposited in areas that were spatially or temporally removed from most of the depositional activity. Where the base of a turbidite sandstone bed overlies the fine-grained top of its predecessor, it is usually sharp and clearly defined (Plate 6) though groove and flute casts are common on many of the bed bases (Plate 7). If the bed is imagined as being rotated back to the horizontal, the orientations of the groove and flute casts indicate the flow direction of the original turbidity current (the palaeocurrent), which also influenced the geometry of the ripples that show up in section as cross-lamination. This becomes an important interpretational tool when linked with the distinctive compositions of the sandstones forming some tracts, particularly those of the Leadhills Supergroup, which indicate derivation from a range of different provenance regions. Additionally, within the Leadhills Supergroup, the compositional differences are not only on a tract scale, but also appear within tracts as the interbedding of differently sourced sandstone units. This complexity has led to an extra level in the Ordovician stratigraphical hierarchy, relative to that applied to the Silurian rocks to the south, but that should not be taken to imply any fundamental change in tectonic and sedimentary processes at the Ordovician–Silurian boundary. However, to ensure clarity, the different levels of lithological variation do require slightly different structures to be adopted for the descriptive sections in the following account.

On a regional scale, compositional contrasts between the different tracts are sufficiently strong to be reflected in the geochemistry of the stream sediment eroded from their sandstones. Contoured maps of element concentration in stream sediment commonly show steep gradients coincident with or parallel to the tract boundary faults: two examples are shown in Figure 16. The dominant feature in the distribution of strontium (Figure 16a) is a marked decrease in its abundance across a narrow zone coincident with the Laurieston–Moffat Valley Fault. The change probably reflects variation in the plagioclase to K-feldspar ratio in the detrital grain population: more detrital plagioclase in the sandstones to the north-west of the fault, more detrital K-feldspar in the sandstones to the south-east. The distribution of chromium (Figure 16b) is dominated firstly by the effects of the mafic and ultramafic igneous rocks forming the Ballantrae Complex, and secondly by a broad zone of very high values running

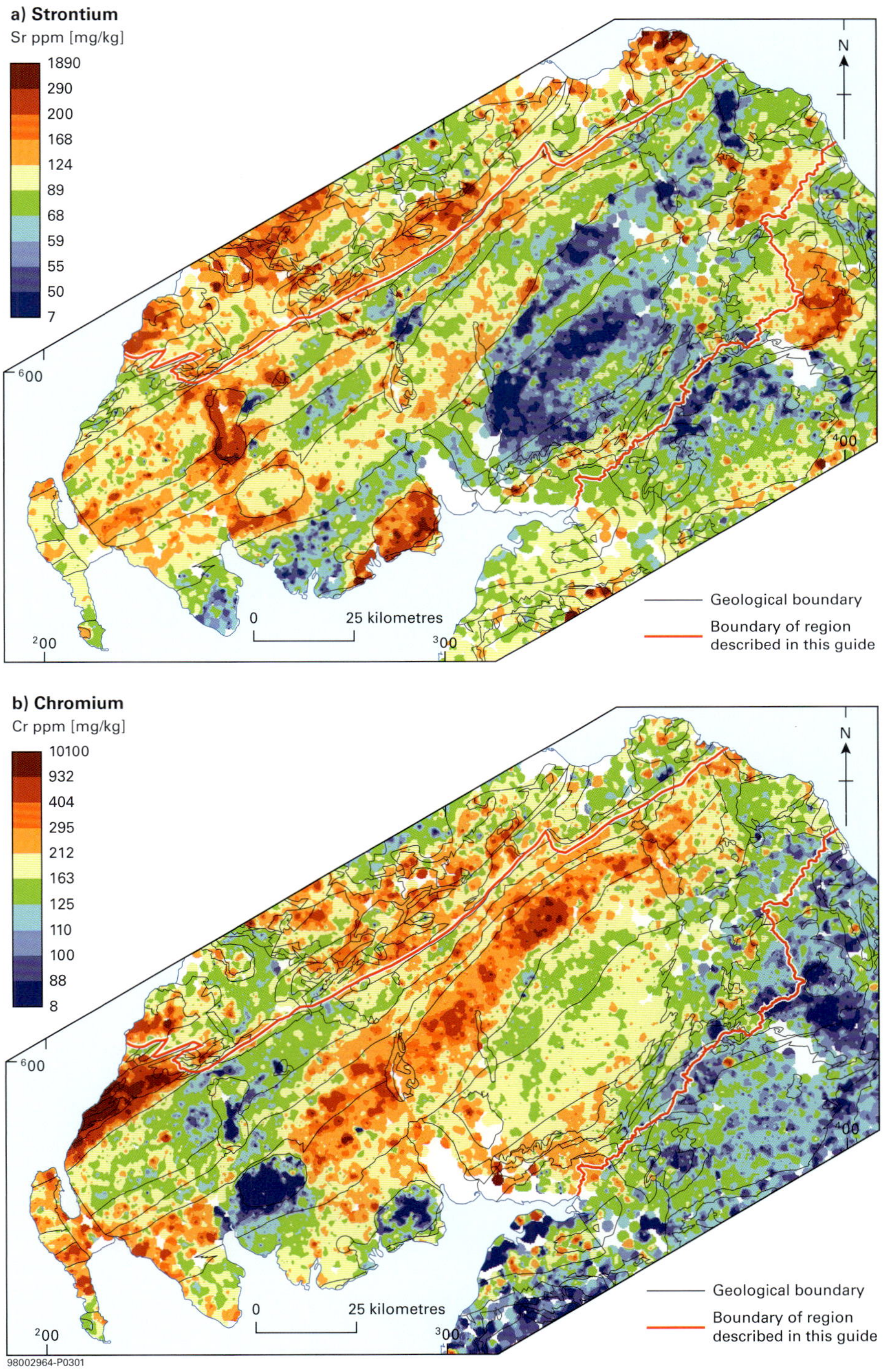

Figure 16 *Regional geochemical images showing the distribution of (a) strontium (Sr) and (b) chromium (Cr) in the fine-grained fraction of stream sediment across the Southern Uplands terrane and adjacent areas. Values are given in parts per million (ppm). No data is available for those areas left uncoloured.*

parallel to, but north-west of, the Laurieston–Moffat Valley Fault. Detrital chrome-spinel in the sandstones underlying this part of the Southern Uplands is the likely cause of this second anomalous zone. In detail, and from multi-element integrations, the regional, stream sediment geochemistry suggests an intensity of cryptic imbrication that has not been fully resolved by field studies. Another useful property that has been exploited in the differentiation of the Leadhills Supergroup tracts is the magnetic susceptibility of the different sandstones, which varies with the proportion of iron-rich minerals that they contain.

CRAWFORD GROUP: ORDOVICIAN

The oldest rocks found in the Southern Uplands belong to the Crawford Group and are of late Arenig to Llanvirn age. They comprise the dominantly volcanic and chert successions of the Raven Gill and Kirkton formations, which underlie the Moffat Shale Group in the Northern and Central belts (Figure 15).

The **Raven Gill Formation** (55–70 m) consists of an interbedded succession of basaltic pillow lavas, blue-grey radiolarian cherts and fossiliferous brown mudstones, with sheet-like dolerite intrusions or lava flows. The stratigraphical limits of the formation are unclear since its southern margin (presumed base) is faulted and no overlying unit is seen, though it is possible that it may be enclosed within the Kirkton Formation (see below) as an olistostrome unit. The brown mudstones at the Raven Gill type locality [NS 9204 1989] have yielded fragmentary graptolites and a conodont assemblage of late Arenig age. Only three proven outcrops are known, all within a strike length of about 2 km in the Abington area. There they form part of an extensive linear belt of black shale and chert (mostly Moffat Shale Group) inliers, up to 1.8 km wide across strike, which crop out within the imbricate Leadhills Fault Zone. These inliers are interpreted as the thrust-faulted repetition of a single, thin basal succession that may have originally been in conformable contact with the (overlying) Kirkcolm Formation.

Pillow lavas and chert are exposed at several other localities within the Leadhills Fault Zone, as well as in association with the analogous Fardingmullach Fault farther south, and have previously been assumed to correlate with those at Raven Gill. However, recent studies have shown that there is considerable variation in the geochemical characteristics of the lavas. For example, samples from the Leadhills Fault Zone at Abington [NS 9295 2141] exhibit the characteristics of mid-ocean-ridge basalts (MORB), whereas those from the Fardingmullach Fault Zone at Gabsnout Burn [NX 2020 6116] near Glenluce are chemically more like island-arc basalts. It seems likely that the lavas underlying the Moffat Shale Group were erupted during several distinct and unrelated volcanic episodes.

The **Kirkton Formation** (>120 m) is a sequence of red and grey cherts and red and green siliceous mudstones with chert nodules that appears to lie between the Raven Gill Formation and the Glenkiln Shale (Moffat Shale Group) within the Leadhills Fault Zone. Similar cherts crop out at other localities throughout the Northern Belt along the tract-bounding faults including, for example, at Normangill Burn [NS 971 241] north-east of Abington, at numerous localities in the Leadhills area, and at Morroch Bay [NX 017 524] south of Portpatrick on the Rhins of Galloway. North of Portpatrick, adjacent to the Glaik Fault [NW 976 588] the chert is associated with hyaloclastite containing broken lava pillows. At many of these localities the cherts yield conodont faunas of late Llanvirn and early Caradoc age. Farther south, the Kirkton Formation is represented by volcaniclastic rocks and chert, underlying *apiculatus-ziczac* Subzone mudstone, that crop out adjacent to the Laurieston Fault in a linear zone extending for about 7 km north-eastward from Crossmichael [NX 735 670].

Curiously, there is a complete absence from the Crawford Group of any evidence for the well-established conodont zones intervening between the upper Arenig and the upper Llanvirn. This absence, together with the restricted occurrence of the Raven Gill Formation and its uncertain relationship with the Kirkton Formation, suggests that the Raven Gill Formation may simply be one or more large olistostromes within the Kirkton Formation.

MOFFAT SHALE GROUP: ORDOVICIAN – SILURIAN

The classical black mudstone succession of the **Moffat Shale Group** was first examined in detail by Charles Lapworth as a part of the extensive and far-sighted stratigraphical work that he undertook around Moffatdale during the early 1870s. Graptolites provide a basis for the biostratigraphical subdivision of the sequence and the faunal biozones established by Lapworth remain the foundation of the modern scheme. Lapworth's Glenkiln, Lower and Upper Hartfell and Birkhill Shale divisions are retained as formations in modern terminology, with the first three being entirely Ordovician in age. Although previously regarded as wholly Silurian, the basal part of the Birkhill Shale is now referred to the Ordovician following the assignment of the *Glyptograptus persculptus* Biozone to that system. The Birkhill Shale therefore spans the Ordovician – Silurian system boundary and indeed includes, at Dob's Linn [NT 196 158], 15 km north-east of Moffat, the Global Stratotype Section and Point (GSSP) for the base of the Silurian (Frontispiece). A selection of the characteristic graptolites that help define the stratigraphy is shown in Figures 5 (Ordovician examples) and 6 (Silurian examples).

The Moffat Shale Group is best developed in its type area of Moffatdale where it spans the Upper Ordovician and Lower Silurian beneath the overlying Llandovery sandstones. Traced towards the north-west, the Moffat Shales are replaced by the wacke sandstone turbidites at progressively lower stratigraphical levels in successive fault-bounded tracts (Figure 15). In the most northerly tract, occupied by the Llanvirn to Caradoc Tappins Group, no true Moffat Shale is seen, the lithofacies being represented instead by black laminae within a turbidite sandstone succession that rests directly on red chert and lava, likely correlatives of the Kirkton Formation.

The **Glenkiln Shale Formation** (about 25 m) is the oldest and least well-known formation of the Moffat Shale Group and consists of at least two units of cherty, black graptolitic mudstone, separated and succeeded by grey mudstones and siltstones. The graptolites establish the lower Caradoc, *apiculatus-ziczac* Subzone, the lower subzone of the *bicornis* Biozone. Though originally named after exposures in the Glenkiln Burn [NY 007 898], near Ae, a type section for the base of the Glenkiln Shale Formation has recently been proposed to the north of the Leadhills area where a stratigraphical contact with underlying grey chert (Kirkton Formation) is exposed in Gripps Cleuch [NS 883 171]; the chert contains conodonts indicative of a late Llanvirn to early Caradoc age.

The overlying **Lower Hartfell Shale Formation** (22 m) consists of dark grey and black pelagic mudstone with an abundant graptolite fauna. As with the Glenkiln Shale, and for the same reasons of limited exposure, more than one section is required to illustrate the different parts of the formation. At Hartfell, soft black mudstone with grey cherty ribs is referred to the *wilsoni* Subzone, the upper subzone of the *bicornis* Biozone, and is overlain by black siliceous mudstone of the *clingani* and *linearis* biozones. The younger part of the Lower Hartfell Shale Formation is also well preserved at Dob's Linn where mudstone from the upper part of the *clingani* Biozone is overlain by softer black mudstones containing *linearis* Biozone graptolites. White beds of metabentonite, representing devitrified volcanic ash, first appear at Hartfell as thin beds and laminae (<2 cm) in the *clingani* Biozone of the Lower Hartfell Shale, but become thicker and more numerous in the succeeding Upper Hartfell Shales and Birkhill Shales (Plate 13).

In contrast to the dark and richly fossiliferous beds of the Lower Hartfell Shale Formation, the mudstones of the **Upper Hartfell Shale Formation** (28 m) are pale grey and virtually unfossiliferous apart from a few isolated black bands, from 1–50 mm thick, containing graptolites. Its type section at Dob's Linn ranges from the *complanatus* Biozone up to the *extraordinarius* Biozone (Figure 15). Although metabentonite layers are present at Dob's Linn in the *complanatus* Biozone, they only start to become common in the succeeding *anceps* Biozone where at least 18 have been recorded (Figure 17). They range up to 20 cm, form 4 per cent of the total thickness and originated as fine-grained volcanic ash, composed of glass shards, fragmented crystals and lithic clasts (up to 0.7 mm), altered in contact with seawater prior to low grade metamorphism. Zircon crystals from one of the *anceps* Biozone bentonites at Dob's Linn have given a U-Pb radiometric age of 445.7 ± 2.4 Ma.

Plate 13 *Thin, pale-coloured layers of volcanic ash (metabentonite) interbedded with mudstone of the Moffat Shale Group at Dob's Linn, Moffatdale (P220194).*

An unusual feature is seen in the Upper Hartfell Shale Formation in its outcrop at Ettrickbridge [NT 390 243], where one of the southernmost inliers of Moffat Shale Group strata in the Central Belt underlies Ettrick Group (Kirkhope Formation) strata. There, about 55 m of coarse, wacke sandstone are interbedded within the mudstone sequence at about the level of the *anceps* Biozone. This sandstone unit is now referred to the **Ettrickbridge Formation** (Figure 15a). As a sandstone turbidite unit, the Ettrickbridge Formation has obvious links with the Leadhills Supergroup and it has generally been included within that division. Alternatively, but less commonly, it is regarded as a local arenaceous member of the Upper Hartfell Shale and so forms part of the Moffat Shale Group. The sandstone is texturally a wacke and though quartz-rich it also contains numerous grains of lava, metamorphic rocks and feldspar, but no ferromagnesian minerals. The most unusual feature of the sandstones is that many of the larger quartz grains (those >0.3 mm) are well rounded, some even being spherical, in marked contrast to the irregular angular quartz grains seen in every other formation in the Leadhills Supergroup. In comparison with the latter, the Ettrickbridge quartz grains have obviously been through at least one additional cycle of mechanical abrasion/erosion. Curiously, it is only the larger quartz grains that are well rounded, all other clasts, including the smaller quartz grains, being quite angular.

Only the basal 1.6 m of the **Birkhill Shale Formation** (40 m), that part lying within the *persculptus* Biozone, is of Ordovician age. At Dob's Linn these beds are black pyritous mudstones with several white bentonite interbeds and a thin development of grey green mudstone at the base. Above this, the Silurian part of the Birkhill Shale Formation is composed predominantly of grey to black, variably laminated mudstone that yields graptolites representative of every Silurian biozone from *acuminatus* to *halli*. In the more southerly outcrops the mudstone

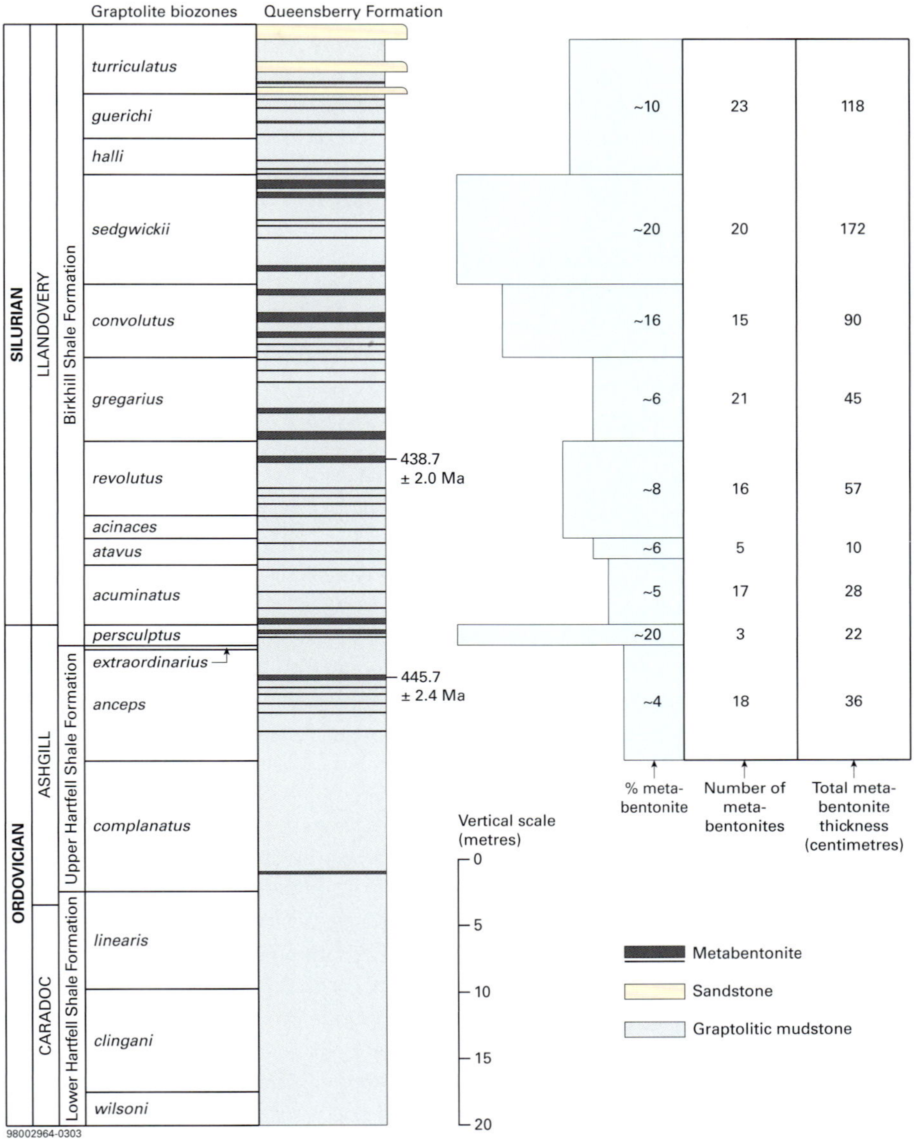

Figure 17 *Distribution of volcanic ash (metabentonite) bands in the Moffat Shale Group at Dob's Linn, Moffatdale. After Merriman and Roberts (1990).*

passes up into a few metres of grey siltstone that is unfossiliferous at Dob's Linn, but farther south includes black mudstone laminae that yield graptolite faunas of the *guerichi* Biozone. Pale metabentonite layers are conspicuous throughout the succession, generally as laminae to thin beds but with a few reaching 50 cm in thickness. They typically comprise 5–10 per cent of the total succession, but this proportion rises to 20 per cent in the *convolutus* to *sedgwickii* biozones (Figure 17). Zircon crystals from one of the *revolutus* Biozone bentonites at Dob's Linn have given a U-Pb radiometric age of 438.7 ± 2.0 Ma. The geochemistry of the bentonites shows a range of ash types, from subalkaline to peralkaline, that were generated in an ensialic volcanic-arc and, possibly, in a backarc setting.

Sandstone interbeds are rare in the Birkhill Shale Formation but occur at a few localities and at various levels, for example: in Wigtownshire, at Culroy [NX 255 540] a few metres of succession at the local top of the Moffat Shale Group (where it underlies the Gala 4 tract — see Figure 15b) spans the *acuminatus*, *atavus* and *acinaces* biozones and part of the *revolutus* Biozone, and is composed of sandstone with mudstone interbeds; at Dob's Linn, thin sandstone beds occur interbedded with the *revolutus* and *convolutus* Biozone mudstones (in the succession underlying the southernmost Queensberry Formation tract — see Figure 15a). These sandstone beds, and the Ettrickbridge Formation described above, probably originated when distal turbidite deposition extended unusually far out across the oceanic sequence.

LEADHILLS SUPERGROUP: ORDOVICIAN

The lithostratigraphical characterisation of the sedimentary tracts forming the Leadhills Supergroup is based on the marked variation in the compositions of the detrital material forming the constituent grains of their sandstones, a phenomenon reflecting differences in the provenance of the original sediment. So, for example, some formations prove to be rich in quartzo-feldspathic material in contrast to others that contain abundant volcanic detritus (Table 2). These compositional variations may be accompanied by differences in the directions of the palaeocurrents deduced from the sedimentary features of individual beds, emphasising the geographical separation of the different provenance regions. Some tracts feature the interbedding of differently sourced sandstone units.

The Ordovician sandstone groups and associated volcanic rocks that make up the Leadhills Supergroup mostly crop out between the Orlock Bridge Fault and the line of the Stinchar Valley Fault/Southern Upland Fault. The one exception is the Ettrickbridge Formation, which occupies a slightly anomalous position farther south, interbedded with mudstone of the Moffat Shale Group. The Leadhills Supergroup incorporates the volcanic Downan Point Lava Formation and other lavas in the Tappins Group, as well as the substantial pile of lower Caradoc lavas and pyroclastic rocks of the Bail Hill Volcanic Group.

Tappins Group

The Tappins Group encompasses the most northerly structural tracts in the Southern Uplands. It lies immediately south of the Southern Upland Fault, and is the oldest division in the Leadhills Supergroup (Figures 14 and 15). It comprises the Traboyack, Downan Point Lava, Currarie, Dalreoch, Corsewall and Marchburn formations, each of which is a discrete tract, separated from its neighbours by major strike faults. The southern boundary of the group is formed by the Glen App and Carcow faults. Limited biostratigraphical (graptolite and conodont) evidence indicates an early Caradoc age (*gracilis* to *apiculatus-ziczac* biozones) for all of the component formations but, unlike the other units in the Northern Belt, they do not rest on Moffat Shale Group strata. Hence, no formational bases can be defined. Most of the sandstones were derived from the north and north-west and are immature; they contain a high proportion of igneous detritus, much of it likely to have been derived from ophiolitic rocks similar to those forming the Ballantrae Complex (Table 2).

Red and purple mudstone and sandstone characterise the **Traboyack Formation** (c.1000 m?), which forms the northernmost sedimentary tract in the Southern Uplands terrane and so most probably contains the oldest strata. The formation crops out on the south side of the Stinchar valley to the west of Barr, between the Stinchar Valley Fault to the north and the Dove Cove Fault to the south, but westwards it is structurally imbricated with the Downan Point Lava Formation. All of the margins to the outcrop are faulted so that no base or top to the formation is known. The Traboyack Formation is rich in ophiolitic detritus and as a result is the most highly magnetic of the sedimentary tracts in the Southern Uplands.

Basaltic pillow lavas of the **Downan Point Lava Formation** (c.1000 m?) crop out on the coast at Downan Point, south of Ballantrae, where the pillow structures, confirmation of submarine eruption, are strikingly well developed (Plate 14). At the coast, the formation occupies all of the ground between the Stinchar Valley and Dove Cove faults, but eastwards the outcrop narrows and the lavas are structurally intercalated with the Traboyack Formation. Although formerly assumed to equate with the superficially similar lavas of the Ballantrae Complex, the Downan Point lavas are now regarded as a separate unit within the Southern

Table 2 *Provenance characteristics of turbidite sandstones from the Ordovician Leadhills Supergroup.*

Group	Formation	Detrital lithic material	Mineral grains
Scaur	Glenlee (andesitic)	Andesite, basalt, rhyolite, quartzite, chert	Quartz (24%), feldspar, augite, hornblende
Scaur	Glenlee (quartzose)	Basalt, rhyolite, quartzite, chert	Quartz (37%), feldspar
Scaur	Shinnel	Granodiorite, granophyre, rhyolite, microdiorite, basalt, quartzite, mudstone	Quartz (57%), feldspar, apatite, zircon
Scaur	Glenwhargan	Quartzite, mica-schist, microdiorite, basalt, chert	Quartz (67%), feldspar, zircon
Scaur	Portpatrick	Pyroxene/hornblende andesite, dacite, rhyolite, basalt, crossite- and garnet-schists, gabbbro, diorite, granite, mudstone	Quartz (15%), feldspar, augite, hornblende, garnet, crossite, spinel, zircon
Barrhill	Blackcraig	Granite, gabbro, rhyolite, basalt, quartzite, chert	Quartz (33%), feldspar, epidote, augite, hornblende
Barrhill	Galdenoch	Pyroxene/hornblende andesite, dacite, rhyolite, basalt, chlorite-schist	Quartz (18%), feldspar, augite, hornblende, apatite
Barrhill	Kirkcolm	Quartzite, basalt, garnet- and mica-schists	Quartz (45%), feldspar, garnet, zircon
Tappins	Corsewall	Granodiorite, quartz-porphyrite, chlorite-schist, microgranite, andesite, gabbro, basalt, serpentinite, chert	Quartz (14%), feldspar, apatite, augite, hornblende, spinel, epidote, biotite
Tappins	Marchburn	Basalt, serpentinite, chert	Quartz (15%), feldspar, hornblende
Tappins	Dalreoch	Serpentinite, basalt, chert, rhyolite	Quartz (26%), feldspar
Tappins	Traboyack	Serpentinite, basalt, chert, andesite, dolerite, gabbro	Quartz (2%), feldspar, epidote, chlorite

98002964-T0301

Uplands Terrane. Apart from a rather imprecise Sm-Nd age date of 468 ± 22 Ma there is little direct evidence for the age of the Downan Point succession. The geochemistry of the lavas suggests their eruption at a within-plate, ocean island volcano. Lavas at Currarie Port and Portandea, previously correlated with those at Downan Point, are of slightly different composition and so could represent an entirely separate phase of volcanic activity.

The **Currarie Formation** (90–300 m) has a restricted outcrop on the coast north of Currarie Port [NX 056 780]. It consists of red and grey-green mudstone and chert with interbedded masses of lava breccia, whilst irregular blocks of pillow lava set in a red siliceous mudstone matrix, that occur at several localities, have been interpreted as olistostromes. Graptolites obtained from the green mudstone are referred to a level near the base of the *gracilis* Biozone. The formation rests on lavas, previously correlated with the Downan Point Lava Formation, but of apparently different composition as mentioned above. Sandstone turbidites overlying the formation on the coast near Currarie Farm are thought to correlate with the Dalreoch Formation (see below). Although the presence of red mudstone in the Currarie Formation is reminiscent of parts of the Traboyack Formation, the magnetic signatures of the two formations are quite different.

The **Dalreoch Formation** (c.1000 m) is a major unit of dark green, turbiditic sandstone with its type section in the Water of Gregg near Barr [NX 294 930]. Its outcrop lies to the south of the Dove Cove Fault and to the north of the Glen App Fault, and it has a magnetic signature intermediate between the adjacent Traboyack (to the north) and Corsewall (to the south, see below) formations. Poorly exposed red mudstone and basalt lava which crop out along the southern margin of the formation may correlate with the lithologically similar Currarie Formation and its underlying lavas.

Formerly known as the the Corsewall Group, the **Corsewall Formation** (c. 1700 m) has its type section at the northern tip of the Rhins of Galloway [NW 982 728] where the spectacular Corsewall Point Conglomerate Member is well exposed on the coast (Plate 15).

Plate 14 *Basaltic pillow lavas from the Ordovician Downan Point Lava Formation (Tappins Group) exposed on the coast at Sgavoch Rock, to the south of Ballantrae (P005827).*

It contains boulders up to 1.5 m across and was derived from the north. The clast suite in the conglomerates is very varied. Granitic lithologies are the most abundant, with porphyritic microdiorite, spilitic lava, sandstone and chert also well represented. Various ages have been obtained from radiometric dating of the granitic boulders, some as old as Mesoproterozoic, but with the most reliable suggesting intrusion in the Early Ordovician. This would require rapid uplift and erosion in the provenance area, to the north, and invites comparison with the situation seen in the slightly older Benan Conglomerate Formation of the Girvan–Ballantrae succession (see Chapter 2). Elsewhere, much of the Corsewall Formation consists of sandstone turbidites, but there are numerous other coarse rudite horizons, including the named Finnarts and Glen App Conglomerate members. The latter unit forms the north-eastern limit of the formation at the Water of Tig, 5 km east of Ballantrae, beyond which the tract pinches out so that eastwards the Dalreoch Formation is faulted against the northernmost tract of the Barrhill Group. Graptolites collected from the Corsewall Formation on the coast north of Finnarts Bay [NX 048 728] indicate that the strata lie within the *gracilis* or *apiculatus-ziczac* biozones.

The **Marchburn Formation** (c. 1300 m), with a type area about 5 km east of New Cumnock [NS 674 130] mostly comprises dark green quartz-poor sandstone turbidites, but with a heterogeneous mixture of other, associated lithologies including red and grey bedded chert, lava and the distinctive 'Haggis Rock'. The latter consists of a green sandy microconglomerate containing a colourful mix of red, grey, green and black lithoclasts of chert, mudstone and dacitic and basaltic lavas. At Ruddenleys [NT 2025 5067] near Leadburn, red cherty mudstone within the Marchburn Formation has yielded early Caradoc conodonts, whilst closely associated dark mudstone yields *gracilis* to *apiculatus-ziczac* Biozone graptolites of about the same age. The red cherty beds almost certainly correlate with the lithologically similar cherts of the Kirkton Formation within the Leadhills Fault Zone.

Plate 15 *The spectacular boulder conglomerate that forms the younger part of the Corsewall Formation (Tappins Group) as exposed at Corsewall Point [NW 982 727],at the northern end of the Rhins of Galloway (P008481).*

In the Leadburn area and along strike to the south-west, pillow lavas are interbedded with the red cherts and cherty mudstones of the Marchburn Formation. Although it is rarely possible to demonstrate clear relationships, these lavas have been defined as the **Noblehouse Lava Member** (c.100 m), with a type section at Grassfield Quarry [NT 1947 4990], 7 km south-west of Leadburn. Geochemically, the lavas have the characteristics of within-plate volcanic rocks, showing affinity to both alkaline and tholeiitic oceanic island basalts.

Barrhill Group

The Barrhill Group consists of the turbidite sandstone units that crop out in a series of tracts between the Glen App/Carcow and Leadhills faults: the Kirkcolm, Galdenoch and Blackcraig formations. Together with differences in age, there are marked differences in the compositions (and hence provenance) of the three formations (Figure 18). The Kirkcolm Formation sandstones consist mainly of quartzo-feldspathic grains, with the Proterozoic and Archaean ages of accessory zircon grains suggesting their original derivation from an ancient Laurentian basement; garnet is another important detrital accessory. The Galdenoch and Blackcraig formations contain much additional igneous detritus, but with a background of quartzo-feldspathic material similar to that forming the Kirkcolm Formation, and with a similar Proterozoic–Archaean population of detrital zircons. Palaeocurrent indicators from the Kirkcolm Formation show that turbidity flow was highly variable (Plate 7) but with a weak overall bias for derivation from the northern quadrant; the evidence is more ambiguous for the Galdenoch and Blackcraig formations.

The thick and extensive **Kirkcolm Formation** (c. 3000 m) occupies an across-strike outcrop width of up to 16 km in the western part of the Southern Uplands (Figure 14). It was originally erected as the Kirkcolm Group in its type area on the Rhins of Galloway and can thence be correlated north-eastwards along strike for almost 170 km as far as the Leadburn area. The formation is dominantly composed of well-bedded quartzose sandstone turbidites (Plate 16) that are sporadically interbedded with thick units of grey laminated siltstone; the sandstone becomes conglomeratic in a few places. Two distinct tracts comprise the formation, separated by

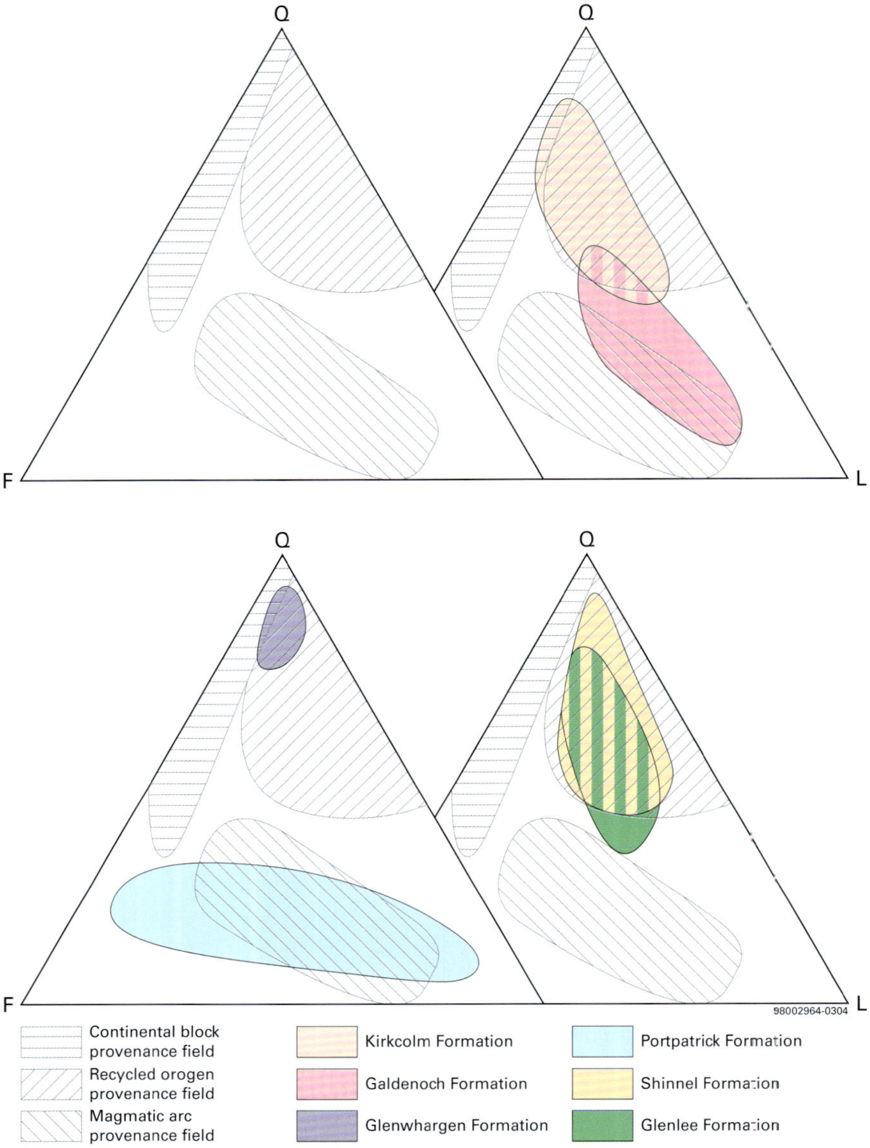

Figure 18 *Compositional variation of sandstones within the Ordovician Barrhill and Scaur groups, Leadhills Supergroup. The triangular plots illustrate the relative abundance of quartzose grains (Q), feldspathic grains (F) and unstable (labile) lithic or mineral grains (L). The corner points of the triangles correspond to 100 per cent Q, F or L, as indicated. Provenance fields from Dickinson and Suczek (1979) Plate tectonics and sandstone compositions.* Bulletin of the American Association of Petroleum Geologists, *Vol. 63, 2164–2182.*

the Glaik Fault. There is a considerable difference in age between the north-western and south-eastern tracts, ranging from the *apiculatus-ziczac* Subzone in the north-west to the *clingani* Biozone in the south-east (Figure 15). However, though the major change in age occurs across the Glaik Fault, in the absence of Moffat Shale Group inliers this structure cannot be accurately delineated along the full length of the Kirkcolm Formation's outcrop. The resulting uncertainty is compounded by the composition and general lithology of the sandstone being very similar in both the north-western and south-eastern tracts despite the difference in their ages.

Plate 16 *Steeply inclined (slightly overturned) turbidite strata of the Ordovician Kirkcolm Formation, as exposed on the west coast of the Rhins of Galloway at Salt Pans Bay [NW 962 620] (P008483).*

Where inliers of Moffat Shale Group strata are seen in association with the Glaik Fault, graptolite faunas prove only the *apiculatus-ziczac* Subzone, i.e. the Glenkiln Shale: examples occur at Knockingarroch [NX 5574 9666] near Carsphairn and at Cowieslinn [NT 2354 5126], 4 km south of Leadburn. At several localities in the 12 km wide zone between the Glaik and Glen App faults, black laminae within the grey siltstone units of the Kirkcolm Formation also yield a graptolite fauna referable to the *bicornis* Biozone, though at Dounan Bay [NW 965 688], at the northern end of the Kirkcolm Formation's type section on the Rhins of Galloway, the fauna in graptolitic interbeds near the top of the Kirkcolm Formation may be from its *wilsoni* Subzone. In contrast, at the southern end of the Rhins of Galloway section (also in Nithsdale and the Peebles area), the Kirkcolm Formation rests on mudstone of the Lower Hartfell Shale Formation ranging in age up to the *clingani* Biozone. The mudstone and underlying Moffat Shale Group strata form inliers in the hanging wall of the Leadhills Fault Zone, which forms their southern margin. No graptolitic interbeds are known in the overlying turbidite succession, which is truncated to the north by the Glaik Fault. However, conglomerates within the Kirkcolm Formation in the vicinity of Kilbucho [NT 056 336] contain derived shelly faunas with ages equivalent to a range from the upper *apiculatus-ziczac* to the lower *clingani* graptolite biozones. The fossils — trilobites, brachiopods, gastropods and corals (Plate 11g) — are shallow-water species carried down from the continental shelf in masses of slumped sediment. Similar forms are seen in situ to the north of the Southern Upland Fault in Upper Ordovician successions at Girvan (see Chapter 2) and in Northern Ireland, which might imply that any lateral movement on the fault — and substantial sinistral displacement is a much-discussed possibility — amounted to no more than a few hundred kilometres.

On the north-east side of Bail Hill, the northern, older outcrop of the Kirkcolm Formation includes two unusual variations which have been separately defined as members: the conglomeratic Spothfore Member and the partly volcaniclastic Stoodfold Member. The conglomeratic **Spothfore Member** (300–800 m) can be traced for about 5 km along strike. All clasts of pebble size (4 mm) and upwards are of intrabasinal origin, with sedimentary rock types constituting 95 per cent of the total and volcanic rocks, including porphyritic lavas, the remaining 5 per cent. The largest proven clasts are boulders of sandstone and siltstone up to 2 m in diameter, though large discontinuous bodies of chert, possibly up to tens of metres across, may represent gravity-emplaced olistoliths. The Spothfore Member conformably overlies laminated siltstone of the Kirkcolm Formation but no upper contact is seen since the member is truncated to the north by a strike-parallel fault. The **Stoodfold Member** consists of at least four units of lithic and crystal-rich volcaniclastic sandstone related to the Bail Hill Volcanic Group (described below) but interbedded with sandstones and siltstones of the Kirkcolm Formation. Whilst there is little doubt that their origin was volcanic, they are interbedded within a turbidite sequence and themselves exhibit the classical suite of turbidite characteristics such as graded bedding, scoured and eroded bases, mudstone rip-up clasts and parallel and cross lamination. The volcaniclastic beds probably aggregate to no more than a few tens of metres in total thickness.

Another conglomeratic facies variation of the northern Kirkcolm Formation is the substantial **Carsphairn Conglomerate Member** (40–450 m). It has a matrix that is petrographically similar to the enclosing Kirkcolm Formation and crops out in the vicinity of Carsphairn as a series of lenticular bodies that represent cross-sections through original channel-fill sequences. Angular pebbles of vein quartz, up to 6 cm in diameter, are the most abundant clasts in the conglomerates, but most of the larger blocks consist of calcareous siltstone or sandstone, which weather to give a carious 'rottenstone' appearance. Well-rounded rhyolite clasts are also quite common and although most are less than 10 cm across, the largest boulder seen (38 x 20 cm) is of this lithology. Local development of similar, albeit less-coarse, conglomerate is seen at approximately the same stratigraphical level elsewhere in the Kirkcolm Formation outcrop.

The **Galdenoch Formation** (up to about 500 m) is a turbidite unit, the component sandstone of which is characterised by a significant content of andesitic detritus. The andesitic material occurs both as lithoclasts of pyroxene and/or hornblende andesite, and as individual detrital grains of pyroxene and hornblende. The combined volcaniclastic contribution may form as much as 30 per cent of the total grains, the remainder being mainly quartzo-feldspathic. Magnetic susceptibility is uniformly high relative to that of the Kirkcolm Formation sandstone. The Galdenoch Formation interdigitates with the older, northern division of the Kirkcolm Formation to form three main members each consisting of several sequences of volcaniclastic beds ranging up to an aggregate thickness of several hundred metres. One such sequence, though only about 25 m thick, can be reliably traced for over 10 km along strike in the Barrhill area. Although most extensively developed to the west of the River Nith, the characteristic andesite-rich sandstones of the Galdenoch Formation also crop out near Abington [NS 9182 2203], and south of Leadburn [NT 2491 5296]. These eastern outcrops confirm that the formation is represented across the full width of the region.

The **Blackcraig Formation** (c. 1500 m) forms a lenticular wedge interfingering with the older, northern division of the Kirkcolm Formation and is best seen in Glen Afton, south of New Cumnock [NS 628 055]. There, spectacular massive gritty sandstone and boulder conglomerate form one of the most distinctive turbidite units in the Southern Uplands. Although a conformable junction with the underlying Kirkcolm Formation sandstone can be seen in Craig Burn [NS 6398 0562], the upper boundary to the Blackcraig Formation is obscured by faulting. The sandstone and the sandy matrix of the conglomerate have a distinct green colour due to abundant detrital epidote and other ferromagnesian minerals. In the conglomerate, boulders of granite, gabbro and both acid and basic lavas are supported in a sandy matrix and the formation probably represents a large channel-fill deposit on the upper reaches of a submarine fan.

Bail Hill Volcanic Group

In the area north of Sanquhar, the Bail Hill Volcanic Group crops out as a heterogeneous sequence of lavas, pyroclastic and intrusive rocks. It forms the largest single area of volcanic rocks in the Northern Belt (c. 4 km^2) and probably represents the remains of an oceanic seamount that built up on a floor of Glenkiln Shale Formation strata in early Caradoc times. Laterally, the group interfingers with sandstones of the Kirkcolm Formation.

Although it superficially resembles a pyroclastic rock, the **Cat Cleugh Formation** (c. 150 m) consists of autobrecciated basaltic lavas containing characteristic large euhedral pyroxenes. It represents the basal unit of the volcanic complex and appears to be in conformable contact with underlying black mudstone of the Moffat Shale Group that contains an *apiculatus-ziczac* Subzone graptolite fauna. The **Peat Rig Formation** (c. 1000 m) is lithologically diverse and forms the larger part of the complex. It includes highly porphyritic autobrecciated lava, lithic tuff and agglomerate of hawaiite/mugearite composition. Phenocrysts in the lava include feldspar (oligoclase/andesine), amphibole (pargasite), apatite and biotite/phlogopite. The Cat Cleugh Formation and the lower part of the Peat Rig Formation are cut by the **Bught Craig Breccia**, which is interpreted as the fill of a small volcanic vent that fed into the upper part of the Peat Rig Formation.

The volcanic rocks of the Bail Hill Volcanic Group are alkaline in character, ranging from alkali basalt to trachyandesite, with the whole-rock geochemical characteristics and enrichment patterns of oceanic, within-plate basalt. Although geochemically distinct and the largest single component, the group forms part of a mixed assemblage of tholeiitic and alkaline, within-plate lavas that are distributed through the Northern Belt of the Southern Uplands, in places intercalated with the sandstone sequences. This assemblage is thought to record a period of extension and within-plate volcanism coincident with the early stages of development of the Southern Uplands accretionary complex.

Scaur Group

The Scaur Group includes all of the sandstone turbidite units that crop out in the Northern Belt tracts between the Leadhills and Orlock Bridge faults. These are the Portpatrick, Glenwhargen, Shinnel and Glenlee formations. The sandstones from the Glenwhargan and Shinnel formations are mostly quartzo-feldspathic, with a similar Proterozoic–Archaean population of likely Laurentian detrital zircons to that found in the Barrhill Group. The Portpatrick Formation sandstones are quite different (Figure 18), as described below, with a large volcanic component and an accessory population of Neoproterozoic detrital zircon grains. The Glenlee Formation is a rather more mixed assemblage. Palaeocurrent indicators show that turbidity flow was predominantly from the north and north-east, except in the Portpatrick Formation where they show the opposite trend, indicating flow from the south and south-west.

The dominantly volcaniclastic **Portpatrick Formation** (c. 2000 m) is one of the most distinctive formations in the Southern Uplands with its characteristic lithology of medium- to thick-bedded, commonly massive, dark blue-grey sandstone turbidite. Detrital andesitic grains are very common throughout the sandstone sequence, usually accompanied by remarkably fresh detrital pyroxene and hornblende; these 'mafic' components may form as much as 20 per cent of the grain population, most of the remainder being quartzo-feldspathic. Despite the apparent freshness of the andesitic detritus, implying penecontemporaneous volcanism, erosion and deposition, radiometric dating of detrital minerals has given Neoproterozoic results. Ages derived from Ar-Ar dating of detrital hornblende, and U-Pb dating of zircon grains, cluster around 560 Ma with no sign of a younger component. This age is closer to that of widespread magmatism on the southern, Avalonian margin of the Iapetus Ocean, than it is to the ages of large-scale magmatism on the Laurentian margin to the north. Coupled with the consistent palaeocurrent evidence for derivation of the Portpatrick Formation sandstones from the south and south-west, this would imply a very different provenance to that from which most other Southern Uplands sandstones were sourced. The presence of detrital blue amphibole is another unusual feature, though the detrital garnet grains have the same characteristics as those seen in the Kirkcolm Formation.

The base of the Portpatrick Formation is defined by the inliers of Moffat Shale Group mudstone along the Fardingmullach Fault, which underlie the turbidite sandstone. The age of the mudstone generally ranges up to the *linearis* Biozone, although there is evidence of a somewhat older (*clingani* Biozone) base to the unit in the Rhins of Galloway. Graptolite-bearing mudstone interbeds occur sparingly, for example at Killantringan Bay where the *linearis* Biozone has been established. There is no defined top to the formation, which is truncated along its northern margin by the Leadhills Fault Zone.

Interfingering within the Portpatrick Formation, the **Glenwhargen Formation** (0–500 m) is a unit of highly quartzose, wacke and arenite sandstones, locally conglomeratic, which is restricted to the western half of the Southern Uplands. At many localities, the Glenwhargan Formation is represented by only a few thin beds of pale grey, quartz-rich sandstone sandwiched between much thicker sequences of darker, partly volcaniclastic Portpatrick Formation sandstone. Several excellent sections display the relationships between the two formations: near Killantringan [NW 985 555] and at Port of Spittal Bay [NX019 521], respectively 1.5 km north-west and 2 km south-east of Portpartick on the Rhins of Galloway; and at Knockville Moor [NX 355 732], 9 km north-west of Newton Stewart, in an area where the Glenwhargen Formation reaches its thickest development and is also conglomeratic in places.

The **Shinnel Formation** (c. 2000 m) crops out between the Fardingmullach and Glen Fumart/Orlock Bridge faults, with its type section along the Scaur Water [NS 787 003] about 10 km north of Moniaive. Although generally dominated by quartz-rich, turbidite sandstone, the Shinnel Formation is characterised throughout its entire outcrop length by thick sequences of laminated grey siltstone. In the Leadhills area where the siltstone is locally the dominant lithology, the informal name 'Lowther Shales' has been applied. A distinct bed-parallel cleavage is commonly developed in these siltstones and they have consequently been worked as an inferior roofing 'slate', for example at the now disused Stobo Quarry [NT 158 365], about 7 km west-south-west from Peebles. Graptolite faunas from mudstone interbeds at several localities within the Shinnel Formation outcrop are referable to the *anceps* Biozone, whilst the base of the formation rests on Lower Hartfell Shale Formation strata containing faunas up to the *linearis* Biozone.

In the Tweeddale area, the Wrae Limestone and the 'Tweeddale Lavas' are exotic breccias of, respectively, limestone and lava clasts that are interbedded within the Shinnel Formation. They are included within the **Tweeddale Member**, which extends for over 20 km along strike south-west of Peebles. It is about 35 m thick in its type section at Wrae [NT 1175 3240] but appears to thicken to over 400 m in the area north of Peebles. The limestone contains a variety of shelly fossils and conodonts that suggest an early Caradoc age yet the member is enclosed within the Shinnel Formation of *linearis* Biozone age or younger, a substantial difference in age that supports interpretation as a submarine slide deposit. The clasts of lava are of peralkaline rhyolite, an unusual lithology most likely to have been erupted from an oceanic island volcano. The shallow-water limestone may have formed in fringing reefs around the volcano, with masses of both lava and limestone intermittently breaking away and slumping down into deeper water.

The **Glenlee Formation** (c. 2000 m) crops out between Thornhill and New Galloway and is the youngest Ordovician turbidite formation of the Northern Belt. It is bounded to north and south by the Glen Fumart and Orlock Bridge faults respectively. Much of the formation consists of sandstone, although a substantial laminated siltstone member, with an outcrop up to 1 km wide, can also be traced for more than 20 km along strike. The sandstone-dominated parts of the formation include sequences rich in detrital pyroxene and/or hornblende as well as others in which ferromagnesian minerals are entirely absent and the sandstones are quartzo-feldspathic. The pyroxenous beds are restricted in distribution and crop out only to the south of (so probably beneath) the siltstone member. The latter contains sporadic black laminae containing graptolite faunas referable to the upper part of the *anceps* Biozone and possibly the *persculptus* Biozone.

GALA GROUP: SILURIAN

The Gala Group is composed of the sandstone-dominated strata that crop out between the Orlock Bridge Fault and, to the south, the faulted boundary with the Ettrick Group. The outcrop is divided by strike-parallel faults, marked by discontinuous outcrops of the underlying Moffat Shale Group, into tracts that range from a few hundreds of metres to a few kilometres in outcrop width. In the Southern Uplands of south-west Scotland the age of the sandstone becomes progressively younger in successive tracts southwards (Figure 15), with an overall range spanning the lower to middle Llandovery *acuminatus* to *guerichi* biozones. In the central and north-eastern parts of the Southern Uplands the earlier and later parts of the group are repeated by faulting but the middle part is missing. The Orlock Bridge Fault has traditionally been seen as one of the more important Southern Uplands strike faults since it was thought to separate sandstone of Ordovician and Silurian age. This view is partly an artifact of the long-established 'Northern Belt versus Central Belt' terminology and, at best, the biostratigraphical break across the Orlock Bridge Fault appears to be no greater than that across most of the other tract boundary faults (Figure 15).

The orientation of palaeocurrent indicators suggests that sediment transport during deposition of Gala Group sandstones was predominantly towards the south-west. Sandstone composition is typically quartzo-feldspathic, with quartz forming up to 55 per cent and feldspar (plagioclase and K-feldspar in varying proportions) 20–30 per cent of the rock. Mica generally forms a minor component but is more abundant in fine-grained sandstone. Pyroxene and amphibole occur sporadically in the older formations where, together with andesitic and basaltic lithic debris, they may very rarely comprise as much as 20 per cent of the sandstone. Other lithic grains comprise clastic sedimentary, felsic igneous, and spilitic volcanic rocks, the latter becoming more abundant in the younger part of the group. A particular feature brought out by the regional geochemical distribution of chromium in stream sediment (Figure 16b) is the abundance of detrital chrome-spinel in the sandstones that make up the older Gala Group tracts. Grains of garnet, tourmaline, zircon and epidote are present in the sandstones forming all of the tracts.

Mudstone–siltstone units are commonly interbedded with the Gala Group sandstones and represent fine-grained elements of the submarine fan succession deposited in areas sheltered from sandstone deposition. The mudstone and siltstone may dominate the succession locally. In the older tracts, dark grey graptolite-bearing mudstone commonly forms units up to 10 m thick, and locally up to 80 m, which are effectively interbedded, lateral equivalents of the pelagic Birkhill Shale Formation. In younger tracts, massive or laminated siltstone dominates units that are tens, and locally hundreds, of metres thick and which contain only a small proportion of medium- to coarse-grained sandstone in thin beds.

As in the Leadhills Supergroup, the variation in age of adjacent Gala Group tracts (albeit based on weak biostratigraphical control in some cases), coupled with subtle variations in the character of their component sandstone successions, has encouraged treatment of each fault-bounded tract as a separate stratigraphical unit. Formation names have been applied in some parts of the outcrop where an adequate degree of lithological distinction can be recognised, but these are difficult to extend with any confidence through the large areas with little exposure. The tops of units thus defined are taken at tract-bounding faults, their formal bases at their contact with the underlying Moffat Shale Group, though at outcrop both top and bottom of 'formations' are commonly faulted with no Moffat Shale Group strata preserved. Consequently, a 'tectonostratigraphical' numbering scheme (Gala 1–8) was adopted in the south-west of the Southern Uplands (Table 3) and extrapolated north-eastward into the area east of the Cairnsmore of Fleet pluton and the Thornhill Basin. Farther north-east, in the central

Table 3 Correlation between Gala Group tract numbers and lithostratigraphy, and their likely relationship to the Ettrick Group.

	SOUTH-WEST SOUTHERN UPLANDS: Tract numbers Wigtownshire and Rhins of Galloway	Formation names	CENTRAL AND NORTH-EAST SOUTHERN UPLANDS	
GALA GROUP	8	Port Logan	ETTRICK GROUP	
GALA GROUP	7	Mull of Logan	ETTRICK GROUP	
GALA GROUP	6	Grennan Point	Queensberry Formation	GALA GROUP
GALA GROUP	5	Stinking Bight/Garheugh	Garheugh Formation / Not present	GALA GROUP
GALA GROUP	4	Float Bay/Sinniness	Not present	GALA GROUP
GALA GROUP	3	Money Head / Not present in Wigtownshire	Not present	GALA GROUP
GALA GROUP	2	Mindork	Mindork Formation	GALA GROUP
GALA GROUP	1	Kilfillan	Not present	GALA GROUP

98002964-T0302

and north-eastern Southern Uplands, repetition of lithologically and biostratigraphically similar rocks across a number of tracts led to the use there of a different suite of formation names, but again the top boundaries are invariably faulted. Overall then, the difficulties of correlation are compounded by the loss of some tracts as faults merge, whilst other parts of the succession expand across several, relatively narrow tracts as faults bifurcate. Hence, though eight units with formational status are recognised on the Rhins of Galloway (Figure 15b), in central to north-eastern parts of the Southern Uplands the Gala Group is divided into only two formations (Figure 15a); correlation is shown in Table 3.

Stratigraphical framework

The older parts of the Gala Group are best preserved in south-west Scotland (Figure 15b) where much of the succession is exceptionally well exposed on the west coast of the Rhins of Galloway and to the east of Luce Bay. The **Gala 1** tract is composed of massive and bedded sandstone facies locally named the **Kilfillan Formation** (up to 2000 m). Sparse units of dark grey to black mudstone interbedded with the sandstone facies range up to 10 m thick. Slumped units up to 80 m thick, comprising blocks of wacke and mudstone in a mudstone or siltstone matrix, are exposed in the north of Luce Bay [NX 199 153] and Penkiln Burn [NX 413 669]. A restricted graptolite fauna suggests a level in the *acuminatus* Biozone. Farther east, the Gala 1 tract is lost as its boundary faults merge.

The outcrop of the **Mindork Formation** (up to 1000 m) extends from Luce Bay eastwards across the full length of the Southern Uplands into the north of the Lammermuir Hills. In Wigtownshire it forms the **Gala 2** tract, but east of the Cairnsmore of Fleet pluton it is the northernmost of the Gala Group units preserved and locally spans at least two structural tracts. The component sandstone is generally well bedded and is petrographically distinguished by pyroxene and amphibole crystal debris together with andesitic and basaltic lithic clasts. Locally the mafic volcanic material forms up to 20 per cent of the rock (e.g. at Mindork Fell [NX 321 583]), but its proportion varies widely and the sandstone commonly contains no discernible volcanic material. There is an increase in lithological variation north-eastward with mudstone, and locally conglomerate, units appearing laterally along strike. The conglomerates, for example the **Raeshaw Conglomerate** near New Channelkirk [NT 484 555], contain clasts of igneous rock and Ordovician Moffat Shale Group mudstone. Graptolite faunas from the interbedded mudstones are mostly long ranging and of low diversity but are broadly indicative of the *acuminatus* Biozone, with one fauna indicative of the middle part of that zone.

The Mindork Formation also forms the northernmost preserved tract on the Rhins of Galloway, although there, andesitic detritus is additionally seen in the lower part of the **Money Head Formation** in the **Gala 3** tract (about 900 m). This latter division is restricted to the western side of the Rhins and generally comprises thickly bedded (up to 5 m) sandstone with intervals of laminated siltstone up to 8 m and several beds up to 3 m thick of intraclast slump breccia. A sedimentary contact with the underlying Moffat Shale Group is exposed at Cairnweil Burn [NX 085 495] and Strandfoot [NX 052 482] where graptolites suggestive of the *acinaces* Biozone provide a maximum age for the Money Head Formation.

The **Gala 4** (up to 1500 m) tract, preserved only to the west of the Cairnsmore of Fleet pluton, is composed of interbedded sandstone and siltstone, with intervals of silty mudstone up to 30 m thick occurring in a few places. The tract is locally termed the **Sinniness Formation** in Wigtownshire and the **Float Bay Formation** in the Rhins of Galloway. The uppermost 200 m of the Float Bay Formation is dominated by laminated siltstone. Soft sediment deformation is evident in a number of places ranging from slump folding of thin wacke interbeds in silty mudstone to complete disruption of bedding, locally forming mélange units from a few metres up to 70 m thick. The age of the succession in the Gala 4 tract is well constrained by graptolite faunas. The basal sedimentary transition from the Moffat Shale Group exposed at Culroy [NX 255 540] yields *revolutus* Biozone faunas, and mudstone beds within the turbidite sequence contain faunas that indicate a range from the *revolutus* Biozone into the *triangulatus* Biozone.

The **Gala 5** tract (up to 900 m) is well exposed in the Rhins of Galloway and east of Luce Bay in Wigtownshire where it is locally named the **Stinking Bight beds** and the **Garheugh Formation** respectively. It is composed of alternating thinly bedded and more massive turbidite sandstone units (Plate 17), locally with interbedded laminated siltstone and mudstone. In Wigtownshire, one such interbedded unit near Rocks of Garheugh [NX 268 501] is 10 m thick and includes very thin beds of red mudstone and bentonite. Sedimentary breccia, composed of sandstone clasts in a coarse-grained sandstone matrix, occurs sporadically in

Plate 17 *Thickly bedded turbidite sandstone beds from the Silurian (Llandovery) Garheugh Formation (Gala Group) as exposed at Rocks of Garheugh [NX 263 507] on the north-east side of Luce Bay (P008401).*

the Garheugh Formation in units up to a few metres thick, and in Wigtownshire is associated with a unit of matrix- and clast-supported conglomerate up to 30 m thick, intermittently exposed for 6 km along strike. Individual clasts are up to 30 cm in diameter and are mostly well-rounded cobbles of crystalline quartz or quartz-arenite. Other clast lithologies are black silty mudstone (one example yielding Ordovician brachiopod fragments), laminated sandstone, rare granodiorite and a highly ferruginous rock. These accessory clasts are up to a few centimetres in size and vary from angular to well rounded. Deposition of the conglomerate in a system of anastomosing channels seems most likely.

Between the Cairnsmore of Fleet granite and the Permian Thornhill Basin, a medium- to very thick-bedded sandstone sequence assigned to the Gala 5 tract includes units of matrix- and clast-supported, cobble to boulder conglomerate. The conglomerate beds are up to 300 m thick and are lenticular over 1–2 km of strike length, e.g. at Castramon Hill [NX 780 835]. Composed of generally well-rounded clasts of quartz, quartz arenite, andesite, limestone, chert and mudstone, these conglomerate units are probably equivalent to that in the Garheugh Formation in Wigtownshire.

The Gala 5 turbidite succession in the south-west of the Southern Uplands is largely unfossiliferous, though one fauna from an interbedded mudstone in the Dalmacallan Forest [NX 816 885] indicates the *triangulatus* Biozone or younger. However, farther west, the maximum age is constrained by a *magnus* Biozone fauna from underlying mudstone of the Moffat Shale Group at The Hooies [NX 068 446] on the Rhins of Galloway. In the north-east of its outcrop, the Garheugh Formation includes an exceptional conglomeratic unit that crops out north-east of Peebles near Fountainhall [NT 423 501]. There, a sequence of thickly bedded sandstones includes a laterally discontinuous body, up to 500 m thick, of intraformational pebbly sandstone known locally as the **Dyker Law Conglomerate Member**; a *triangulatus* Biozone age has again been determined from graptolites in a rare mudstone interbed. It may be that the Garheugh Formation becomes diachronously younger to the south-west.

In the Rhins of Galloway, the **Grennan Point Formation (Gala 6)** (300–600 m) is well-exposed above several tectonic repetitions of the basal part of the succession and the top of the underlying Moffat Shale Group at Drumbreddan Bay [NX 075 437]. A few metres of grey mudstone and laminated siltstone occur at the transition, overlain by a succession that mostly comprises sandstone but includes thick units of laminated siltstone, the latter sporadically including thin beds of pale grey and red mudstone. A *convolutus* Biozone fauna has been obtained from the base of the formation at Grennan Point [NX 075 439] and in Drumbreddan Bay. In Wigtownshire, a lenticular sandstone tract in an analogous structural position extends south of the Cairnsmore of Fleet granite. Farther east the stratigraphical relationships become more complicated, with anastomosing tract-boundary faults resulting in multiple tracts of similar composition and overlapping, mid Llandovery age; this has previously led to interpretations involving an apparent structural reversal of the Gala 5 and 6 tracts. All of these mid Llandovery tracts are probably equivalent to the Queensberry Formation as exposed to the east of Thornhill (see below and Table 3).

In the south-west of the Southern Uplands, the Gala Group outcrop extends southward with the **Gala 7** tract (which locally bifurcates) comprising the **Mull of Logan Formation** (about 1800 m). The southern part of the outcrop is largely composed of well-bedded wacke sandstone with mudstone intervals varying from thin laminae to units 40 cm thick. Very thick (<2 m) sandstone beds occur dispersed throughout the succession and locally, for example north-east of Elrig [NX 331 484], even thicker units of massive sandstone are evident. The northern part of the outcrop, well exposed north from the Mull of Logan on the west coast of the Rhins of Galloway and inland on the western side of the Wigtown peninsula, is composed of well-bedded sandstone with interbedded units of laminated siltstone, generally less than 15 m thick but rarely up to 100 m. Sparse graptolite faunas from the mudstone range across the *guerichi*, *turriculatus* and *crispus* biozones. Beds of laminated red mudstone up to

3 m thick occur locally towards Dumfries, but only thin laminae of red mudstone are present in the south-western part of the tract.

The northern part of the Gala 7 tract includes sporadic and laterally discontinuous units of massive sandstone up to several hundred metres thick, with individual beds recognisable up to 8 m thick, and spectacular intraformational breccia deposits most probably emplaced as debris flows. The breccia is composed of angular to subrounded clasts of siltstone and sandstone, variably matrix- or clast-supported in a sandstone matrix. In the Rhins of Galloway, breccia clasts range up to 10 m across in a 550 m-thick breccia-dominated succession. East of Luce Bay, breccia with clasts from a few centimetres up to 50 cm in diameter forms two units, 150 m and 200 m thick, separated by about 350 m of thickly bedded sandstone. In both areas the breccia units die out laterally within a few kilometres.

As originally conceived, the tract sequence of the Gala Group extended to a **Gala 8** tract, or **Port Logan Formation** (about 800 m), with an outcrop restricted to the Rhins of Galloway and western Wigtownshire (Figure 14). It is largely composed of wacke-type sandstone beds with mudstone intervals ranging from thin laminae to beds 40 cm thick. On the Rhins of Galloway, several thicker mudstone units, up to 135 m thick, include laminated siltstone and thin-bedded sandstone and, locally, chert layers with carbonate laminae. The transition from the Moffat Shale Group into the sandstone succession is exposed at Clanyard Bay [NX 101 380] with faunas from the mudstone ranging up to the *sedgwickii* Biozone. Graptolites from mudstone interbeds within the sandstone succession range from the *guerichi* to the *crispus* biozones.

The regional relationships of the southern part of the Mull of Logan Formation in the Gala 7 tract, and also of the Port Logan Formation in the Gala 8 tract, both became ambiguous following the establishment of the Ettrick Group in the central Southern Uplands. There is an overlap in age between the Gala 7 and 8 tracts and the Ettrick Group, whilst all three divisions share a distinctive sandstone composition, shown most clearly by the whole rock geochemistry. Likely correlation between the Ettrick Group and the previously defined Gala Group tracts is summarised in Table 3 and discussed further, below, as part of the Ettrick Group account.

Eastward from the Cairnsmore of Fleet pluton, an increasingly complex array of anastomosing faults is indicated by discontinuous outcrops of Moffat Shale Group until, east of Moffat, ten or more relatively narrow, laterally discontinuous tracts can be recognised, all of similar lithological character and overlapping in age. The succession repeated in these tracts (uncertain but probably around 1000 m) comprises the **Queensberry Formation** (Figure 19). At its base, a few metres of unfossiliferous grey siltstone overlies the Moffat Shale Group (e.g. at Dob's Linn [NT 196 159]) and is succeeded by alternations of thickly and thinly bedded sandstone sequences; interbedded units of siltstone range upward from a few metres in thickness, commonly to 50 m and more rarely to 200 m. Units of conglomerate and intraformational breccia occur locally. These are generally up to a few metres thick but rarely 100 m of breccia is seen, associated with massive sandstone. The breccia is variably matrix- to clast-supported and is mostly composed of angular to rounded, fine-grained sandstone and siltstone clasts, generally a few centimetres in diameter but locally up to 50 cm, in a coarse sandstone matrix. In the north-east of the Southern Uplands, the Queensberry Formation continues through the Lammermuir Hills from Oxton to the limit of the Silurian outcrop south-west of Dunbar.

The maximum age of the Queensberry Formation is constrained by the youngest faunas in the numerous outcrops of the underlying Moffat Shale Group. Faunas indicative of the *convolutus* Biozone and, in more southerly tracts, the *sedgwickii* Biozone are common; at Dob's Linn, at the southern edge of the outcrop, the youngest fauna proves the *halli* Biozone. Graptolite faunas from interbeds within the formation itself range through the *sedgwickii*, *halli* and *guerichi* biozones.

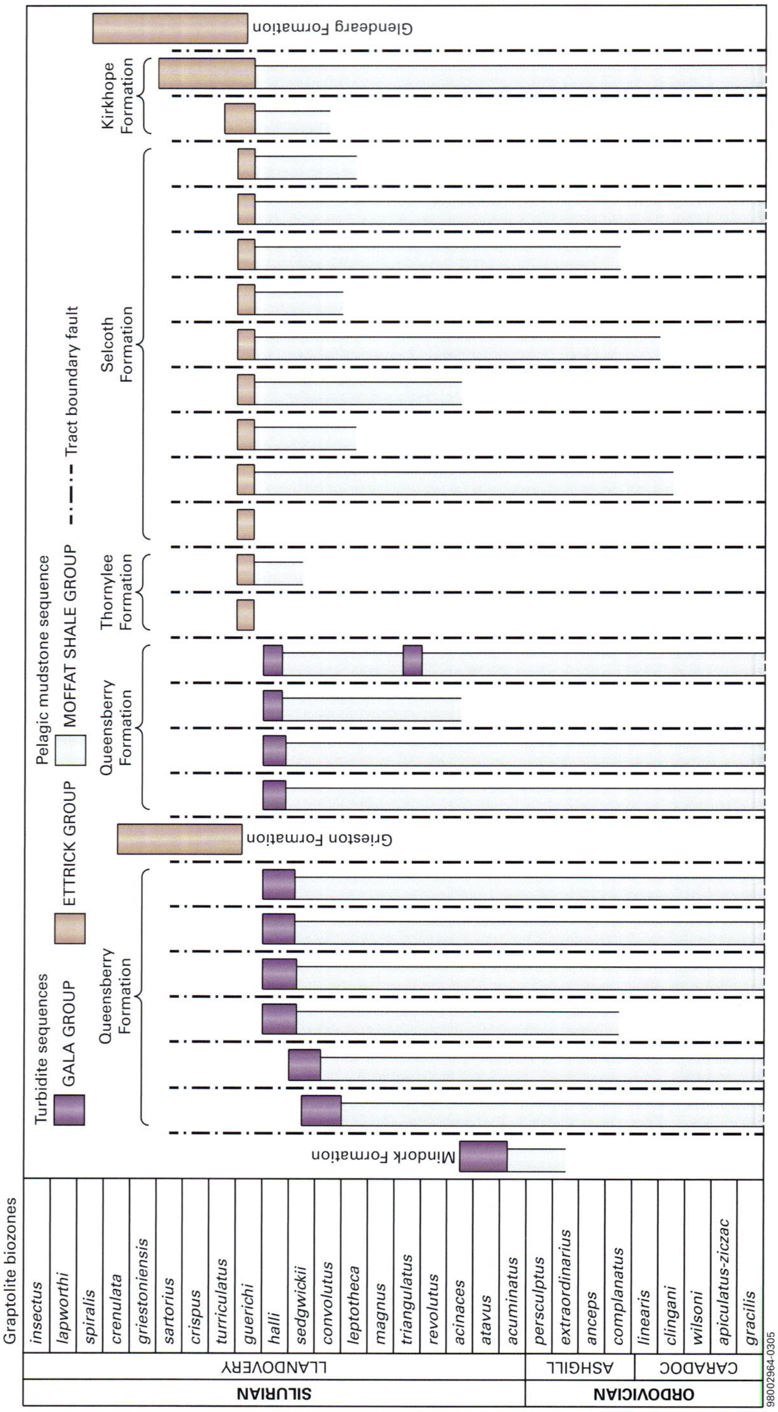

Figure 19 *Representative structural-stratigraphical profile for the Queensberry Formation and the Ettrick Group in the Ettrick area.*

ETTRICK GROUP: SILURIAN

The Ettrick Group was first formally recognised in 2009 and utilised for the BGS 1:50k map sheets for Moffat and Ettrick (Scotland 16 W & E). In the Moffat–Ettrick area, the Ettrick Group incorporates strata lying to the south of the Moffat Valley Fault that had been formerly assigned to the Gala Group, and some strata lying to the south of the Laurieston Fault that had been previously regarded as lying within the Hawick Group. Farther south-west, it is likely that strata included within the Gala 8 tract, and the southern part of the Gala 7 tract, should now be regarded as forming part of the Ettrick Group. In the eastern part of the Southern Uplands, around Innerleithen, the Ettrick Group incorporates strata sometimes described in the older literature as either the 'Garnetiferous Group' or the 'Buckholm Formation'.

The Ettrick Group is characterised by a turbidite lithofacies of well-bedded sandstone, locally coarse-grained, with mudstone interbeds up to about 40 cm thick. The individual sandstone beds are typically massive or poorly graded for most of the bed thickness, with grading to finer-grained sandstone only seen in the uppermost part, where cross-lamination may also be developed. Very thick sandstone beds occur sporadically, either singly or in multiples up to 20 m thick; thick massive siltstone units also occur locally. One division, the Grieston Formation, forms narrow outliers separate from the main outcrop and is largely composed of laminated to thin-bedded alternations of fine-grained sandstone, siltstone and mudstone. Sole marks are common on the base of the turbidite sandstone beds with linear grooves and flute casts establishing the flow direction of the depositing palaeocurrent. Once corrected for rotation during folding, the orientation of these structures suggests that the sediment transport direction was dominantly towards the south-west.

Petrographically, the sandstone is typically composed of poorly sorted angular grains set in a variable amount of fine matrix material. Quartz and feldspar (plagioclase and subordinate K-feldspar) dominate the grain population. Mica is generally a minor component but may be abundant in fine-grained sandstone. Garnet is common locally in coarse-grained sandstone; otherwise, tourmaline, zircon and epidote occur widely as minor accessory mineral grains. A variable proportion of lithic detritus includes grains of clastic sedimentary and metasedimentary rock, fine- and coarse-grained acidic igneous lithologies and conspicuous fine-grained, basic volcanic rocks that show a range of spilitic textures. Some carbonate (CaO up to 7 per cent) is usually present as replacement of matrix or framework grains. Whole rock geochemical analyses of the Ettrick Group sandstones show them to have a relatively homogeneous compositional character and allow them to be readily distinguished from both the Gala Group and Hawick Group sandstones (Figure 20).

Unusually for the Southern Uplands, the biostratigraphical age of the Ettrick Group is demonstrated by graptolite faunas from mudstone interbedded with the turbidite succession. It spans the upper Llandovery *guerichi* to *spiralis* biozones (Figure 15a), and so overlaps with the range established for the youngest part of the Gala Group, and the oldest part of the Hawick Group.

In the central Southern Uplands, abundant outcrops of the Moffat Shale Group indicate relatively closely spaced tract-bounding faults and define numerous anastomosing tracts in parts of the Ettrick Group (Figure 19). Farther north-east, Moffat Shale outcrops become relatively rare and the tracts that can be recognised appear correspondingly broader, although a similar level of cryptic tectonic imbrication may well be present. Tract width also increases to several kilometres in the southern part of the Ettrick Group. However, the tract arrangement of the Ettrick Group is in some respects unusual, since the oldest strata in every formation is of *guerichi* or early *turriculatus* Biozone age, and the same base level is repeated across several adjacent tracts in the closely faulted areas; the sandstone succession in individual tracts may thence expand across several biozones. Uniquely, the Grieston Formation, also biostratigraphically wide ranging, forms outliers within the Gala Group, to the north of the main Ettrick Group outcrop.

Stratigraphical framework

Four formations are recognised in the main outcrop of the Ettrick Group in the central to north-eastern Southern Uplands, distinguished primarily by their sedimentological and petrographical character. All of the top formation boundaries are taken at faults, but the basal transition from the Moffat Shale Group is clearly preserved in all but the Glendearg Formation. The Thornylee Formation spans two or three tracts, the Selcoth Formation is made up of about 9 or 10 tracts, the Kirkhope Formation bifurcates at its western end, whilst current information allows only a single tract to be assigned to the Glendearg Formation (Figure 19). To the north

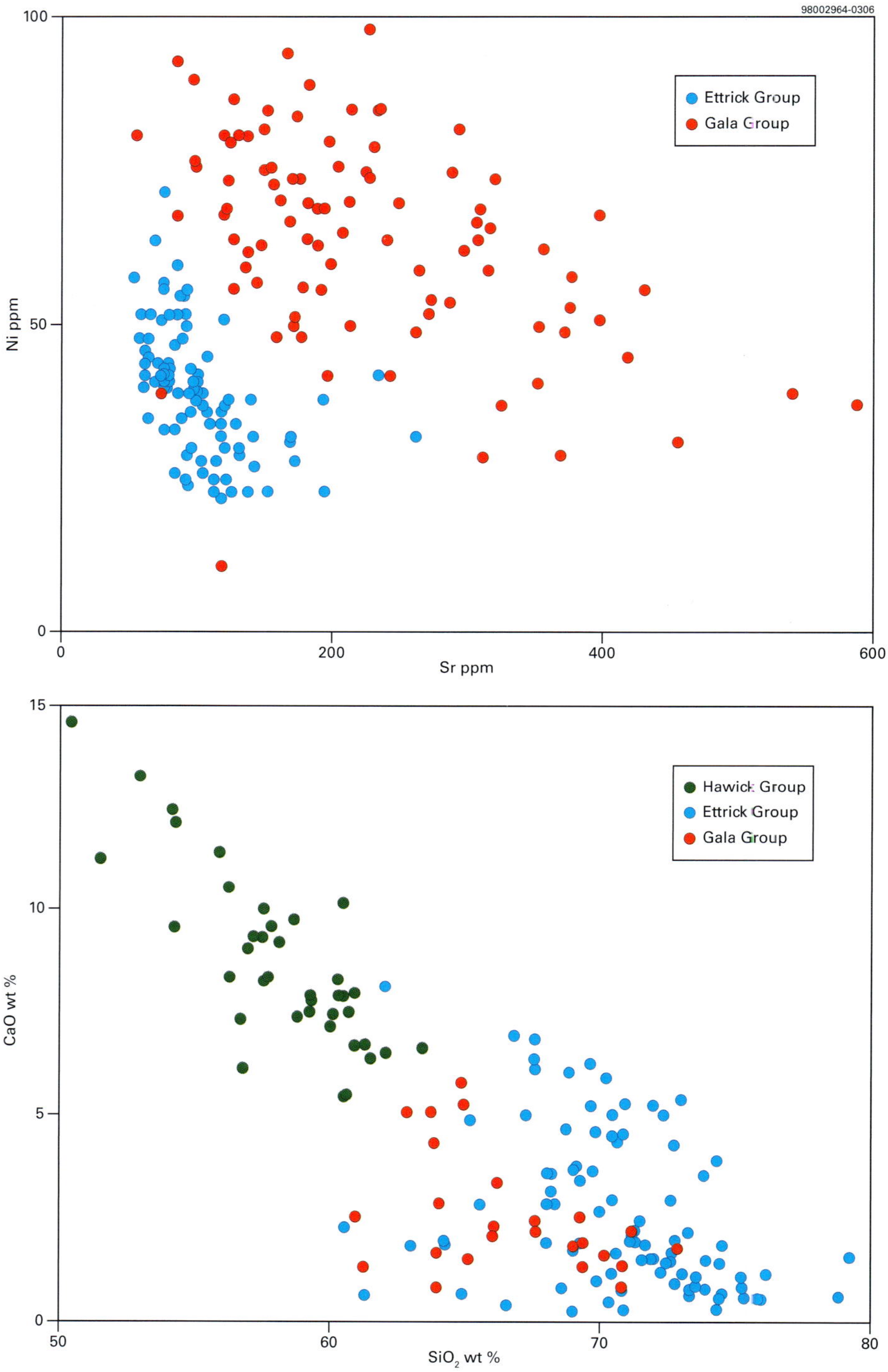

Figure 20 *Geochemical variation shown by sandstones from the Ettrick, Gala and Hawick groups. The Ni-Sr plot differentiates between sandstones of the Gala and Ettrick groups, the CaO-SiO_2 plot differentiates between these and sandstones from the Hawick Group. Compare this figure with the regional geochemical patterns shown in Figure 16; the agreement in Sr distribution is striking.*

of the group's main outcrop, the Grieston Formation occupies narrow, faulted outliers within the outcrop of the Gala Group.

The **Thornylee Formation** (at least 500 m) was distinguished originally as the 'Garnetiferous Group' since the coarse-grained massive sandstone contains significant garnet in fresh, angular and commonly large grains. The formation is well exposed at Thornylee Craigs [NT 402 370] where it is dominated by very thick-bedded or massive, coarse-grained sandstone, and in old railway cutting east of Thornielee House [NT 419 364] where tens of metres of medium- to thick-bedded, fine- to medium-grained sandstone and massive siltstone are interbedded with the massive coarse-grained sandstone. Brown mudstone is present locally and may be extensively bioturbated by the meandering feeding burrow *Dictyodora*. The outcrop of the formation narrows towards the south-west and is structurally terminated at the south side of Dob's Linn. Near there, in the vicinity of St Mary's Loch, the massive coarse-grained sandstone apparently forms lenticular bodies tens or hundreds of metres in thickness and 1–3 km in length within a background of medium- to thick-bedded sandstone. To the north-east, it is likely that the Thornylee Formation tracts broaden to occupy parts of the Lammermuir Hills and the Coldingham Moor inlier of Lower Palaeozoic strata (Figure 21). Throughout its outcrop, graptolite faunas from thin mudstone interbeds demonstrate

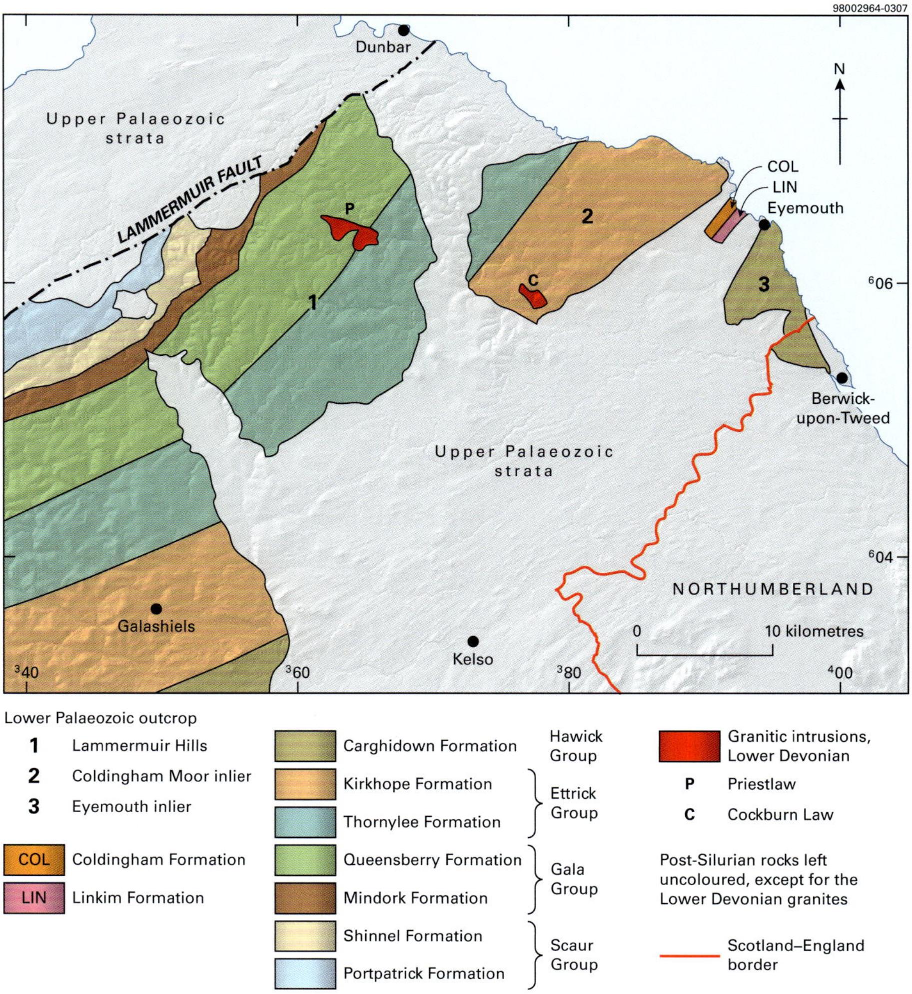

Figure 21 *Outline geology for the north-eastern extremity of the Southern Uplands terrane.*

the *guerichi* Biozone, perhaps extending up into the lower part of the succeeding *turriculatus* Biozone.

To the south-east of the Thornylee Formation, but extending farther to the south-west (Figure 14), the **Selcoth Formation** (500–2000 m) forms several narrow tracts within a demonstrably much imbricated outcrop (Figure 19) that pinches out between converging strike-parallel faults both north-eastward, near Clovenfords [NT 449 364], and south-westward, near Ae [NX 985 890]. The transition from the underlying Moffat Shale Group is well exposed at a number of localities including those in the Selcoth Burn [NT 159 062] and Pot Burn [NT 180 090]. At the transition, graptolitic mudstone of the Birkhill Shale Formation is overlain by about 8 m of greenish grey siltstone with thin interbeds of pale-coloured bentonite and, towards the top, dark mudstone laminae containing *guerichi* Biozone graptolites. Overlying the siltstone, the Selcoth Formation is consistently composed of medium- to thick-bedded sandstone with silty mudstone interbeds, interspersed with sequences of thinly interbedded sandstone and mudstone with sporadic thicker (1–3 m) sandstone beds. A typical section is exposed in the Selcoth Burn stream section [south-east from NT 137 072].

The **Kirkhope Formation** (at least 500 m) forms another south-westerly tapering outcrop, pinching-out between converging faults to the south-east of Moffat (Figure 14). In its south-western extent, the Kirkhope Formation lies to the south-east of the Selcoth Formation, but farther north-east, where the latter is absent (from Clovenfords eastwards), the Kirkhope Formation abuts the Thornylee Formation. There is evidence from Moffat Shale Group outcrops for closely spaced faulting in the south-west of the outcrop, but subdivision cannot be carried north-eastward, where a single tract broadens to over 7 km in width in the Galashiels–Melrose area; large-scale folding may be responsible. In general, the Kirkhope Formation is composed of medium- to thick-bedded sandstone with thin mudstone interbeds, interspersed with sporadic, more thinly bedded units, 1–2 m thick. Massive, medium- to coarse-grained sandstone beds up to 2 m thick are common, locally forming sequences that amalgamate into laterally impersistent, coarse-grained units, tens, perhaps hundreds of metres thick. In common with the Thornylee Formation, the massive coarse-grained sandstone beds are commonly conspicuously garnetiferous. The typical lithology forming the lower part of the Kirkhope Formation is well exposed in the Ettrick Water around Ettrickbridge [NT 390 243], whilst characteristic higher parts of the formation are exposed to the north on Kirkhope Hill [NT 385 255]. It is likely that the Kirkhope Formation continues north-eastward to form much of the Coldingham Moor Lower Palaeozoic inlier, where it is exposed in coastal sections from St Abbs Head towards Siccar Point (Figure 21). In this part of the outcrop, the distribution of numerous graptolite faunas from mudstone interbeds indicate that at least several hundred metres of strata in the *turriculatus* Biozone overlie a much thinner, condensed turbidite succession that here appears to span the *sedgwickii* to *guerichi* biozones, equivalent to the youngest part of the Moffat Shale Group as seen to underlie the Kirkhope Formation elsewhere. The youngest Kirkhope Formation strata, within the *crispus* Biozone, occupy a broad area in the north of the Coldingham Moor inlier that is characterised by large-scale folds, as clearly visible in the coastal section.

The Moffat Shale Group underlying the Kirkhope Formation crops out in narrow lenticular inliers in the Ettrick Water at Ettrickbridge and south of Melrose between Lindean Glen [NT 490 315] and Rhymers Glen [NT 527 327]. At these localities the age of the transition up into the Kirkhope Formation lies within the *guerichi* Biozone, with graptolite faunas from interbedded mudstones higher in the Kirkhope succession ranging up to the *crispus* or *griestoniensis* biozones.

The **Glendearg Formation** (500–2000 m) forms the most south-easterly of the Ettrick Group tracts. It has a lenticular outcrop, attaining a maximum outcrop width of 4.5 km but pinching-out between converging faults near Glenkiln [NY 016 897] in the south-west and towards Selkirk in the north-east (Figure 14). The outcrop lies to the south of that of the Kirkhope Formation, but the Glendearg Formation extends a few kilometres farther to the south-west so that its northern margin abuts the Selcoth Formation. The Glendearg Formation is dominated by thickly bedded sandstone with generally thin mudstone interbeds, interspersed with sporadic thinly bedded sequences that only rarely exceed 2 m in thickness. Isolated, thicker sandstone beds (up to 2 m) are relatively common and may locally combine to form units up to 20 m thick. Red mudstone occurs sporadically, usually in beds a few centimetres thick that have gradational margins with grey mudstone. No Moffat Shale Group strata are preserved at the base of the Glendearg Formation, but graptolite faunas from the interbedded mudstones establish an age range from the *turriculatus* Biozone up to the *spiralis* Biozone. A characteristic section through the formation is exposed in the Glendearg Burn stream section [north from NT 230 062].

The boundary between the Glendearg Formation and the Hawick Group has been defined by a combination of geochemical data derived from sandstone samples (Figure 20) and from stream sediment. From the latter, of particular value is the distribution of CaO, which reflects an abrupt southward increase in carbonate detritus in the Hawick Group sandstones. The steep gradient in CaO abundance in the stream sediment is coincident with a strike-parallel fault zone exposed in Buck Cleuch [NT 334 145] that is now taken as the boundary between the Ettrick and Hawick groups.

North of the main outcrop of the Ettrick Group, the **Grieston Formation** (up to 500 m) occupies narrow, lenticular, fault-bounded outliers within the outcrop of the Queensberry Formation; these extend from the Megget reservoir [NT 209 223] north-east to Innerleithen and possibly as far as Lauder. The formation is composed of siltstone with a variable proportion of fine-grained sandstone in very thin to medium beds. Best known from the eponymous quarry at Grieston Hill [NT 313 361], these fine-grained deposits yield graptolite faunas ranging in age from the *guerichi* Biozone up to the *griestoniensis* Biozone, in contrast to the surrounding Queensberry Formation siltstone which is of *convolutus* to *sedgwickii* Biozone age. The 'Grieston Shales' have previously been regarded as part of the Gala Group, but seem more likely to have been deposited above Gala Group strata with a low angle unconformity intervening. The Grieston Formation can best be regarded as a distal component of the Ettrick Group, deposited in small basins that formed on top of the accretionary complex soon after the Gala Group tracts were incorporated therein.

As discussed above when considering the Gala 7 and 8 tracts, a case can be made for including within the Ettrick Group much of the strata previously assigned to those tracts in south-west Scotland. The evidence is drawn from whole-rock geochemistry, the presence of garnetiferous, coarse-grained sandstone beds, and the ages defined by graptolite faunas, all of which demonstrate considerable overlap. The graptolites from the Moffat Shale Group, seen locally at the base of the successions forming the Gala 7 and 8 and the Ettrick Group tracts, range up to the *sedgwickii* Biozone in all cases, whilst faunas from mudstone interbeds in the sandstones overlying the Moffat Shale in all of those tracts range from the *guerichi* Biozone up to the *crispus* Biozone. The principal sandstone-dominated facies association of the Gala 7 and 8 tracts is sedimentologically most similar to the Selcoth or Glendearg formations, but the thicker, mudstone-rich parts of the Gala 7 and 8 tracts are more akin to the Grieston Formation. However, the level of current information does not permit confident correlation at formational level.

HAWICK GROUP: SILURIAN

The Hawick Group comprises the upper Llandovery to lower Wenlock, sandstone-dominated turbidite strata that crop out between the Ettrick Group to the north and the Riccarton Group to the south. The steeply-dipping fault-bounded tracts that characterise the structural pattern in the older Southern Uplands divisions are clearly present only in the oldest part of the Hawick Group in south-west Scotland, where narrow slivers of the underlying Moffat Shale Group are locally preserved. Elsewhere, whilst it is possible that a cryptic tract structure is present, it is evident only that the succession is much folded.

All of the formations in the Hawick Group are composed of alternating turbidite lithofacies (Plate 18). The succession is dominated by well-bedded turbidite sandstone (bed thickness

Plate 18 *Variable bed thickness in Hawick Group turbidite strata exposed on the north-east coast of Wigtown Bay at Low Auchenlarie [NX 539 518] (P008453).*

most commonly in the 20–60 cm range) with interbeds of silty mudstone up to about 30 cm thick. Interspersed with this dominant facies are units of thinly interbedded sandstone and silty mudstone up to a few metres thick, and thickly bedded, massive sandstone either in single beds or forming units up to 10 m thick. Even the thinner sandstone beds have marked lateral continuity. Thin red mudstone beds are common in the Llandovery part of the succession, but are replaced in the upper part by beds of hemipelagic, laminated and carbonaceous siltstone that are characteristic of the Wenlock. The siltstone beds are commonly fossiliferous, with graptolite faunas proving the *centrifugus* to *riccartonensis* biozones. In the earlier, Llandovery part of the succession, graptolite faunas are very sparse, ranging through the *guerichi*, *crenulata* and *insectus* biozones.

The sandstone is usually fine- to medium-grained although locally includes coarse-grained detritus and mudstone intraclasts. Beds are usually massive through much of their thickness, grading only in the upper part to siltstone that may be cross-laminated, though thinner beds may be laminated or cross-laminated throughout. Sandstone beds have parallel, sharply defined tops and bases, the latter commonly carrying sole marks; these are typically linear grooves and flute casts that generally indicate axial sediment transport towards the south-west. Rippled top surfaces of beds (Plate 19) are also widely seen and the palaeocurrent orientations determined from them are commonly oblique or perpendicular to those derived from sole marks. Trace fossils, usually small burrows or feeding traces, are abundant in places.

Petrographically, the sandstone is a lithic wacke, with angular to subrounded sand grains and up to 40 per cent silt-grade matrix. The sand fraction is dominated by quartz with significant amounts of feldspar (c.10 per cent, mainly plagioclase) and mica (3–15 per cent depending on grain size) with reddened, haematite-coated mica grains conspicuous in younger parts of the group. Carbonate is a significant component, forming up to 15 per cent of the rock, and though now extensively recrystallised it probably originated largely as detrital material. The high carbonate content of the rocks is shown by the whole-rock sandstone geochemistry (Figure 20), which also demonstrates the relatively homogeneous compositional character of the group, and its contrast with sandstones from both the Gala and Ettrick groups. The lithic grain assemblage in the sandstone (variable around 15 per cent) consists mainly of mafic-volcanic and sedimentary rocks, polycrystalline quartz and granitic material. Accessory minerals include green tourmaline, zircon and garnet.

Stratigraphical framework

The difficulties that beset stratigraphical definition in the more northerly parts of the Southern Uplands are less of an impediment in the Hawick Group, which is divided into conventional formations based on subtle variations in the lithological character of the turbidite succession and the presence within it of interbedded red mudstone or laminated hemipelagite. Four formations are identified in the south-west of the Southern Uplands — Cairnharrow, Kirkmaiden, Carghidown and Ross — but only the Carghidown and Ross formations can be recognised north-east of Dumfries.

The **Cairnharrow Formation** (up to 1000 m) forms the northern part of the Hawick Group in south-west Scotland. Poorly preserved slivers of Moffat Shale Group strata thought to underlie the formation indicate that it forms at least two fault-bounded tracts. The northern tract is best exposed in the hills north-west of Gatehouse of Fleet, the southern tract is well exposed in coastal sections on either side of the Wigtown peninsula and the eastern side of Wigtown Bay. In both tracts, the Cairnharrow Formation is composed mostly of thickly bedded sandstone characterised by a rough, parallel lamination throughout most of the bed thickness, grading to a thin cross-laminated layer near the top; there is very little interbedded mudstone. Sporadic thinly bedded units range up to 40 m in thickness, whilst very thick-bedded, massive sandstone units occur locally. Despite the paucity of mudstone interbeds, sparse graptolite faunas have been recovered from several of them and all indicate the *guerichi* Biozone. The Cairnharrow Formation is, therefore, contemporaneous with the older part of the Ettrick Group.

The **Kirkmaiden Formation** (1000–1500 m) is well exposed in coastal sections in Luce Bay and Wigtown Bay whence it is transitional — north-eastwards, laterally along strike;

Plate 19 *Rippled bedding surface in steeply inclined turdidite strata from the Kirkmaiden Formation (Hawick Group) exposed at Carrick Point [NX 575 506] on the south-east side of Fleet Bay (P220405).*

south-eastwards, in stratigraphical sequence across strike — with the Carghidown Formation as that unit's characteristic red mudstone beds appear in the succession. It is dominated by medium to thickly bedded sandstone with thin silty mudstone interbeds that together form units up to tens of metres in thickness, separated by more thinly bedded sandstone units less than 2 m thick. Very thick sandstone beds occur sporadically and may amalgamate into units up to 10 m thick. Biostratigraphical evidence is not good, with only sparse and ill-preserved graptolite faunas in a few mudstone beds, but does confirm a *crenulata* Biozone age. Hence the Kirkmaiden Formation may be in part coeval with the Glendearg Formation of the Ettrick Group, from which it can be distinguished by the carbonate-rich nature of its sandstone.

The **Carghidown Formation** (1000–1500 m) is well exposed in long coastal sections: in south-west Scotland, on the east side of Luce Bay and around Wigtown Bay and in Berwickshire, in the north-east of the Southern Uplands, to the south of Eyemouth. Like the Kirkmaiden Formation, much of the Carghidown Formation is composed of strata displaying classical turbidite facies features (Plate 20). Mudstone interbeds are a little more abundant than in other formations, particularly in the area north-west of Hawick where mudstone, commonly well laminated, locally forms up to 40 per cent of the succession. The Carghidown Formation is characterised by red mudstone interbedded with the ubiquitous grey-green silty mudstone, and by the common presence of distinctive red, haematite-coated mica flakes in the sandstone. The red mudstone beds are thin and rare in the northern part of the outcrop, but they become more common southwards with their maximum thickness increasing up to several metres. A primary origin for the red colouration is indicated by the occurrence of patches of red mudstone mixed with green mudstone in the matrix of slumped units.

Evidence for soft sediment deformation is widespread in the southern part of the Carghidown Formation, where disrupted sequences up to several tens of metres thick are common. Disruption varies from incipient pinch and swell of otherwise continuous bedding to mélange

Plate 20 *Steeply inclined, thinly bedded turbidite strata of the Carghidown Formation (Hawick Group) exposed at Brighouse Bay [NX 632 453] to the south of Borgue (P220426). Note the well-developed cleavage.*

zones, in which bedding has completely disaggregated into lenses and irregular blocks of sandstone of varying size and density in a silty mudstone matrix. Small slump folds in the mélange zones characteristically have a wide range of axial orientations. The deformation in many of the disrupted zones has been accentuated by tectonic effects, but early, soft sediment disruption is favoured by gradational boundaries with coherent strata, blocks lying across the matrix foliation, and variations in matrix lithology.

Much of the outcrop of the Carghidown Formation is devoid of macrofossils. None are known from the south-west of the Southern Uplands, where the formation's age at the northern margin of its outcrop is constrained to *crenulata* Biozone or older by its stratigraphical position beneath the Kirkmaiden Formation, whilst to the south the junction between the Carghidown and Ross formations (see below) also appears to be a stratigraphical transition in which the Carghidown Formation passes up into the lower Wenlock Ross Formation. However, in the central part of the Southern Uplands, several sparse graptolite faunas recovered close to the northern boundary of the outcrop suggest a range from the *crenulata* Biozone, through the upper Llandovery, and into the *murchisoni* Biozone of the lower Wenlock. The lower part of the Carghidown Formation overlaps in age with the Kirkmaiden Formation (in which there is no red mudstone). The upper part of the formation was deposited contemporaneously with the Ross Formation (see below) but lacks the latter's defining interbeds of pelagic laminated siltstone. There was also a depositional overlap with the upper part of the Ettrick Group, demonstrating the contemporaneous deposition of distinctive carbonate-poor and carbonate-rich wacke sandstones. The implication of these relationships is that the late Llandovery and early Wenlock was a period during which several distinct lithofacies were deposited with interfingering relationships, whilst the accretionary process that had previously dominated development of the terrane was relatively inactive.

The youngest division of the Hawick Group, the **Ross Formation** (up to 2000 m) is best exposed in coastal sections at Burrow Head and south of Kirkcudbright. The formation is dominated by sequences up to 50 m thick comprising turbidite sandstone units with sporadic mudstone interbeds, which alternate with sequences up to 4 m thick with more thinly bedded sandstone. The formation is characterised by distinctive dark grey, finely laminated carbonaceous siltstone in beds ranging from a few millimetres up to 1.5 m in thickness, which forms up to 10 per cent of the succession. Red mudstone beds, varying from a few centimetres to 1.2 m in thickness, occur together with the laminated siltstone in a transitional zone from the Carghidown Formation. In south-west Scotland, the transition is a few tens of metres in thickness, but south of Hawick it forms an outcrop about 2 km in width and has been separately named the Stobbs Castle Beds. Also in the Hawick area, the Ross Formation includes a unit comprising thick beds of very coarse-grained, pebbly sandstone (known as the Penchrise Burn Beds and best exposed south-east of Berryfell Farm [NT 524 074]) that is lithologically very similar to the Gypsy Point Member of the Raeberry Castle Formation (see below). Some of these coarse-grained sandstone beds contain small shelly fossils such as crinoid ossicles, brachiopods and solitary corals. Graptolites are relatively common in the carbonaceous laminated siltstone, with faunas ranging from the *insectus* Biozone, near the top of the Llandovery, up to the *riccartonensis* Biozone of the lower Wenlock. The distribution of the graptolites allows four biostratigraphically distinct structural tracts to be identified (Figure 22), albeit they show much internal imbrication.

The Coldingham–Linkim inlier

A small inlier of Silurian strata crops out on the North Sea coast of the Southern Uplands between Eyemouth and St Abbs (Figure 21). The inlier contains two distinct divisions, separated by a fault. These have been widely described as the Coldingham and Linkim formations, either with no group assignation or with an implied association with the Hawick Group, but have also been reduced to member status within the Carghidown Formation.

The northern division, the Coldingham Formation/Member, is a sequence of thinly bedded, quartzose sandstone and siltstone with a high proportion of carbonate in the matrix

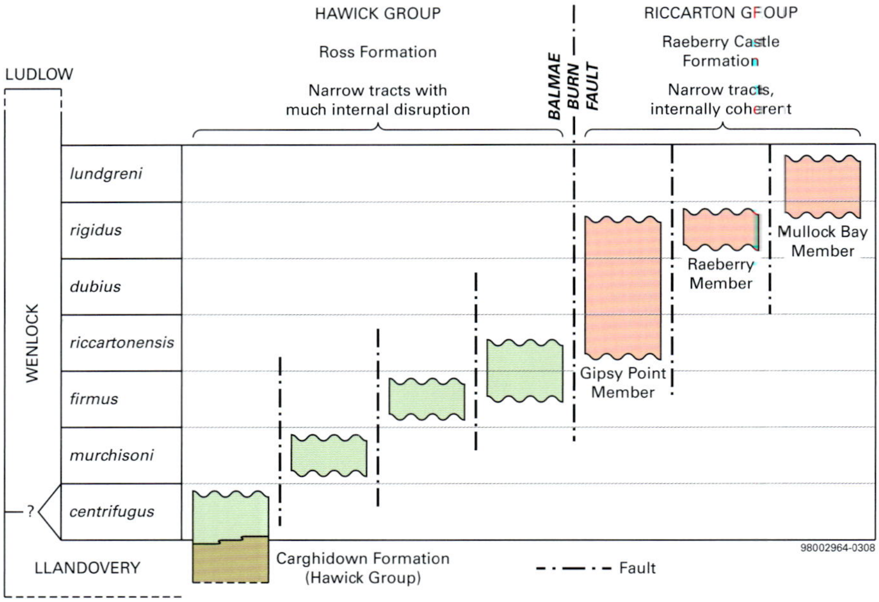

Figure 22 *Representative structural-stratigraphical profile for the Ross and Raeberry Castle formations in the Kirkcudbright area.*

(hence the association with the Hawick Group). The sequence has been thoroughly and chaotically deformed by slumping. The southern division, the Linkim Formation/Member, is a well-bedded succession of turbidite sandstone and mudstone that has been pervasively reddened; but despite a low dip and apparent lack of deformation, it is inverted. The only biostratigraphical control is provided by a poor graptolite fauna and a microflora assemblage from Linkim Kip [NT 928 654]. The graptolite fauna is either late Llandovery or early Wenlock in age. The microfloral assemblage is probably early Wenlock in age and shows similarity with a microflora obtained from the Carghidown Formation in Brighouse Bay, south-west Scotland. This floral comparison, and the pervasive reddening, encourages the association of the Linkim succession with the Carghidown Formation.

There is no evidence for a tectonic cause of the disturbances seen in the Coldingham–Linkim inlier. Soft-sediment deformation is widespread in the Carghidown Formation and it may be that Coldingham–Linkim examples are exceptionally extensive manifestations of this common phenomenon. Alternatively, it has been proposed that the two divisions originated separately as the fill of small, isolated trench-slope basins that formed on the flank of the accretionary thrust complex. Deformation and inversion might then have been caused by large-scale, gravity-induced sliding soon after deposition. Such an origin would weaken the direct association with the Carghidown Formation.

RICCARTON GROUP: SILURIAN

The Riccarton Group, the southernmost exposed part of the Silurian succession in the Southern Uplands, is in fault contact with the Hawick Group to the north but is unconformably overlain by lower Carboniferous strata to the south. It is composed of a range of turbidite lithofacies interbedded with finely laminated carbonaceous siltstone, most probably a hemipelagic deposit. Wenlock graptolite faunas are common in the latter, ranging in age from the *riccartonensis* Biozone to the *lundgreni* Biozone in the Kirkcudbright coastal outcrop (Raeberry Castle Formation), but ranging no higher than the *rigidus* Biozone farther north-east between Langholm and Hawick (Caddroun Burn Formation). The component turbidite sandstone, variably fine- to coarse-grained, is quartz-rich (up to 67 per cent of the grains) but can also contain abundant carbonate, largely as an alteration product but also as rock fragments and bioclastic debris. Minor grain contributions include 5–10 per cent feldspar (predominantly K-feldspar), mica (particularly in fine-grained sandstone) and lithic debris. The latter forms a significant component of coarser-grained sandstone and includes quartzite, chert, limestone and acid and basic (spilitic) igneous material along with sedimentary material likely to be of intrabasinal origin.

In Kirkcudbrightshire, the **Raeberry Castle Formation** (about 1500 m) is divided into three, partly coeval structural tracts that are accorded member status (Figure 22). In contrast to the adjacent Ross Formation tracts, the Raeberry Castle Formation members are internally coherent, with very little disruption of the bedding. The **Gipsy Point Member** is composed of thin- to medium-bedded turbidites overlain by channel-fill accumulations that include slump deposits, and which are interspersed with overbank or levee deposits of thinly interbedded sandstone and mudstone with well-developed dewatering structures. The **Raeberry Member** consists of variable but well-bedded turbidites with isolated thick sandstone and hemipelagite beds. The **Mullock Bay Member** features channelised sandstone, some very coarse, alternating with thinly interbedded, fine-grained sandstone and mudstone; this assemblage is overlain by upward thickening and coarsening units that pass from hemipelagite through thin- to medium-bedded to thick-bedded sandstone.

The equivalent strata in the Langholm to Riccarton area make up the **Caddroun Burn Formation** (about 1500 m). This comprises thin-bedded fine-grained sandstone and mudstone with intercalations, tens of metres in thickness, of more thickly bedded, coarser sandstone with thin mudstone interbeds. Laminated carbonaceous siltstone occurs throughout, generally as relatively thin beds but locally much thicker (8 m at one locality near Saughtreegrain [NY 568 998]). Palaeocurrent data indicate that most sediment transport was towards the south-east quadrant, with a subsidiary flow towards the south-west. Some thin, fine-grained beds are rich in plant and crustacean remains.

four

Caledonian structure and magmatism

As the Iapetus Ocean closed, collisional events at its margins caused episodes of deformation and metamorphism. Two main events can be identified. The Grampian Event (which peaked at about 470 Ma) was caused by the arrival at the Laurentian margin of arc terranes represented, in small part, by the Upper Ordovician Ballantrae Complex ophiolite (Figure 4b). The Scandian Event (about 430 Ma) resulted from the collision of Laurentia with Baltica but its effects are largely restricted to northern Scotland. The slightly later convergence of Laurentia with Avalonia and the final elimination of the Iapetus Ocean was a much less dramatic affair that is barely identifiable in the tectonic record. Though it might be anticipated to have had a profound effect in the Southern Uplands terrane, the leading edge of Laurentia, the deformation seen therein is mostly the result of the Late Ordovician to mid Silurian, diachronous accretionary process (Figure 4c,d). One possible effect of the collision, implicit in Figure 7, was the north-directed thrust emplacement of the accretionary complex onto Laurentian crystalline basement. Farther to the north, more demonstrable north-directed thrust movements probably instigated by the collision produced the large-scale, Wenlock (about 425 Ma) thrust reactivation of faults affecting the Girvan sequence. The coming together of Laurentia and Avalonia was then followed during the Early Devonian (at about 400 Ma) by the Acadian orogenic event, which was initiated much farther south by the arrival of another Gondwanan continental fragment at the southern margin of Avalonia as the Rheic Ocean closed (Figure 3c). Acadian effects are most pronounced in northern England and Wales but some tectonic features of the Southern Uplands were influenced by that event. Of particular importance was the transtensional regime that, between about 410 Ma and 397 Ma, focused the intrusion of dyke swarms and granitic plutons.

OBDUCTION OF THE BALLANTRAE COMPLEX

As described in Chapter 2, the Ballantrae Complex is an assemblage of mantle and crustal rocks formed in a variety of geotectonic settings: island arc, marginal basin and oceanic island. The various components were dismembered and juxtaposed during the obduction of the Ballantrae Complex onto the margin of Laurentia, which can be regarded as part of the Grampian Event. The initiation of obduction is dated by the radiometric (K-Ar) age of 478±8 Ma obtained from amphibolite within the dynamothermal metamorphic aureole at the base of the northern serpentinite belt. The graptolite fauna from mudstone associated with a serpentinite conglomerate at North Ballaird is likely to date 'obduction-in-progress' to the late Arenig. Obduction would then have been complete prior to the deposition of the Llanvirn basal beds of the Barr Group (the Kirkland Conglomerate Formation) on top of the assembled Ballantrae Complex. Detailed structure within the obducted ophiolite is difficult to unravel, but some indication of the structural complexity is shown by the tectonic repetition of a thin, fossiliferous mudstone–chert–lava succession between Bennane Head and Balcreuchan Port (Figure 11). The structural contacts between the different components of the complex are generally steep, and there is evidence for widespread strike-slip movement along the major faults.

Though there is no proof of the subduction polarity, it is conceptually easier to envisage southward subduction of marginal basin crust beneath an Iapetus island arc, since this would make arc–continent collision inevitable (Figure 10). To continue the closure of the Iapetus Ocean, collision and ophiolite obduction must then have been followed by a reversal in subduction direction (a well-established phenomenon elsewhere in the geological record), so that the ocean crust was lost northward beneath the Laurentian continental margin, now endowed with an attached ophiolite complex. Once northward subduction was established, the build-up of the Southern Uplands accretionary terrane could commence.

DEFORMATION OF THE GIRVAN SUCCESSION

Whilst the Southern Uplands accretionary complex was building up at the subducting margin of the Iapetus Ocean, the Ordovician to Silurian Girvan succession was accumulating on the subsiding margin of the Laurentian continent. The depositional setting was essentially a forearc basin between the Midland Valley volcanic arc, then rising to the north, and the developing accretionary complex to the south (Figure 4c). However, a note of caution must be added. Though this overall association of geological regimes fits the present-day outcrop pattern, it is highly unlikely that the original spatial relationships are preserved. The arc, forearc basin and accretionary complex all extended for considerable distances along strike, and substantial, sinistral strike-slip movement, post-Ordovician and particularly focused on the Southern Upland Fault, has brought about the juxtaposition currently seen.

For much of its history (and as described in Chapter 2) deposition of the Girvan succession was controlled by a series of major faults, throwing down to the south and sequentially stepping back into the Laurentian hinterland (Figure 13). There would undoubtedly have been some rotation of the early-deposited beds during subsequent fault growth, but there is only one stratigraphical break preserved in the Girvan succession. This coincides, approximately, with the Ordovician–Silurian boundary where the basal Silurian strata overlie the Ordovician at an unconformity with low-angle structural discordance.

The major structural disruption occurred in the mid to late Silurian, when the faults that had controlled sedimentation were reactivated as north-directed thrusts. The cause of this reversal is uncertain, but was most probably related to the collision, farther south, of Laurentia and Avalonia. The effect was to carry parts of the Girvan succession northward, over other parts of the succession and the underlying Ballantrae Complex. In the process, strata were rotated and folded so that steep dips are now commonly seen at outcrop. The thrust reactivation could have begun quite early in the Wenlock, with the youngest strata affected (and now preserved) being the Straiton Grits Formation, with an age equivalent to the *murchisoni* Biozone (Figure 12).

SOUTHERN UPLANDS ACCRETIONARY COMPLEX

A stratigraphical paradox was identified in the Southern Uplands at an early stage of its geological exploration. At outcrop, most of the steeply inclined beds face northward so that the sequence becomes younger in that direction, yet the oldest strata are found in the north of the terrane and the youngest in the south. To explain this apparent contradiction various models of folding and faulting were proposed, but it is now generally accepted that the Southern Uplands terrane formed as an accretionary thrust complex above the northward-dipping subduction zone that carried Iapetus Ocean crust beneath the Laurentian continental margin. It was built by a series of southward-propagating, imbricate thrusts that sequentially stripped the oceanic sedimentary cover from the descending plate and stacked-up the ensuing, fault-bounded strips, each strip being inserted at the base of a stack of previously accreted strips (Figure 4c, d). The original thrust faults would have advanced at a relatively low angle. As the complex developed, the faults and their enveloped strata were structurally rotated to become near vertical, but at depth they remained listric, merging into a subhorizontal décollement surface. Their near-vertical attitude is now exposed across the Southern Uplands and defines the tracts as described in Chapter 3.

Only one phase of ductile deformation (D_1) is evident throughout the Southern Uplands. It is related to thrust propagation and the folds produced are hence diachronous, becoming younger southwards (Figure 23). Later phases of deformation are apparent locally, associated either with accommodation in the thrust hinterland commensurate with D_1 deformation at the thrust front (and so likely to be equally diachronous) or with intermittent sinistral shear imposed across the entire belt but focused into major strike-fault zones. These post-D_1 deformation phases have been referred respectively to D_2 (co-axial with the gently plunging D_1 folds) and D_3 (sinistral, steeply plunging folds) but their relationship may not be the same everywhere.

One caveat should be attached to the interpretation of the entire Southern Uplands terrane as an accretionary complex. There is much evidence for a change in structural style in

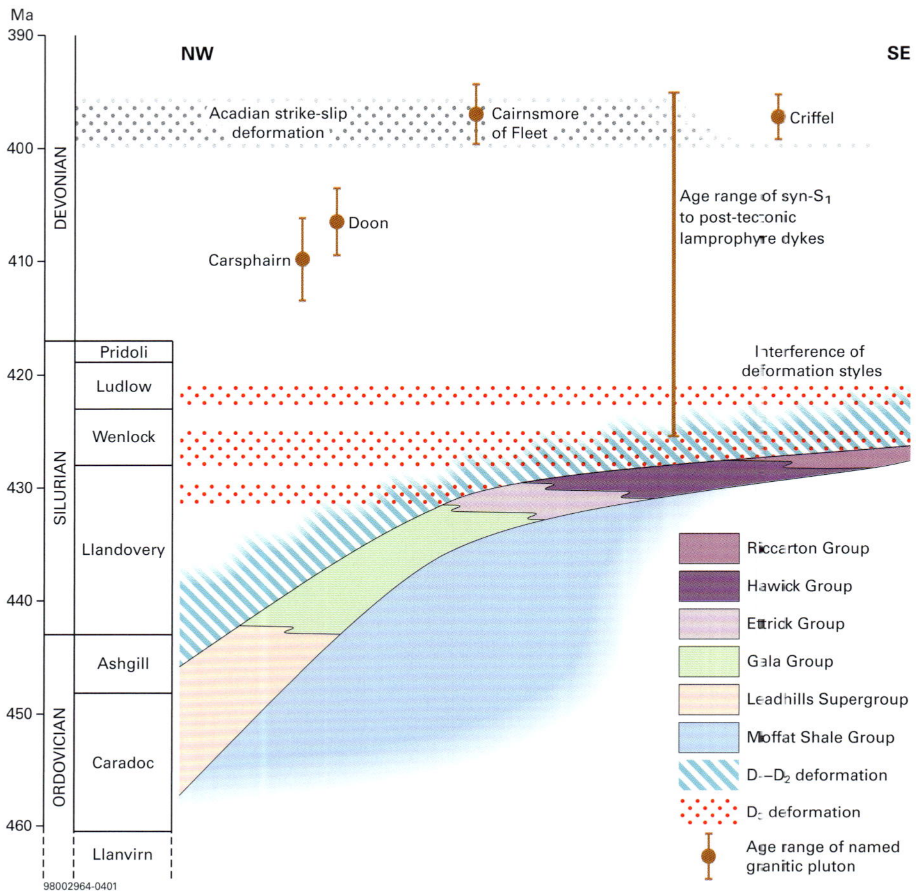

Figure 23 *Summary of the regional controls on the timing of deformation across the Southern Uplands accretionary complex.*

the late Llandovery, as manifest in the Ettrick and Hawick groups. The sequential southward younging of the structural tracts, established through the Barrhill and Scaur groups and the lower part of the Gala Group, is replaced in the Ettrick Group by the repetition of narrow tracts all with similar ages. The Gala 8 tract, the likely equivalent on the Rhins of Galloway of the Ettrick Group farther east, and the Hawick Group tracts to the south of it, show a reversal of the structural pattern with north-directed thrusts imbricating a succession that becomes younger southward. Elsewhere, and more generally, the Hawick Group tracts are much folded and affected by a stronger cleavage than is apparent to the north. These changes are most probably linked to the effective closure of the Iapetus Ocean with the younger Llandovery to Wenlock turbidite successions deposited in a relict basin above the suture (Figure 4d).

Some convergence of Laurentia and Avalonia continued after the elimination of the Iapetus Ocean, with the Laurentian margin overriding that of Avalonia. In this situation, deformation of the Ettrick–Hawick–Riccarton successions occurred in a foreland fold and thrust belt setting, with the deformation front migrating southward towards the Avalonian hinterland. Sedimentary links, particularly clear for the hemipelagite facies of the Ross and Raeberry Castle formations (Hawick and Riccarton groups repectively), suggest that in time this basin extended to the Isle of Man and into the south of the English Lake District (Figure 24). The final stage of this process is seen in the southern Lake District, where loading by the advancing Southern

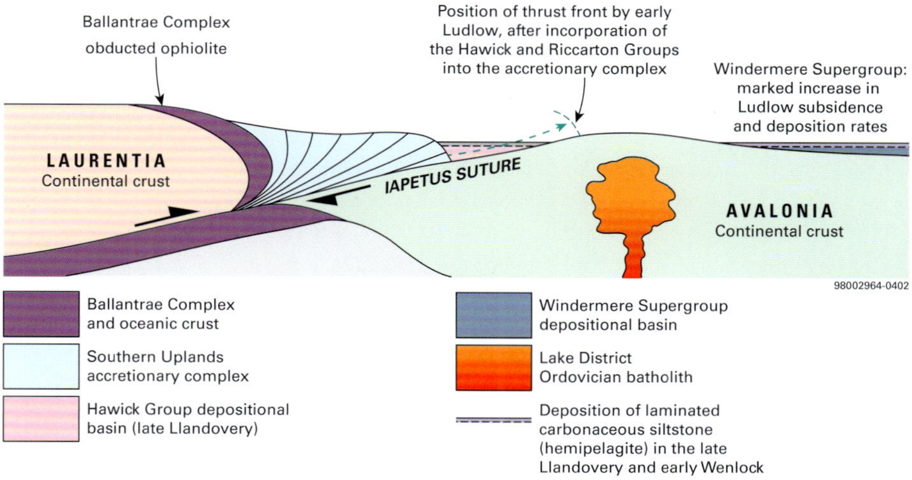

Figure 24 *Migration of the Southern Uplands accretionary complex onto the Avalonian continental margin as a foreland fold and thrust belt.*

Uplands thrust belt caused an acceleration of subsidence and the deposition of the thick, Ludlow turbidite sequence of the Coniston Group. Convergence stalled at this point and the Coniston Group basin had stabilised and filled by Pridoli times. Acadian deformation did not follow for about another 10 million years.

A possible tectonic outcome of the Laurentia–Avalonia collision at the northern margin of the accretionary complex was its northward emplacement onto Laurentian crystalline basement (Figure 7). Such large-scale northward thrusting of the accretionary complex perhaps accompanied the demonstrable north-directed thrusting of the Girvan succession during the mid to late Silurian. The considerable horizontal shortening of the accretionary complex that would have been likely in such circumstances could have been accommodated by the widespread rotation of bedding towards the vertical, an attitude hard to attain across the whole terrane only by accretionary activity.

Major faults

The northern boundary of the Southern Uplands terrane is formed by the Southern Upland Fault (= the Stinchar Valley Fault in the west), a long-lived and much reactivated structure marked by a gouge and breccia zone that may reach several tens of metres in width. Significant sinistral movement has occurred along the Southern Upland Fault and this process may have truncated the older part of the Southern Uplands accretionary complex as it was juxtaposed against the Midland Valley terrane. Up to about 1500 km of lateral movement has been proposed, based on the correlation of boulders in the Corsewall Conglomerate with a provenance in Newfoundland. This correlation now seems tenuous and a more likely limit of several hundred kilometres of post-Caradoc movement is suggested by the similarities between the derived fauna in slump conglomerates within the Kirkcolm Formation and the in situ fauna in Upper Ordovician limestone in the Midland Valley terrane, at Girvan and at Pomeroy in Northern Ireland. In this situation, ophiolitic detritus in the northernmost of the Southern Uplands tracts may have been derived from obducted masses now represented by the Ballantrae Complex, though the provenance is not definitive.

In northern and central parts of the Southern Uplands, the thin, discontinuous and lenticular inliers of Moffat Shale Group mudstone define lineaments that divide the turbidite succession into the strike-parallel tracts. Individual tracts so established range from a few tens of metres to several kilometres in outcrop width. The mudstone sequence is commonly disrupted but in places a stratigraphical transition from mudstone to turbidite sandstone is preserved at its northern margin. The southern mudstone–sandstone junction in each inlier, typically juxtaposing older mudstone against younger sandstone, is interpreted as a major, steeply dipping, strike-parallel fault that originated as an accretionary thrust, as discussed above. In some cases, closely spaced repetitions of the mudstone at the base of the sandstone tracts show these bounding structures to be anastomosing fault zones. There is rarely much associated fault rock preserved but breccia and gouge may locally reach a few metres in width.

The recognition of faults is largely dependent on the preservation of the inliers of the Moffat Shale Group. Apparent fault density is highest in the central Southern Uplands, in a zone of about 10 km across-strike width spanning the Gala and Ettrick groups in the Moffat–Ettrick area (Figure 19), where fifteen or more faults occur spaced from 100 m to 1.5 km, commonly anastomosing to define laterally discontinuous, lenticular tracts. Elsewhere, the recognised faults appear as subparallel structures spaced up to about 5 km, although in the north-west the tract containing the Kirkcolm and Galdenoch formations is up to 16 km wide. It is very likely that these broader tracts are compound, with internal strike-parallel faults that are not currently recognisable.

The southernmost inliers of the Moffat Shale Group lie within the outcrops of the Ettrick or Hawick groups or at the boundary between them. Only a few strike-parallel faults are recognised south of this limit, substantiated either by extrapolation from limited local exposure of fault-related fabrics or by abrupt changes in sandstone composition. So, for example, east of Moffat the faulted boundary between the Ettrick and Hawick groups is identified by the marked increase in carbonate in the Hawick Group sandstone (Figure 20). Coincident with this 'carbonate lineament' is an outcrop of north-east-trending fault gouge and breccia at Buck Cleuch [NT 334 145]. Tract boundaries within the Hawick Group are particularly problematical, as discussed in Chapter 3, due to the highly folded nature of the succession and the biostratigraphical overlap between adjacent tracts. Hence, whilst the boundary between the Carghidown and Ross formations might appear to be faulted due to the conundrum of apparently younger strata appearing southwards in a northward-younging succession, there is some conflicting biostratigraphical evidence to suggest that the two units are partially coeval; locally (at Burrow Head [NX 457 340], Ross Bay [NX 647 448] and south of Hawick) the boundary appears to be a stratigraphical transition. The fault thought to separate the Hawick Group from the younger Riccarton Group to the south is deduced largely from an abrupt decrease in deformation into the latter but, in the Kirkcudbrightshire coastal section, abundant biostratigraphical data within the Riccarton Group does establish a sequence of narrow tracts repeated by closely spaced faults (Figure 22).

Although the major strike-parallel faults are thought to have developed as thrusts with a 'top to the south' sense of displacement, many have experienced different senses of reactivation with widespread evidence for significant sinistral lateral displacement during the later stages of ductile deformation of the terrane (the D_3 episode, see below). Many of the strike-parallel faults were then reactivated during later tectonic episodes, with variable amount and sense of displacement along their lengths due to their segmentation by cross-strike faults. These disrupt the tract fault system and are typically associated with narrow zones of gouge that may weather recessively, forming marked features. The cross faults are concentrated on two trends: north-west to south-east and north-north-east to south-south-west. It is evident from their close relationship with minor intrusions that many of them pre-existed, or were active during, the approximately 400 Ma intrusive episode.

It is generally difficult to quantify vertical displacements on most of the cross-strike faults, although some larger examples were reactivated as major bounding structures to the post-Silurian basins that cut across the Southern Uplands terrane. Offsets in metamorphic grade, such as that affecting the thermal aureole of the Cairnsmore of Fleet granite pluton, or abrupt changes in the low-grade metamorphic pattern (see below) also point to vertical movement on cross faults transferring onto reactivated segments of the strike-parallel fault system. Lateral offsets are more commonly identifiable with north-west-trending structures, with most proving to have a dextral sense; where north-north-east-trending faults show evidence for lateral movement it is usually sinistral. In general, any lateral movements on the cross faults are relatively small, but in a few cases they appear to reach several kilometres.

Early, thrust-related deformation (D_1)

D_1 folds were developed variably throughout the Southern Uplands. Many occur as gently plunging, south-verging anticline-syncline pairs compatible with the 'top to the south' movement on the thrusts, with which the folds may be intimately connected via hanging-wall detachments (Figure 25). Fold structures occur on all scales, with congruous minor fold pairs in the limbs of the larger structures (Plate 21). Across-strike, highly folded zones occur interspersed with long homoclinal sections, usually of steeply inclined, north-younging strata. This variation in structural style is, at least in part, related to the nature of the strata, with thickly bedded, massive greywacke less likely to be intensely folded than more thinly bedded strata. Slickensides or slickenfibres in thin veins along bedding surfaces, perpendicular to the fold axial orientation, demonstrate early fold growth by flexural slip, although the veins were themselves folded in the later stages of deformation. Individual tracts are commonly marked by subtle variations in the style, orientation and intensity of D_1 folding. The D_1 deformation was particularly intense in the generally finer-grained rocks of the Hawick Group (Plate 22), in the southern part of which many fold hinges become strongly curved to steeply plunging and are locally downward facing. This phenomenon suggests that a significant component of sinistral shear was involved in D_1 as the Hawick Group tracts were deformed, and this may well have coincided with the onset of D_3 in the thrust hinterland (see below).

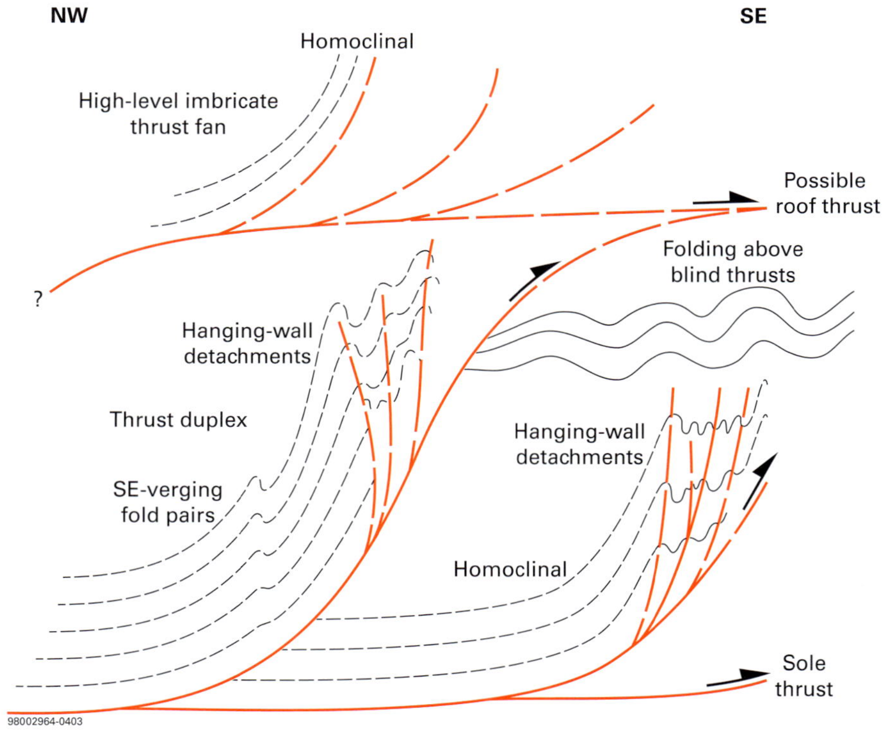

Figure 25 *Variation in structural style seen in the Southern Uplands diagrammatically illustrated as a consequence of position and thrust geometry within a developing accretionary complex.*

Plate 21 *Examples of D_1 folding in the Kirkcolm Formation as seen on the west coast of the Rhins of Galloway: a) A large-scale overturned anticline exposed to the north of Salt Pans Bay [NW 964 617] (P008502); b) An anticline–syncline fold pair developed in the short, flat-lying limb of a major fold pair such as that seen in Plate 21a. This example is exposed at Portobello [NW 960 662] (P008464).*

Plate 22 *Upright, tight, D_1 anticline–syncline fold pair affecting thinly bedded strata of the Carghidown Formation (Hawick Group) at Hare Glen [NX 622 445] to the south of Borgue (P220420).*

A widespread, regional manifestation of D_1 is the penetrative cleavage (S_1). It is commonly developed as an axial planar fabric to the major folds, but in cross-section may be slightly fanned (Plate 23a). In northern and central parts of the Southern Uplands, S_1 is best developed in the fine-grained, silt- and mud-rich lithologies, though even there it can be quite weak. It is much less apparent macroscopically in sandstone, and when present may be strongly refracted across the sandstone-mudstone contact (Plate 23b) and also with grain size in the graded sandstones. It is more widely seen in thin sections of sandstone as an irregular anastomosing fabric. In parts of the Hawick Group, however, the foliation is more pervasive and is well developed in sandstone, where it is strongly refracted through graded beds. Significantly for the timing of deformation, the cleavage is also commonly developed in felsic and lamprophyre dykes intruded into the Hawick Group. Cleavage is only very weakly developed in strata of the Riccarton Group.

The S_1 cleavage is typically axial planar to the folds in the north-western tracts of the Southern Uplands (Figure 26a) but in the south-eastern tracts, and in particular those containing Hawick Group strata, it may locally show a clockwise transection in plan view of up to 20° relative to the fold axial traces (Plate 24) (Figure 26b). Where folds are overturned, and such overturning is generally towards the north-west, the cleavage may cut across the inverted fold limb (Figure 26c) so that there the bedding faces downward on cleavage whilst in the opposing limb of the same fold the bedding is normal relative to the same cleavage. As a consequence of this confusing effect, the assessment of way-up or vergence from bedding-cleavage relationships alone is generally unreliable in the south-eastern tracts of the Southern Uplands. The transecting cleavage is undoubtedly a result of the D_1 deformation and can be explained as resulting from rotation of the D_1 stress field between

Plate 23 Examples of D_1 cleavage development: a) An anticline with an axial cleavage fan from the Kirkmaiden Formation (Hawick Group) near Barlocco [NX 581 490] (P220411); b) Cleavage refracted by variation in lithology and grain size in Portpatrick Formation (Scaur Group) strata near Dunskey Castle, to the south of Portpatrick [NX 005 529] (P220143).

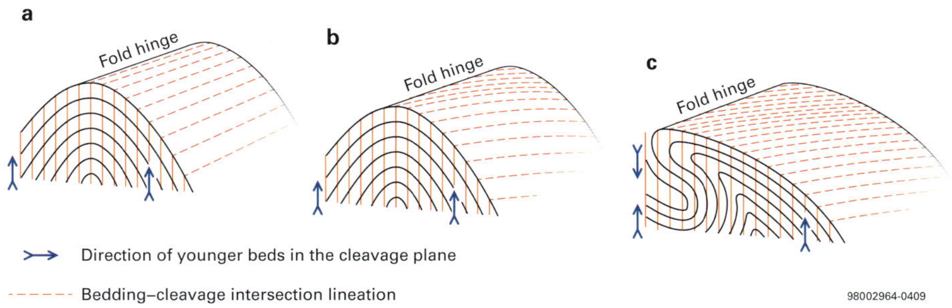

Figure 26 Variations in the relationship between folding and cleavage developed during the D_1 phase of accretionary deformation (adapted from Anderson, 2001): a) an axial planar cleavage in an upright fold, b) a clockwise transecting cleavage in an upright fold, c) a clockwise transecting cleavage in an overturned fold resulting in opposite facing directions of bedding in the cleavage plane on the limbs of the fold.

initiation of folding and imposition of cleavage. An added complication in some sections is the alternation of zones of axial planar and transecting cleavage, apparently partitioned by major north-east-trending faults and providing further evidence for a variable stress regime during accretion.

Plate 24 Transecting cleavage (aligned with hammer handle) in a D_1 anticline affecting strata of the Kirkmaiden Formation (Hawick Group) at Corseyard Point [NX 591 481] near Borgue (P220414).

a

b

Plate 25 *Examples of D_2 refolding of D_1 structures affecting Kirkmaiden Formation (Hawick Group) strata to the west of Borgue: a) At Craigmore Point [NX 572 517] (P220404); b) At Isle Mouth [NX 576 498] (P220407).*

Plate 26 Steeply plunging D_3 fold affecting Garheugh Formation (Gala Group) strata at The Hooies [NX068 448] on the west coast of the Rhins of Galloway (P008491).

Late, superimposed deformation (D_2 and D_3)

The extent and character of post-D_1 deformation varies widely across the Southern Uplands. In northern and central parts, post-D_1 structures coaxial with D_1 tend to occur only very locally and are difficult to correlate. To the south, in central and southern parts of the Hawick Group outcrop, coaxial post-D_1 deformation is widespread. Gently plunging, minor to mesoscale folds, coaxial with but refolding D_1 structures, occur in two styles, upright to inclined and recumbent (Plate 25). These have conjugate geometry and locally occur together as open box folds suggesting that they formed together; consequently both are classified as D_2. Small recumbent folds, verging down the dip of bedding, are most common and are associated with a widely developed, gently dipping, S_2 crenulation cleavage. The orientation and geometry of the D_2 structures suggests that they formed either through a continuation of 'D_1' stress after locking of the original D_1 folds, or by subsequent renewal of shortening on the tract-bounding faults. Alternatively, the recumbent folds may have been formed by subvertical shortening of bedding in more or less its present attitude, causing down-dip vergence, rather than by a consistent sense of shear on tract-bounding or other faults.

The steeply plunging sinistral folds (D_3) developed locally throughout the Southern Uplands are usually in narrow zones of shearing adjacent to tract-bounding faults and may therefore be associated with reactivation of these structures (Plate 26). The relationships of the putative D_3 folds to D_2 are ambiguous with indications of sinistral shear or refolding co-axial with development of D_1 folds at various times. It also seems likely that there were several episodes of sinistral shear superimposed on the diachronous D_1 and D_2 folding at different times. This means that in some localities in the south of the terrane, folding of 'D_3' style and origin preceded that with 'D_2' character (Figure 23).

Moniaive Shear Zone

One particularly important example of strike-parallel sinistral shear is the Moniaive Shear Zone, named from the area around Moniaive [NX 780 910], north-east of the Cairnsmore of Fleet pluton. It is a zone of high strain that shares structural characteristics with, but is much wider than, the narrow shear zones associated with a few of the tract-bounding faults. It has been recognised over a strike length of about 100 km through the central part of the Southern Uplands, where it is up to 5 km wide, and continues westward across Ireland as the Slieve Glah Shear Zone. In the Southern Uplands, the Moniaive Shear Zone is truncated abruptly to the north-west by the Orlock Bridge Fault but dies out southward within the northern tract of the Gala Group. It is characterised by the intermittent development of a pervasive foliation nearly parallel to bedding, locally with a strong linear component, which commonly transposes all original structure. Strain within the shear zone is very variable but a variety of structural indicators consistently show a sinistral sense of shear. Because the shear zone fabric is inseparable from the relatively weaker S_1 cleavage outwith the shear zone, the two cannot be differentiated and unequivocal relative age relationships are difficult to establish. Cordierite (and some garnet) porphyroblasts, widely distributed throughout the thermal metamorphic aureole of the Early Devonian Cairnsmore of Fleet Pluton (c. 397 Ma, see below), are deformed by the shear zone foliation but the latter is generally overprinted by the biotite hornfelsing and later stages of the thermal metamorphism. These relationships closely constrain the timing of final development for the Moniaive Shear Zone to around 397 Ma, which means that it could have been active during the Acadian deformational event.

It is likely that the Moniaive Shear Zone is a composite feature, representing progressive but intermittent deformation over a long time period from its initiation during D_1 (accretionary) deformation in the early Silurian until the possibly Acadian effects in the Early to Mid Devonian. Despite the possible long duration of deformation there are no grounds for assuming very large lateral displacement from evidence in Scotland. However, substantial movement has been proposed for the Slieve Glah Shear Zone, along-strike to the west in Ireland. In addition, the Moniaive Shear Zone lies above a deep crustal discontinuity (Figure 7) that may have focused movement during and after the collision of Laurentia and Avalonia.

Regional metamorphism

The thermal metamorphic aureoles surrounding the larger plutons are superimposed on low grades of regional metamorphism that were acquired by burial during development of the accretionary complex. The prehnite-pumpellyite facies can be established in some of the volcaniclastic sandstone beds, but elsewhere grades within the deep diagenetic zone and anchizone are shown by the crystal thicknesses of white mica from the mudstone — the illite crystallinity. Only locally does the grade rise to epizone, and mostly that is either in the vicinity of intrusions or where cleavage is unusually well developed. There is no consistent pattern of increasing grade into older strata, which would be expected in normal sedimentary burial, and instead, some of the highest grades (though still mostly anchizonal) affect the Llandovery Hawick Group. In a number of cases the metamorphic grade increases sequentially into younger tracts. If the succession had initially acquired a pattern of normal burial metamorphism whereby grade increases into older strata, and had subsequently been imbricated and rotated, older strata would still be expected to show higher grades than younger strata; this is demonstrably not the case. As an additional complication, in places the grade changes abruptly along strike at major cross faults, with lower grade on the downthrow side. These relationships are most readily explained if a depth-related pattern of metamorphism was imposed on strata that were already steeply inclined. This would have been achieved after

Plate 27 *Intrusion styles of lamprophyre dykes:*
a) Dyke intruded parallel to bedding in the Kirkmaiden Formation (Hawick Group) at Mossyard Bay [NX 552 517] (P104245); b) Offsets in a dyke facilitated by its intrusion synchronous with development of a D_2 fold, Back Bay, near Monreith [NX 372 392] (P008561).

their incorporation in the accretionary complex. Abrupt changes in grade across some of the tract-boundary faults probably arose through their post-metamorphic reactivation. Assuming low to moderate heat flow (<25°C), the depth of burial within the accretionary complex, would have ranged from about 6 km, for those strata now in the deep diagenetic zone, up to about 13 km for some of the epizonal rocks.

LATE CALEDONIAN DYKE SWARMS

Towards the end of accretionary deformation, the Ordovician and Silurian strata of southern Scotland were intruded by two coeval suites of calc-alkaline dykes: one microdiorite to microgranite in composition and the other comprising lamprophyres. The overall abundance of dykes increases southward and though both suites range across the full extent of the Southern Uplands, there are proportionately more microdiorite and microgranite dykes in the north, and more lamprophyre dykes in the south. Both suites of dykes appear to have been intruded over a considerable interval of time, from the late Silurian to the Early Devonian, and in structural terms they are pre-D_2 to post-tectonic. The two suites are broadly coeval but it is likely that the earliest intrusions were mica-bearing lamprophyres, whereas the last dykes emplaced were probably microdiorites.

Most dykes postdate the D_1 deformation (Plate 27a), but in the south of the Southern Uplands a few lamprophyre dykes seem to have been folded by the youngest phase of the diachronous D_1 episode. Rather more dykes, though still only a minority and including some from both suites, carry the S_1 cleavage fabric, particularly at their margins. Some dykes, mostly lamprophyres, are demonstrably folded (or otherwise affected) by D_2 structures (Plate 27b). These early dykes were mostly intruded parallel to the bedding of the host strata and so have a general north-east to south-west orientation. Other dykes, that appear to be entirely post-folding, may also lie north-east to south-west, parallel to the host bedding, but have also been widely intruded into the north-west- and north-north-east-trending cross faults, where the foliated margins of some show their emplacement to have been synfaulting. In a few cases, for example within the Moniaive Shear Zone, north-east-trending dykes have been foliated by sinistral strike-slip shear, either as a D_3 effect, or during subsequent 'Acadian' deformation. Overall, there is no preferential pattern of orientation in either dyke suite relative to the other.

Successful radiometric dating has been restricted to undeformed lamprophyre dykes, from which has been obtained a range of ages from 400 ± 9 Ma to 418 ± 10 Ma (Figure 23). This provides a minimum age for the end of penetrative deformation, but equally suggests that dyke intrusion was protracted and probably polyphase. The youngest dykes are likely to include the microdiorite intrusions into the younger of the granitic plutons of south-west Scotland (described below), which have themselves been dated at about 397 Ma. In contrast, some lamprophyre dykes are metamorphosed by the thermal aureoles surrounding the plutons.

In petrographical terms, each of the two dyke suites contains a range of rock types:

(i) The mafic, lamprophyre suite is the more distinctive of the two, comprising mica-bearing kersantite and hornblende-bearing spessartite as well as coarse-grained hornblende-rich lamprophyric rock informally known as 'appinite'.
(ii) The microdiorite–microgranodiorite–microgranite suite, originally termed the 'porphyrite–porphyry series', is more variable and comprises a broad range of intermediate to felsic rock types.

The **lamprophyre dykes** consist of fine-grained, highly altered, feldspathic rock that is characterised by the presence of large hornblende or mica phenocrysts within a finer grained groundmass of plagioclase, amphibole and biotite with minor quartz and K-feldspar. Clinopyroxene is present in some rocks, with pseudomorphs after olivine being more common in mica-lamprophyres. The lamprophyres may contain felsic segregations and veinlets, as

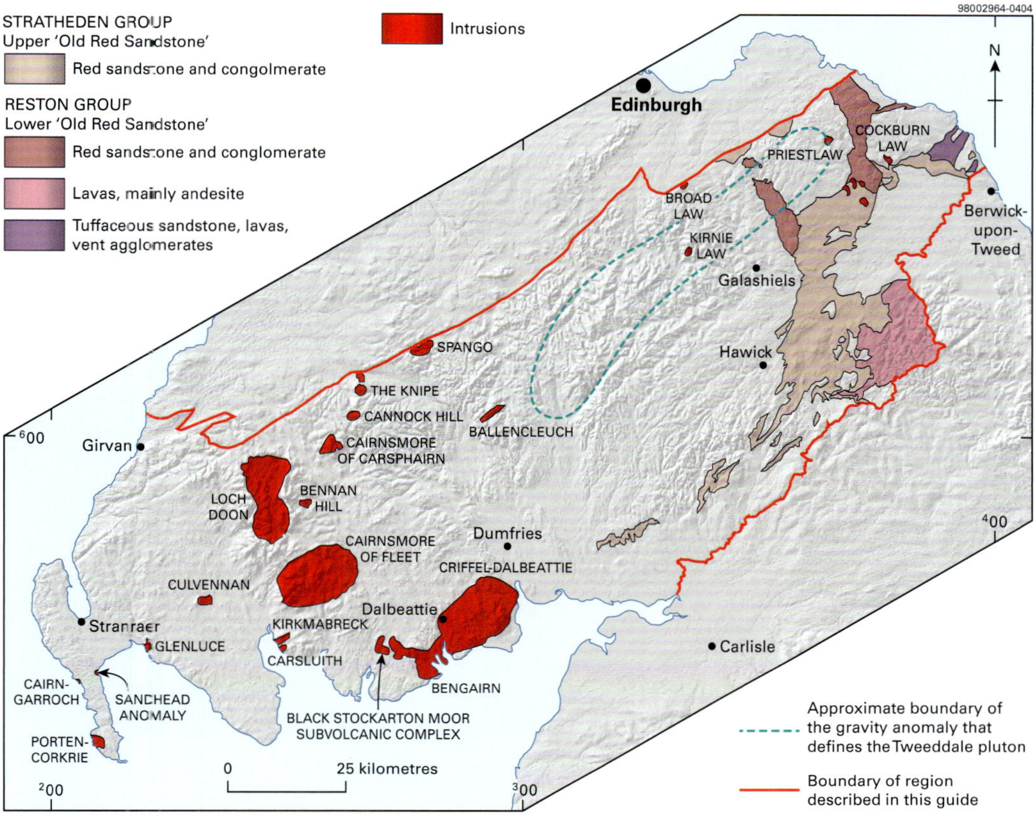

Figure 27 *Principal Caledonian intrusions in the south of Scotland. The large, concealed Tweeddale intrusion is defined from its gravity anomaly — see Figure 28.*

well as locally abundant quartz xenoliths. A dyke exposed near Garheugh [NX 269 501] contains xenoliths of variably foliated and metamorphosed diorite and/or tonalite which have been interpreted as representing either the basement to the Southern Uplands or intrusions emplaced at depth within the sedimentary cover. In general, however, xenoliths are uncommon. The calc-alkaline lamprophyric magmas originated deep within the mantle, with the kersantites and spessartites probably resulting from slight variations in melting conditions and differing degrees of modification in the upper mantle and crust. The lamprophyre dykes in the southern part of the Southern Uplands are mostly mica-bearing kersantites similar in composition to dykes of the same age found to the south in the English Lake District. This similarity has been taken as evidence that Avalonian mantle was underthrust beneath Laurentian Scotland during the final destruction of the Iapetus Ocean.

The **microdiorite, microgranodiorite** and **microgranite dykes** are locally common and are characterised by high plagioclase content and a low proportion of mafic minerals. They are generally highly altered, range from andesitic to rhyolitic in composition and comprise varying proportions of feldspar, amphibole, biotite and quartz phenocrysts set in a fine-grained quartzofeldspathic groundmass. Corroded clinopyroxene phenocrysts may also be present within the microdiorites. Coarser grained dioritic rocks form major dykes (up to 100 m wide) and small irregular shaped intrusions. This dyke suite, although calc-alkaline in character, is petrogenetically unrelated to the lamprophyre suite and was probably produced by melting within the lower crust or upper mantle. Only a few of the pervasively altered rhyolitic dykes may represent felsic differentiates of the lamprophyric magmas. In general, the felsic, late Caledonian minor intrusives are spatially associated with, and have

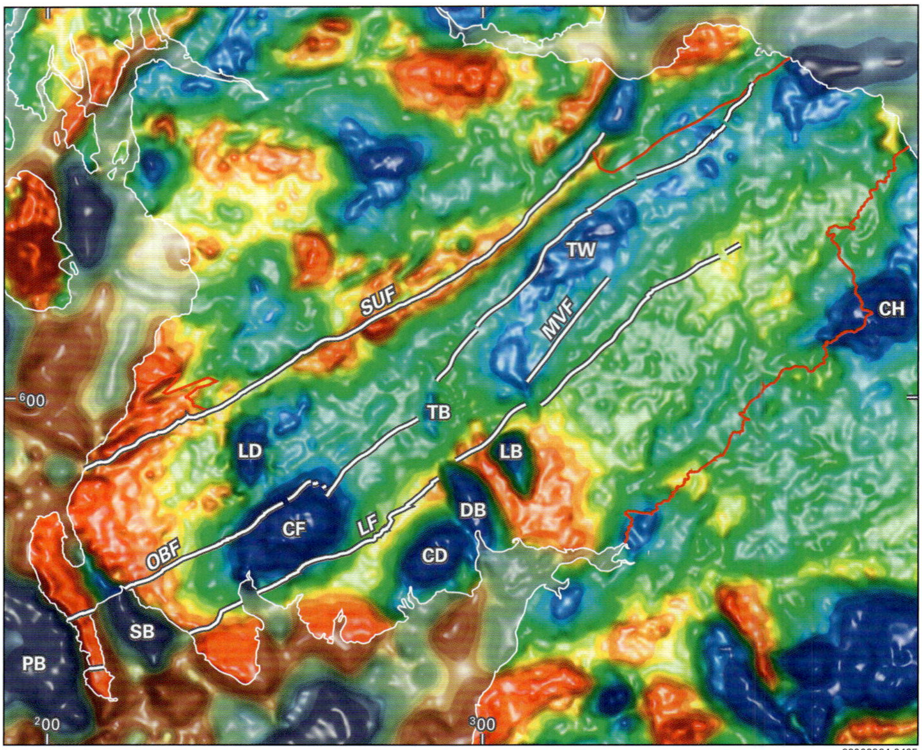

Figure 28 *Residual regional gravity field over the south of Scotland, after removal of a long-wavelength component derived by upward continuation of the observed field by 10 km. Colour shaded-relief image with equal-area colour and vertical illumination. Faults: LF Laurieston Fault, MVF Moffat Valley Fault, OBF Orlock Bridge Fault, SUF Southern Upland Fault. Plutons: CD Criffel–Dalbeattie, CF Cairnsmore of Fleet, CH Cheviot, LD Loch Doon, TW Tweeddale. Carboniferous–Permian basins: DB Dumfries, LB Lochmaben, PB Portpatrick, SB Stranraer, TB Thornhill.*

similar chemical compositions to, the larger granitic plutons, and were probably derived from the same magma chambers that ultimately produced the plutonic rocks.

Associated with the dykes are a number of **volcanic vents** and **breccia pipes**. Most are poorly exposed bodies mainly composed of highly altered sandstone-siltstone agglomerate with diffuse patches of tuffisite. One of the largest examples crops out on the west side of Kirkcudbright Bay at Shoulder O'Craig [NX 663 492] where a vent agglomerate, mostly of sandstone debris, is associated with a number of basaltic minor intrusions and kersantite sheets. Elsewhere, a particular concentration of vents, and of minor intrusions in general, forms the **Black Stockarton Moor subvolcanic complex** to the north-east of Kirkcudbright (Figure 27). The complex comprises a mass of intersecting lamprophyre and microdiorite dykes, granodiorite sheets, small granodiorite stocks and breccia pipes intruded into the Carghidown Formation. The development of the complex involved several phases of intrusion (Table 4) and overlapped with the emplacement of the nearby Bengairn Complex and Criffel–Dalbeattie Pluton (see below). A sygmoidal intrusion pattern and locally developed foliations suggest that intrusion of some dykes was accompanied by movement on the Caledonian fault network, though in this case reactivation of the faults may have been caused by the intrusion process. Compositionally, the subvolcanic complex and the Criffel–Dalbeattie Pluton show similarities that suggest they both arose from the same evolving magma(s).

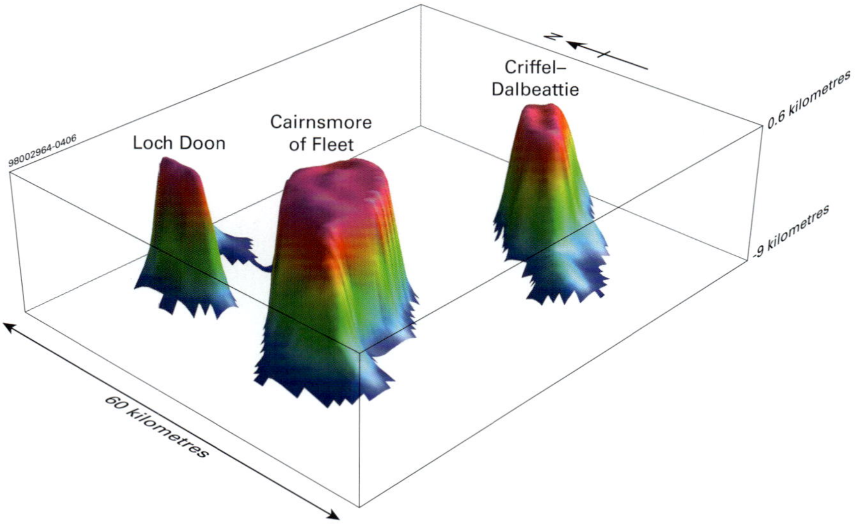

Figure 29 *Perspective view of the subsurface form of the low-density components of the major granitic plutons in Galloway, south-west Scotland.*

LATE CALEDONIAN PLUTONIC ROCKS

The major granitic plutons in south-west Scotland — Loch Doon, Cairnsmore of Fleet, Criffel–Dalbeattie and Bengairn (Figure 27) — are less dense than their Ordovician–Silurian host strata and so cause pronounced gravity lows (CF, LD and CD respectively in Figure 28). It is possible to model the subsurface forms of the plutons from their gravity anomalies and so establish that they are steep sided and extend down to depths of 8–12 km (Figure 29). Farther to the north-east, a gravity low in the Tweeddale area (TW in Figure 28) has been ascribed to a large, concealed Caledonian granitic pluton.

The isotope ratios of the igneous rocks, in particular the isotopic ratios of lead and strontium, provide insight into the origin of the magmas and their source rocks. The granitic plutons in the north of the Southern Uplands (such as Loch Doon) show similarities to igneous rocks in the Midland Valley whilst those in the south (such as Criffel–Dalbeattie) have characteristics more similar to igneous rocks in the English Lake District. This implies that the southern plutons were generated in response to collisional melting of underthrust Avalonian

Table 4 *Sequence of intrusive and structural events in the development of the Black Stockarton Moor subvolcanic complex.*

Phase of activity	Lithology and type of intrusion	Related igneous events
3	Microgranodioritic dyke swarm: minor E–W-trending porphyritic microgranodiorite dykes	
2	Mafic dyke swarm: minor hornblende lamprophyre (spessartite) and ophitic microdiorite	
	Microgranodioritic dyke swarm: sigmoidal, NW–SE-trending porphyritic microgranodiorite dykes	Criffel–Dalbeattie Pluton
1	Granodioritic stocks: elliptical granodiorite bodies	Bengairn Complex
	Granodiorite sheet complexes: NE–SW-trending granodiorite laccoliths with minor microdiorite	
	Mafic dyke swarm: minor hornblende lamprophyre (spessartite) and ophitic microdiorite	Volcanic vents
	Microgranodioritic dyke swarm: sigmoidal, NE–SW-trending porphyritic microgranodiorite dykes	

98002964-T0401

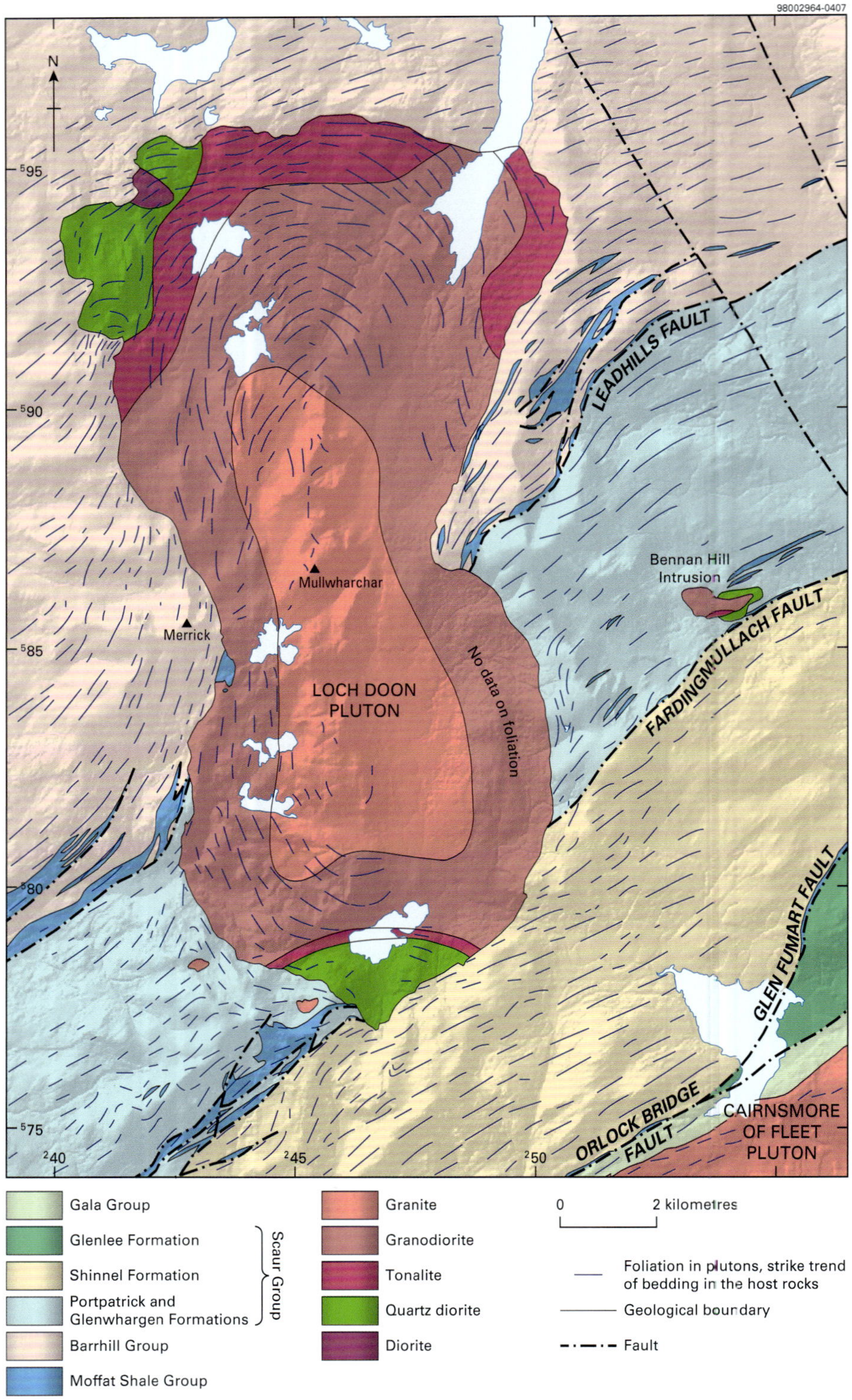

Figure 30 *Loch Doon pluton showing compositional zones, internal foliation and trend of strike within the host Ordovician strata.*

lithosphere, whilst the northern plutons were sourced in Laurentian lithosphere. At the very least, the contrast requires different crustal properties beneath the northern and southern sectors of the Southern Uplands (Figure 7), though as a complicating factor there is no absolute requirement for any one pluton to have been derived from a single source. For example, the isotopic evidence shows that the parental magma of the Criffel–Dalbeattie pluton contained a significant mantle component, although its central part, and the whole of the Cairnsmore of Fleet pluton, has the characteristics of crustal-derived granites. All of the plutons were probably formed in response to partial melting of lower crustal rocks initiated by underplating and intrusion of basic, mantle-derived melts. They are all compositionally zoned and so are likely to have been formed by several successive intrusive pulses, all derived from a single parental magma which was evolving at depth.

Loch Doon Pluton

The Loch Doon Pluton (406 ± 2 Ma, 410 ± 1 Ma, U-Pb, zircon) is an hour-glass-shaped body, elongated north–south, which cuts across the Leadhills and Fardingmullach tract-bounding faults (Figures 14 and 30) and intrudes the sedimentary rocks of the Kirkcolm, Portpatrick and Shinnel formations. Small satellite bodies occur to the east and south-west of the main mass, which the gravity model shows to be steep sided down to about 10 km, with a shallow subsurface plateau on the eastern side linking the main mass to at least one of the satellites. The steep sided nature of the pluton is also indicated by the relatively narrow thermal aureole of biotite hornfels. A steeply dipping igneous foliation is developed parallel to the margin of the pluton and the common xenoliths and autoliths are aligned parallel to the fabric. The presence of this planar fabric within the pluton, coupled with the marked deflection of the regional strike across its central 'waist', argues for forceful emplacement (Figure 30). However, there is less evidence of distortion of the regional strike at the northern and southern ends of the pluton suggesting that there the intrusion was, at least in part, permissive. The large number of xenoliths in those parts of the pluton suggests that stoping was an important process. Overall, intrusion of the Loch Doon Pluton may have been partially facilitated by a transtensional stress regime related to strike-slip movements along the tract-bounding faults.

The main mass of the pluton is compositionally zoned from diorite at the margin, through granodiorite and into granite at the centre (Figure 30). The boundaries between the various phases are gradational. The fine- to coarse-grained diorites and quartz-diorites are mainly composed of plagioclase and pyroxene (including hypersthene) with minor amounts of biotite, K-feldspar and quartz. Relict olivine and actinolitic hornblende after pyroxene are locally present. Passing through a transitional tonalite facies into the granodiorite, plagioclase becomes more calcic and pyroxene decreases, whilst quartz and K-feldspar increase towards the granitic centre of the pluton. The central granite is mainly composed of K-feldspar, plagioclase and quartz, with minor biotite, and locally contains large (up to 25 mm) K-feldspar phenocrysts. The main phases of the pluton are cut by a series of aplite, microgranite, porphyritic granite and pegmatite veins. Petrogenesis is likely to have been complex and polyphase, involving partial melting of basic, lower crustal rock, with variable contamination by assimilated sedimentary material.

Cairnsmore of Carsphairn Pluton

The roughly triangular Cairnsmore of Carsphairn Pluton (410 ± 4 Ma, Rb-Sr, mineral/whole-rock) lies to the north-east of the Loch Doon Pluton, intruded into Kirkcolm Formation strata. The pluton has a well-developed thermal aureole and cross-cuts the strike of the host rocks, although this is locally deflected around the western and northern margins of the intrusion. Though relatively small, the pluton is zoned with a central granite phase enclosed within microgranite, granodiorite and a marginal quartz diorite with lenses of diorite. The boundaries between these zones are gradational, though chilled contacts have been recorded between the microgranite and granodiorite, and between the two granite phases. The diorites and granodiorites at the margin of the pluton contain xenoliths that show variable

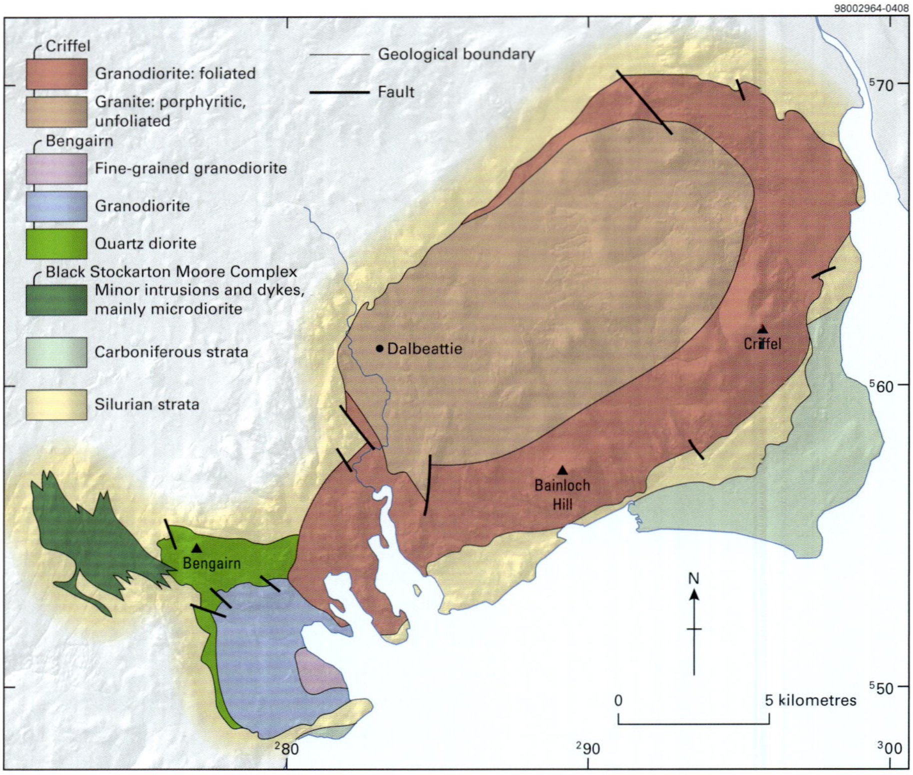

Figure 31 *Compositional zoning within the Criffel–Dalbeattie and Bengairn plutons.*

degrees of melting and assimilation. The pluton was emplaced as a series of magma pulses, ranging from dioritic to granitic in composition and derived from a single parental basaltic magma which fractionated at depth, though it is likely that the diorite melt fractionated in situ to quartz-diorite.

Cairnsmore of Fleet Pluton

The Cairnsmore of Fleet Pluton (397 ± 3 Ma, U-Pb, zircon) is the most evolved of the major granitic plutons and may also be the youngest. It is ovoid in form with a long axis extending east-north-east to west-south-west and is intruded into Gala Group host strata. The gravity model shows steeply dipping margins down to a depth of about 10 km. The pluton lies entirely within the Moniaive Shear Zone, but its intrusion postdates the main phase of deformation within the shear zone. As a result, it distorts the regional strike of the shear zone fabric and has produced a thermal metamorphic aureole of biotite-hornfels, up to several kilometres in width, which overprints the shear fabric. However, garnet and cordierite porphyroblasts, which developed during the earliest stages of thermal metamorphism, are deformed by the shear fabric. This suggests that initial emplacement of the pluton was facilitated by sinistral shear during the transtensional stress regime of the Early to Mid Devonian.

The pluton is compositionally zoned with a broad margin of coarse-grained, commonly microcline-phyric granite concentric to a core of fine- to medium-grained aphyric granite. Blocks of the coarse-grained, marginal granite are present within the aphyric granite, which is therefore likely to be a later intrusive phase. A weak foliation, defined by aligned feldspar and mica crystals, is present within both granite phases and occurs parallel to the margins of the pluton. The overall intensity and dip of this foliation increases towards the margins and are consistent with the fabric having developed during forceful emplacement. The local presence of screens of country rock within the southern part of the pluton suggest that it was, at least in part, emplaced as thick sheets. The absence of veining and only rare occurrence of country rock xenoliths, particularly in the roof zone, argues against significant stoping. Two pulses of intrusion are likely, the parental magmas having originated as partial melts of deep-seated basic rocks that began to fractionate during ascent, whilst at the same time becoming increasingly contaminated by assimilated sedimentary material.

As described above, in the thermal metamorphic aureole around the pluton, early garnet and cordierite porphyroblasts are contained in a broad zone of later biotite hornfels. In addition, skarn alteration occurs at several places within the aureole as calc-silicate lenses up to several metres in length and composed of epidote, amphibole and rare diopsidic pyroxene. These skarn minerals overgrow and replace the earlier-developed biotite.

Criffel–Dalbeattie Pluton

The Criffel–Dalbeattie Pluton (397 ± 2 Ma, Rb-Sr, whole-rock) is the largest of the major granitic plutons within southern Scotland. It is a composite, concentrically zoned granitic body with an ovoid shape, elongated north-east to south-west and intruded into Hawick Group strata. The gravity model shows steeply dipping margins down to a depth of at least 9 km, but the pluton was probably emplaced at a high level in the crust (and was rapidly unroofed) as boulders of granodiorite are found in Upper Devonian to lower Carboniferous conglomerate exposed a few kilometres to the south.

The pluton has an incomplete outer shell of foliated granodiorite containing aligned xenoliths of diorite, which surrounds a core of unfoliated porphyritic granite (Figure 31). The

Plate 28 *Kirkmabreck Quarry [NX 481 565], to the south of Creetown, reveals a microdiorite dyke swarm at the margin of a large granitic body (behind vehicle) (P008440).*

foliation is defined by the preferred orientation of feldspar, hornblende, biotite and quartz crystals. The development of the foliation and preferred alignment of the xenoliths within the granodiorite shell probably occurred during the forceful emplacement of the later granite core, possibly as a result of oblique, north-east-directed diapirism. Thermal metamorphism associated with the emplacement of the pluton resulted in a 2 to 3 km wide zone of biotite-hornfels and hornblende-diopside-hornfels.

The compositional boundary between the foliated shell and inner core is gradational and characterised by steep geochemical gradients. The outer shell comprises unzoned metaluminous hornblende-biotite-granodiorite, locally pyroxene-bearing at the outer margin. This passes progressively inwards through a zone of locally porphyritic biotite-granite, which contains small enclaves of hornblende-biotite-granodiorite, and then into muscovite-biotite-granite towards the core of the pluton. The outer granodiorite was probably derived from a mantle or lower crustal source, whereas the more evolved granitic core is more likely to have been generated by partial melting of middle and upper crustal, immature metasedimentary rocks.

Bengairn Complex

The Bengairn Complex is a composite pluton lying between the Black Stockarton Moor subvolcanic complex, to the north-west, and the Criffle–Dalbeattie Pluton, to the north-east (Figure 31). Its emplacement into the sedimentary rocks of the Hawick Group postdated the first phase of activity within the Black Stockarton Moor complex (Table 4) but the Bengairn Complex was intruded by a suite of north-west-trending lamprophyre and microdiorite dykes during a later phase of subvolcanic activity. Although no intrusive contacts between the Bengairn Complex and the Criffel–Dalbeattie Pluton have been recognised, it is likely that the Bengairn Complex was emplaced prior to, and is intruded by, the Criffel–Dalbeattie Pluton.

The Bengairn Complex is a zoned pluton with an incomplete outer rim of quartz-diorite intruded by a granodiorite core. The medium- to coarse-grained quartz diorite consists of plagioclase, hornblende, biotite, quartz and K-feldspar. It possesses a weakly developed foliation and contains xenoliths of partially assimilated, hornfelsed country rock and basic igneous material. The granodiorite is a fine- to medium-grained rock composed of zoned plagioclase, quartz, K-feldspar, biotite and hornblende. Xenoliths are uncommon within the granodiorite, but where present are aligned parallel to a more widely developed igneous foliation.

Minor intrusive bodies

A number of small intrusive masses cut the Ordovician and Silurian strata in the far west of the Southern Uplands (Figure 27). The **Culvennan diorite** (to the west of Newton Stewart) has an outcrop area of about 1.5 km^2 and is mainly composed of pyroxene-biotite-diorite with some tonalite and granite. The surrounding thermal aureole encloses several satellite pods of diorite, which suggests that its subsurface form may be significantly larger than the outcrop. The **Glenluce diorite** (exposed at the head of Luce Bay, on the eastern side) is mainly composed of a coarse-grained plagioclase-phyric pyroxene-diorite with minor plagioclase-pyroxene-biotite-phyric andesite, quartz diorite and pyroxene-rich melanocratic diorite phases. On the opposite (western) side of Luce Bay, an approximately 2 km^2 area of hornfelsed sandstone at Sandhead coincides with a magnetic anomaly indicative of a probably dioritic body rising to within 50 m of the ground surface. This feature — the **Sandhead anomaly** — straddles the Orlock Bridge Fault, which may well have influenced the site of intrusion. The small **Cairngarroch Complex** (on the west coast of the Rhins of Galloway, south of Portpatrick) lies immediately to the north of the Orlock Bridge Fault, which forms the southern margin of this polyphase, dioritic to granodioritic body. The complex must postdate movement on the fault since the southern part of its metamorphic aureole, which extends across the fault, is not displaced. Nevertheless, it is probable that the Orlock Bridge Fault had a major influence on the site and shape of the intrusion. The older part of the Cairngarroch Complex is composed of a plagioclase-phyric biotite-hornblende-microtonalite which was intruded by a later phase of plagioclase-phyric biotite-hornblende-granodiorite. The southernmost part of the intrusion contains numerous screens of sandstone as well as brecciated, quartz-porphyritic granite, the latter occurring as xenoliths and enclaves within the quartz-microdiorite. Pervasive hydrothermal alteration and widespread Fe-As-Cu-Mo mineralisation are associated with the Cairngarroch Complex. Farther south, the roughly rectangular **Portencorkrie Pluton** crops out over about 5 km^2 on the Rhins of Galloway between Laggantalluch Head and Crammag Head. It is a compound dioritic to granodioritic stock with an outer zone of locally foliated pyroxene-mica-diorite on its northern side and hornblende-mica-diorite to the south. The centre of the intrusion is composed of medium-grained granodiorite, locally containing phenocrysts of K-feldspar. Regular, continuous geochemical trends between the diorites and granodiorites have been used to suggest that the Portencorkrie Complex developed in response to the multiple intrusion of magma that was fractionated at depth.

Two small but significant intrusive bodies crop out on the east side of Wigtown Bay (Figure 27). The **Carsluith** intrusion is a near-vertical body (up to 150 m wide at outcrop) lying parallel to bedding in the steeply dipping country rocks. It is a composite intrusion with an outer carapace of plagioclase-phyric microdiorite enclosing a microgranodiorite core. The presence of screens of microdiorite within the microgranodiorite core, and xenoliths of mudstone that are aligned parallel to bedding within the adjacent country rocks, suggest that the Carsluith intrusion was emplaced as a series of sheets. An irregular dolerite dyke that cuts both phases is most probably of Palaeogene age. The **Kirkmabreck** (or **Creetown**) intrusion is a relatively small, subvertical and sheet-like microgranodiorite body (up to 200 m wide at outcrop) emplaced along the Laurieston Fault. The intrusion has a complex emplacement history with an early phase, represented by a swarm of closely spaced microdiorite dykes, cut by the main microgranodiorite body, which contains xenoliths of the dyke microdiorite (Plate 28). The intrusion is enclosed by a well-developed thermal aureole (up to 100 m wide) of biotite-hornfels with subsequent skarn alteration (grossular + diopside + clinozoisite + actinolite), though these effects appear to have been generated by the early dyke swarm and predate the main intrusion.

A number of small intrusive bodies crop out in the central, northern part of the Southern Uplands (Figure 27). None are well exposed and the **Ballancleuch granodiorite** is known only from a boulder field, the shape of which suggests a dyke-like body adjacent and parallel to the Leadhills Fault; though no more than 300 m wide it extends for about 4 km along strike. The most substantial of these bodies are those at Spango and The Knipe. The **Spango pluton** has an outcrop of about 10 km^2. It is a hornblende-biotite-granodiorite that cuts across the tract-bounding Carcow Fault but is itself cut by the Southern Upland Fault along its north-western flank. The granodiorite becomes more melanocratic towards its margin and contains small xenoliths of comagmatic microdioritic rocks as well as rarer xenoliths of hornfelsed sedimentary rocks. **The Knipe** intrusion is a composite body with a core of hornblende-biotite-granodiorite or quartz-diorite passing outwards into pyroxene-bearing dioritic rocks. The main part of the intrusion has an outcrop area of about 1.5 km^2, but a linear offshoot to the north forms a separate, narrow outcrop about 2 km long. To the south of The Knipe intrusion, several small, dioritic bodies lie in a linear zone between it and the Cairnsmore of Carsphairn Pluton; examples include the intrusion at **Cannock Hill**. In the same generalised zone, areas of hornfelsed Ordovician sandstone indicate the presence of additional small intrusions, concealed but close to surface.

In the north-eastern part of the Southern Uplands, only a few granitic/granodioritic intrusions crop out, the largest of which are those at Priestlaw, Cockburn Law and Broad Law (Figure 27). The **Priestlaw** intrusion in the Lammermuir Hills is an irregularly shaped pluton with an outcrop area of about 3.5 km^2, which is mainly composed of a variably porphyritic biotite- or hornblende-biotite-granodiorite. The granodiorite is enclosed within a marginal phase of slightly porphyritic quartz-augite-biotite-diorite or microdiorite; olivine-norite forms the north-west corner of the intrusion. To the south-east of the Priestlaw intrusion, the **Cockburn Law** intrusion (c. 1.5 km^2) is zoned from an outer shell of quartz-diorite into a granodiorite core. For its size, this relatively small intrusion is enclosed withir a remarkably wide thermal aureole, locally extending up to 1 km from the intrusion. The small intrusive mass at **Broad Law**, in the Moorfoot Hills, is largely made up of biotite-granodiorite with phenocrysts of plagioclase.

five

Devonian: the 'Old Red Sandstone' and associated volcanic rocks

By late Silurian times, the Iapetus Ocean had closed and the more-or-less orthogonal convergence of Avalonia and Laurentia had given way to a sinistral, transtensional tectonic regime across the Iapetus Suture Zone. The intrusion of granitic plutons and dyke swarms spanned the Silurian–Devonian transition, as described in Chapter 4, and by the Early Devonian the south of Scotland region had become an upland area subjected to intensive subaerial erosion. The steeply dipping Lower Palaeozoic strata, underpinned by the granitic plutons, gave rise to a rugged landscape of hills and valleys, the latter controlled by reactivation of Caledonian faults under the influence of a sinistral stress regime orientated north-east to south-west.

The region was subject to an arid climate in low latitudes south of the equator (Figure 3). Terrestrial vegetation was poorly developed at the time and so erosion was rapid and extensive. Screes and piedmont fans were laid down in the areas of high relief, whilst on the lower ground fluvial and lacustrine sediments were deposited by braided (and most probably seasonal) rivers on floodplains and in ephemeral lakes. These deposits are now seen as coarse conglomerates, sandstones and mudstones and have long been known informally as the 'Old Red Sandstone'. They are preserved mostly in the eastern part of the region, where they overlie the deeply eroded Lower Palaeozoic rocks with a marked unconformity. Volcanic activity associated with the Caledonian intrusions continued and was widespread at times. In the south of Scotland it gave rise to the series of andesitic lavas and volcaniclastic rocks which are interbedded with sandstones in the area around Eyemouth and St Abbs, the Eyemouth Volcanic Formation, and the less well-exposed but much more extensive outcrop of similar volcanic rocks around Cheviot, the Cheviot Volcanic Formation. Much of the latter, together with an associated pluton and dyke swarm, crops out to the south-east of the English border in Northumberland (Figure 32).

The historical descriptor 'Old Red Sandstone' now has supergroup status, and is largely supported by formal lithostratigraphy applied on a regional basis. In the south of Scotland, the 'Lower Old Red Sandstone' is represented in the Girvan area by part of the Lanark Group, and to the south-east of the Southern Upland Fault by the Reston Group. In the latter area, the 'Upper Old Red Sandstone' now forms part of the Stratheden Group.

In common with the rest of Scotland south of the Great Glen, the Devonian strata in the Southern Uplands region can be divided into two quite distinct successions separated by an unconformity: the Reston and Lanark groups below, and the Stratheden Group above. The discordance between the upper and lower divisions is markedly angular in places, testifying to some tectonic tilting during the Mid Devonian, perhaps as a peripheral effect of the Caledonian Orogeny's Acadian event (see Chapter 4). There are no known Middle Devonian rocks in the south of Scotland region, this interval of about 20 million years being a period of erosion both there and in the adjacent parts of northern England and the Midland Valley of Scotland. During this episode of basin inversion and uplift, it is likely that much of the Lower Devonian cover was eroded and recycled towards the north and south. The present distribution of the Lower Devonian strata in several isolated and disconnected outliers, originally sediment-filled valleys in the Devonian landscape, is probably a relic of a more extensive presence prior to Mid Devonian erosion.

The conglomerate, sandstone and siltstone of the Upper Devonian Stratheden Group have a more widespread distribution than its subjacent counterpart, and in places oversteps onto the Lower Palaeozoic rocks. The resulting, sharply angular relationship between subvertical Silurian sandstone and the gently-dipping Stratheden Group strata can be demonstrated at two localities, both known as 'Hutton's Unconformity' after the celebrated 18th century savant who first recognised their significance for the age of the Earth: Siccar Point [NT 813 710] near Cockburnspath on the North Sea coast (Plate 29), and Allar's Mill [NT 652 198] on the Jed

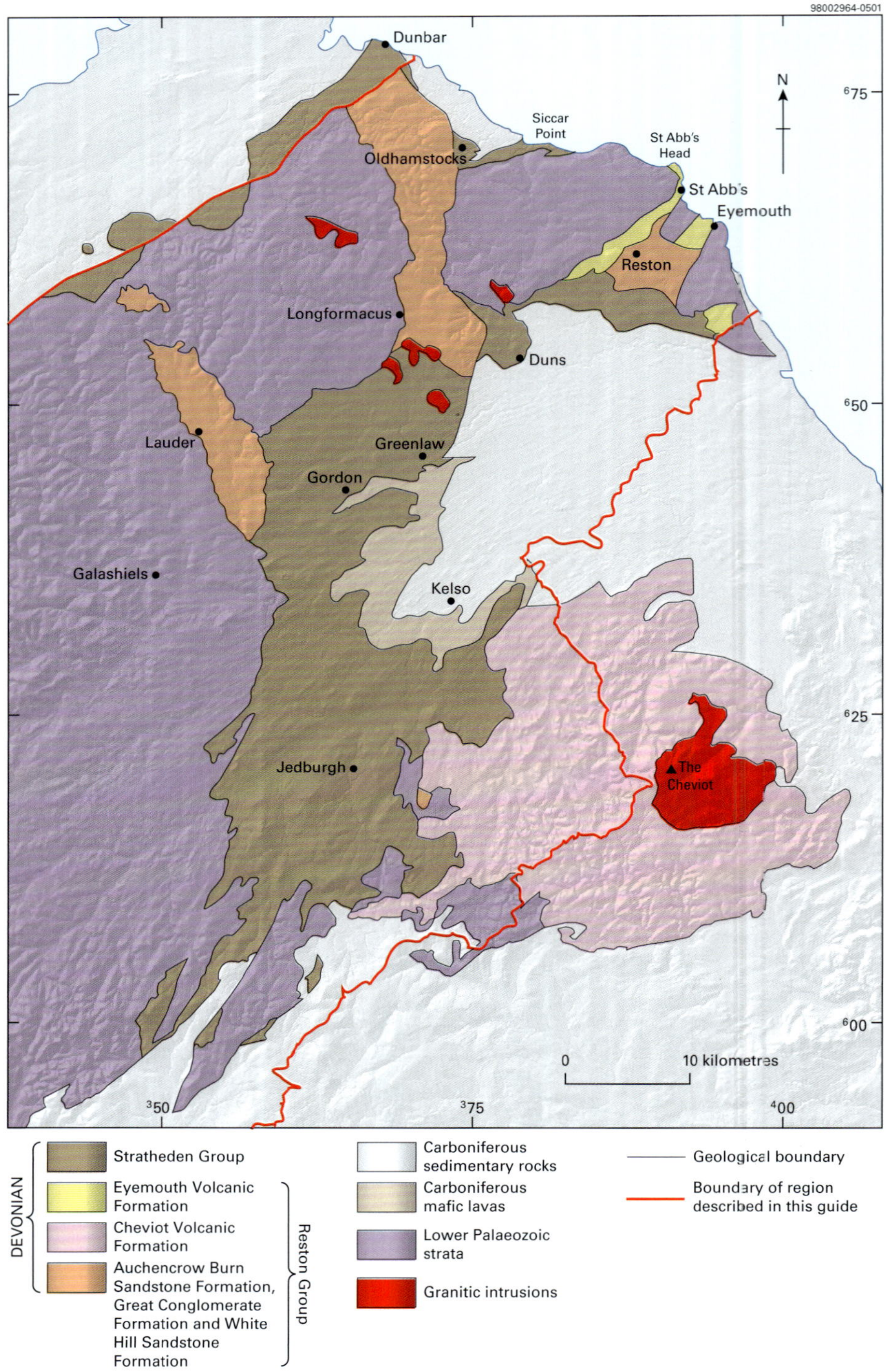

Figure 32 Devonian ('Old Red Sandstone') outcrop in the east of the Southern Uplands, showing the distribution of the Reston and Stratheden groups, Lower and Upper Devonian respectively.

Plate 29 *'Hutton's Unconformity', Siccar Point [NT 812 710]. Steeply inclined Lower Silurian turbidite sandstone beds of the Ettrick Group are unconformably overlain by Upper Devonian pebbly sandstone beds of the Stratheden Group (P774192).*

Water just to the south of Jedburgh. In both these examples, the Stratheden Group strata rest unconformably on steeply inclined, Silurian (Llandovery) sandstones: Ettrick Group at Siccar Point, Hawick Group at Jedburgh. Both of the 'Hutton's Unconformities' have been proposed as 'Global Geosites' on the basis of their historical importance, whilst a full-sized casting of part of the Siccar Point unconformity is housed in the American Museum of Natural History, New York.

LANARK GROUP

Lanark Group strata were deposited in the Lanark Basin, along the south-east side of the Midland Valley terrane. Part of its outcrop extends into the Girvan district, where it unconformably overlies and is faulted against the Lower Palaeozoic rocks. Conglomerate of the **Greywacke Conglomerate Formation**, containing clasts of the eponymous sandstone, unconformably overlies steeply dipping Silurian and Ordovician strata in a much-dissected outcrop between Straiton and Girvan, dominated by Hadyard Hill [NX 271 993] and Daljedburgh Hill [NX 313 973]. There, the conglomerate is succeeded by red-brown sandstone of the **Swanshaw Formation** overlain, in turn, by basaltic lavas of the **Carrick Volcanic Formation** that are correlated with the **Duneaton Volcanic Formation** farther to the north-east. These examples are peripheral to much more extensive outcrop of the formations elsewhere in the Midland Valley of Scotland that are more fully described in the companion volume for that area in the British Regional Geology series.

RESTON GROUP

The strata comprising the Reston Group crop out to the south-east of the Southern Upland Fault. They are mostly red-brown conglomerates and sandstones laid down in a fluvial-terrestrial environment wherein large rivers drained broadly towards the south-west from a catchment in what is now Scandinavia. Reston Group strata are mostly preserved as an infill to pre-existing, approximately north–south valleys in the Early Devonian land surface that were instigated by reactivation of the Caledonian, north-east and north-north-west conjugate fault sets.

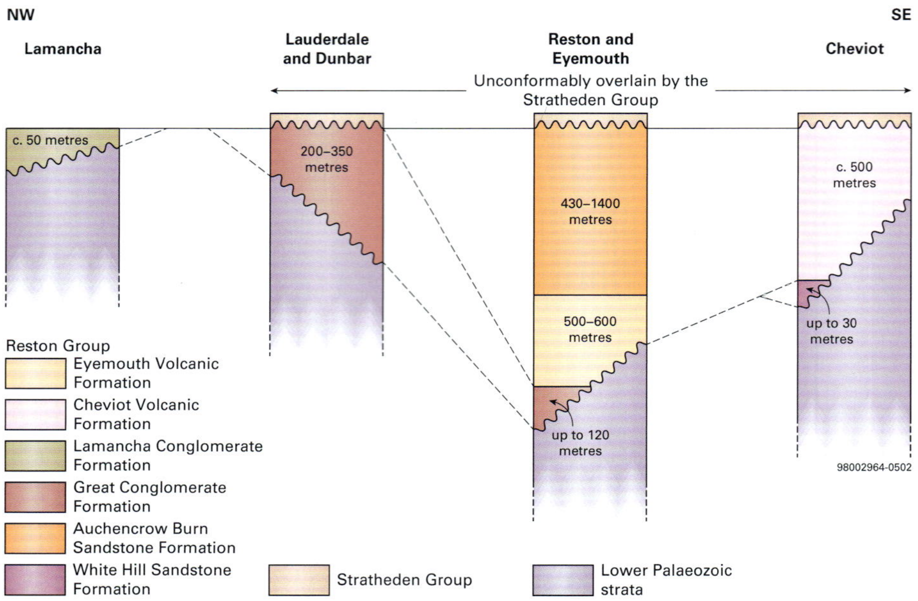

Figure 33 *Comparative stratigraphical columns for the Reston Group (Lower Devonian) in its outcrops in south-east Scotland.*

The most westerly outcrop of the Reston Group is a small isolated patch of poorly exposed red and purple conglomerate, known as the **Lamancha Conglomerate Formation**, located adjacent to the Southern Upland Fault near Leadburn [NT 200 515]. This unit is largely composed of rounded pebbles and boulders of wacke-type sandstone, with some basalt and chert detritus, set in a sandy matrix; it is probably of relatively local origin. Its elevated position on top of a ridge, rather than filling a valley, suggests that it is just the last vestige of a formerly much more extensive deposit.

The westernmost of the three northern outcrops of the Reston Group (Figure 32) occurs in Lauderdale with an associated but disconnected outlier to the north-west at Corfrae Common, near Soutra [NT 470 590]. In the Lauderdale outcrop, red and purple conglomerate of the **Great Conglomerate Formation** is up to 350 m thick (Figure 33). It consists of subrounded pebbles and cobbles of wacke-type sandstone, with some chert and jasper, set in a red-brown sandstone matrix. Farther east, another outcrop of very similar Great Conglomerate Formation strata (Plate 30) extends north–south from the vicinity of Dunbar in the north, to the area between Longformacus and Duns in the south (Figure 32). Some spectacular exposure is provided in the north of this outcrop by a series of deeply incised glacial meltwater channels near Oldhamstocks (Plate 31). There is little direct evidence for the age of the Great Conglomerate and it is assigned to the Lower Devonian principally on lithological grounds. One imprecise constraint is provided by the radiometric (K-Ar, biotite) age of 400±9 Ma from a lamprophyre dyke cutting conglomerates that underlie the Eyemouth Volcanic Formation (described below) near St Abb's Head (Figures 32 and 33).

Inland from Eyemouth and St Abbs, the easternmost outcrop of the Reston Group (around Reston itself) contains about 120 m of red feldspathic sandstone and conglomerate with exclusively wacke-type sandstone clasts; the overall lithofacies is very similar to that of the Great Conglomerate Formation as seen elsewhere. A few thin beds of pedogenic nodular limestone (calcrete or 'cornstone') are also present. Calcrete is more characteristic of the Upper Devonian and lowermost Carboniferous lithofacies in the south of Scotland, but the Reston–Eyemouth sequence is overlain by about 600 m of Lower Devonian lavas with interbedded volcaniclastic

Plate 30 *Great Conglomerate (Reston Group) exposed in Burn Hope [NT 699 699], about 4 km up-stream from Oldham-stocks (P005866).*

rocks and red sandstones, together known as the **Eyemouth Volcanic Formation**. In addition, as described above, conglomerate underlying the Eyemouth lavas near St Abb's Head is cut by a lamprophyre dyke dated at about 400 Ma. To the north-west of Eyemouth beach, in the cliffs beneath Eyemouth Fort [NT 943 648] the purple basaltic-andesite lavas are overlain, in turn, by a younger red conglomerate containing clasts of sandstone and andesite. This may correlate with the upper part of the succession around Reston, the **Auchencrow Burn Sandstone Formation** (Figure 33), in which the sandstone is largely pebbly and/or volcaniclastic, with plentiful detritus derived from the underlying or possibly coeval andesitic lavas and volcanic breccias. Alternatively, the volcaniclastic succession at Eymouth Fort could be younger and a part of the Stratheden Group (see below).

The lavas of the Eyemouth Volcanic Formation may possibly have been erupted from volcanic vents now filled with andesitic agglomerate that crop out locally along the coast between Eyemouth and St Abbs; a steep, brecciated and hydrothermally veined vent margin is particularly well exposed to the north-west of St Abbs harbour. The lavas are mostly fine-grained and aphyric or olivine-phyric (up to 12 per cent olivine), or more rarely orthopyroxene-olivine-phyric. They have a compositional range from basalt to basaltic andesite, though pervasive hydrothermal alteration makes analytical data unreliable. Many of the lavas are block flows with a variety of internal features including flow-jointing, vesiculation, auto- and mass-flow brecciation. Most of the associated volcaniclastic beds (Plate 32) are coarse sandstones that appear to have been derived from the local lavas and to have been deposited under high-energy, flood conditions. In addition, there are local intercalations of breccio-conglomerates that consist mostly of large blocks of lava and volcaniclastic rock up to 2 m across. Many of the blocks are moderately well rounded and in places a rough stratification is discernible.

Plate 31 *Great Conglomerate (Reston Group) exposed in Fairy Castle Dean, Back Water [NT 700 699] (P001091). Erosion has produced a 'badlands' landscape featuring boulder-capped residual pillars.*

Spanning the border with England in south-east Scotland, the andesitic lavas of the **Cheviot Volcanic Formation** are poorly exposed over an area of about 600 km^2 with a preserved thickness of about 500 m (Figures 32 and 33). It is likely that the coeval granite pluton (exposed to the south of the border) once had a substantial cover, and the original thickness of volcanic rocks may have exceeded 2000 m. The volcanic rocks unconformably overlie steeply dipping, tightly folded sandstone and cleaved mudstone of the Wenlock Riccarton Group, though locally the **White Hill Sandstone Formation** intervenes. This unit is up to 30 m thick and comprises red sandstone and mudstone with sporadic beds of conglomerate. The Cheviot lavas are overlain by either Upper Devonian, Stratheden Group, conglomerates containing abundant andesite and granite clasts, or by lower Carboniferous beds, only some of which contain andesite fragments. Thin intercalations of red sandstone within the volcanic outcrop have proved to be unfossiliferous, but a radiometric age of about 396±4 Ma (Rb-Sr, biotite) suggests that volcanism occurred at the end of Early Devonian times, late in the Acadian tectonic episode.

The Cheviot lava succession comprises stacked sheets of trachyandesite and subordinate trachyte that contain phenocrysts of plagioclase, hypersthene, augite, ilmenite and apatite. There are also a few sheets of biotite-feldspar-phyric trachyte (previously described as 'mica-felsites') near the base of the succession that may represent eruptions of more fractionated magma. The sheets are massive to amygdaloidal and scoriaceous, with near-horizontal platy jointing a characteristic feature of many. The more readily weathered tops of the units have been eroded in some areas to form a prominent bench and scarp landscape. The benches reflect the gentle eastward dip of the sequence. The sheets have been interpreted as lavas, erupted in a subaerial setting from a cluster of low-profile volcanoes, though the presence of a significant proportion of sills cannot be ruled out.

The basal unit of the volcanic formation, spanning the border in the south-west of the outcrop, comprises up to 60 m of breccia composed of rubbly, angular to subangular clasts of silicic volcanic rock along with some mudstone fragments. This unit is most likely to have been the product of initial phreatomagmatic eruptions, though a sedimentary origin cannot be entirely discounted. Other intercalations of pyroclastic and volcaniclastic sedimentary rocks are present higher in the succession, though they are sparse. In the uppermost parts of some sheets there are enclaves and fissure-fills of green fine-grained sandstone and siltstone. Fragments of these rocks are commonly seen in streams, suggesting that this lithology is more common in the unexposed parts of the succession.

STRATHEDEN GROUP

The largely sandstone lithofacies forming the 'Upper Old Red Sandstone' in Scotland spans the Devonian–Carboniferous chronostratigraphical boundary. In lithostratigraphical terms, this terrestrial and generally unfossiliferous sequence is divided into a lower unit of red sandstone and conglomerate, the Stratheden Group, and an upper unit characterised by thin beds of pedogenic nodular limestone (calcrete or 'cornstone'), known as the Kinnesswood Formation and assigned to the Carboniferous Inverclyde Group. The presence of the calcrete beds suggests that the climate, though generally hot and semi-arid, included a substantial seasonal rainfall. A Famennian (latest Devonian) age is suggested for at least part of the Stratheden Group on the basis of a few discoveries of fossil fish (Plate 33). The Kinnesswood Formation is generally thin and variably developed and its base, though probably diachronous, is commonly taken to be the local base of the Carboniferous succession. However, sparse fish faunas from a few localities show that the lower part of the Kinnesswood Formation may also be of Famennian age. Despite any resulting anachronism, in this account the description of the entire Kinnesswood Formation is reserved until the Carboniferous succession is discussed in Chapter Six.

The main outcrop of the Stratheden Group lies in the east of the Southern Uplands around Gordon and Jedburgh (Figure 32). There the sequence comprises up to about 200 m of

Plate 32 *Coarse volcaniclastic sandstone interbedded with lava of the Eyemouth Volcanic Formation at Horsecastle Bay [NT 918 685] to the south-west of St Abb's Head (P005893).*

Plate 33 *Plate from the armoured fish* Bothriolepis hicklingi, *BGS GSE 10396. Stratheden Group near Greenheugh Point, East Lothian [NT 800 705] (P693051).*

alluvial sandstone and siltstone, with sparse conglomerate at the base, deposited in a fluviatile environment. In general, and ignoring the basal conglomerate, there is an upward trend to coarser lithologies, accompanying a shift from high to low sinuosity fluviatile deposits, and in the east of the outcrop a sediment provenance in the volcanic rocks of the Cheviot Hills is likely. The Famennian age is confirmed by fish faunas. To the west, a thin sequence comprising about 20 m of red sandstone and conglomerate underlies the lowermost calcrete at the base of the Kinnesswood Formation and extends intermittently, following the margin of that formation's outcrop, past Langholm and Dumfries as far as Kirkbean [NX 975 592] on the Solway Firth. At Kirkbean, the presence of granitic clasts in the conglomerate suggests a provenance in the Galloway Hills. Though palaeocurrent evidence is inconclusive, it is possible that the Late Devonian rivers that laid down the Stratheden Group strata drained to an internal basin in the Jedburgh area.

An aeolian influence is seen particularly in the younger strata, and is most apparent in the north-east, around Siccar Point, site of the more spectacular of Hutton's unconformities (Plate 29). Overall, a topographical relief of around 100 m can be established there at the unconformity between the Stratheden Group and the underlying Silurian rocks Immediately above the unconformity lies the thin basal conglomerate member (up to 6 m thick) of the **Redheugh Mudstone Formation**. The conglomerate member is mostly made up of angular clasts of the subjacent Silurian sandstone, contained in a red sandstone matrix. Above the conglomerate, the greater part of the Redheugh Mudstone Formation, which is about 200 m thick, comprises red-brown, sandy mudstone interbedded with variable brown, grey-green or cream sandstones that become more numerous and massive upwards. Eventually sandstone dominates the succession and the Redheugh Mudstone Formation passes transitionally into the overlying **Greenheugh Sandstone Formation**, about 70 to 100 m of mostly red-brown, coarse-grained sandstone with thin interbeds of red mudstone. Famennian fish faunas have been recovered from both the Redheugh and Greenheugh formations.

six

Carboniferous

At the end of the Devonian Period, Scotland lay in low latitudes just south of the Equator, within the northern part (Laurussia) of the huge Pangaea 'supercontinent'. During Carboniferous times, Pangaea drifted slowly northward, thus moving Scotland from arid, through humid tropical and back to arid climatic zones (Figure 3d). Throughout this time, the Southern Uplands was a terrestrial area of low to moderate relief forming an elongate but discontinuous barrier between depositional basins in the Midland Valley to the north and the Northumberland–Solway Basin to the south; a more open connection between these two basins probably existed to the north-east of the present North Sea coastline and extended onshore as the Tweed Basin (Figure 34). The Northumberland–Solway Basin was itself the northernmost depositional centre within the larger Pennine Basin, which covered much of northern and central England and had sedimentary links westwards into North Wales and Ireland, and eastwards into mainland Europe.

Across the south of Scotland, Carboniferous strata are now preserved along the eastern and southern margins of the Southern Uplands: the Dunbar district, much of Berwickshire and Roxburghshire, the south of the Langholm district (wherein lies the Canonbie Coalfield) and the Solway coastal fringe of Dumfriesshire. Other sequences form outliers that originated as the fill of palaeovalleys within the massif: examples include the outliers at Whitecleuch, Sanquhar, Thornhill and Stranraer. A range of lithofacies and depositional environments are represented.

As described in Chapter 5, the youngest Devonian and oldest Carboniferous rocks form a continuous terrestrial (fluvial and aeolian) sequence containing sporadic miospore floras and rare fish faunas. Accordingly, there has been persistent difficulty in defining a reliable and consistent base for the Carboniferous throughout the region. In south and central Scotland the base of the Kinnesswood Formation is commonly correlated with the base of the Carboniferous, and is taken at the incoming of a pedogenic nodular limestone (calcrete or cornstone)-bearing sequence that probably represents a general and widespread change to a seasonally wetter climate. However, though assigned to the Carboniferous and included within the Inverclyde Group (Table 5), the lower part of the Kinnesswood Formation displays 'Upper Old Red Sandstone' lithofacies and is similar in character to the underlying Stratheden Group. In places, like the higher formations of that group, the Kinnesswood Formation in the south of Scotland contains a probably Famennian (Late Devonian) fish fauna as described in Chapter 5 of this account; elsewhere in the region, Carboniferous miospores have been recorded close to the base of the Kinnesswood Formation. To compound the problem, both Carboniferous and Devonian miospores have been recovered from low in the Kinnesswood Formation in different parts of the Midland Valley. On a regional scale the base of the formation therefore seems likely to be diachronous. In some parts of southern Scotland the Kinnesswood Formation is succeeded by basaltic lavas of earliest Carboniferous age, but the error range on the available radiometric dates precludes a definitive chronostratigraphical allocation.

During the Tournaisian the sea gradually encroached onto the Southern Uplands massif, giving rise to a conformable transition in south-east Scotland, within the Inverclyde Group, from the fluvial, calcrete and red-bed facies of the Kinnesswood Formation up into the lagoonal mudstone with dolostone (cementstone) facies of the Ballagan Formation. The sedimentary transition was interrupted locally by the subaerial eruption of basalt lava around Langholm and Kelso. Some of the larger valleys such as Nithsdale experienced partial marine inundation and became narrow arms of the sea connected to the more open seaways along the Solway Firth and the Midland Valley, but there was never an open marine connection across the massif through these palaeovalleys. Through the late Visean and Namurian, marginal and

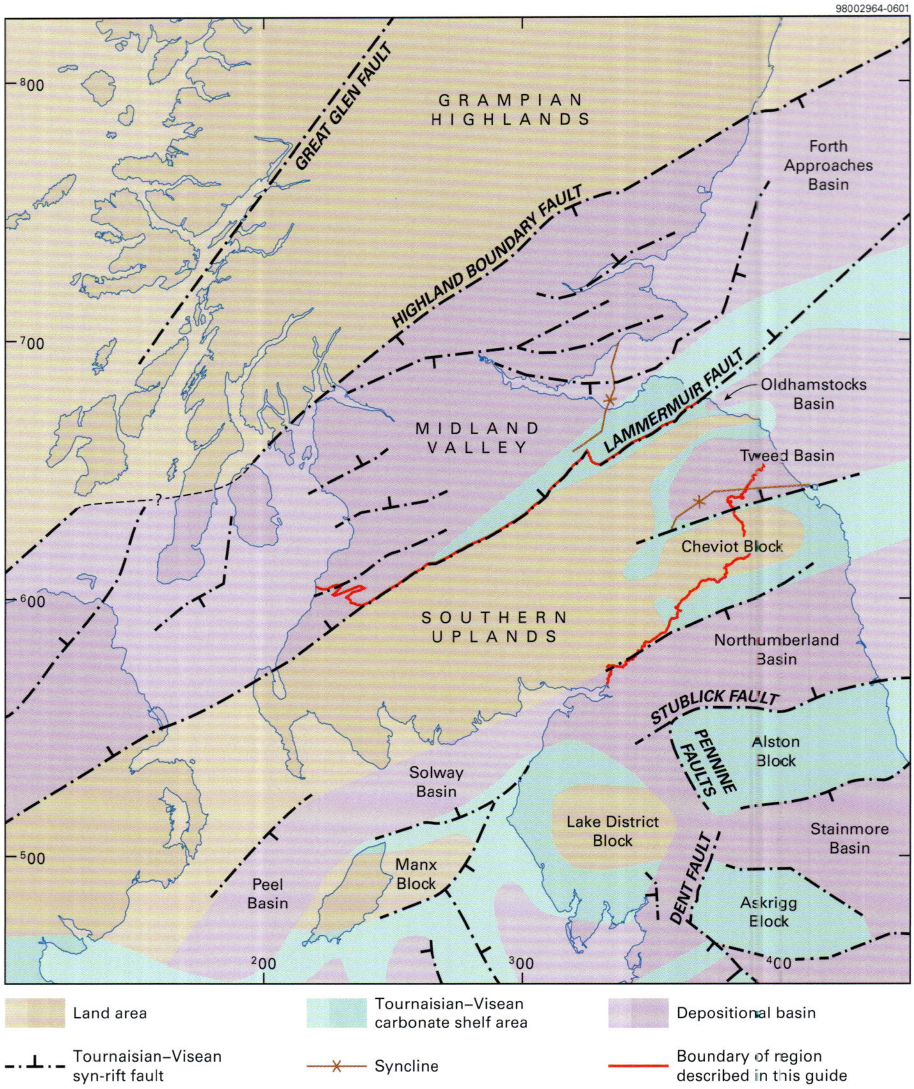

Figure 34 *Structural setting for Carboniferous sedimentation across the south of Scotland and the surrounding regions. (After Waters et al., 2007. BGS Research Report RR/07/01).*

shallow marine conditions prevailed with repeated flooding cycles creating a characteristic succession of 'Yoredale-type' cyclothems. When fully developed, each individual cyclothem has a limestone at the base, which is overlain sequentially by mudstone, sandstone, seatearth and coal. The cycles commonly attain a thickness of around 20 m but may be much thicker. Each one records a marine transgression followed by a progressive shallowing and change to fluvial conditions with subsequent emergence and the growth of delta-top swamp vegetation. The deltaic conditions then became dominant during the Westphalian, with extensive development of the characteristic Coal Measures cyclic sequences in which limestone is generally absent, replaced by a thicker mudstone development followed by sandstone, seatearth and coal; individual cyclothems mostly range up to about 10 m in thickness, some attain 15 m.

As might be expected in narrow basins within a generally stable, upland area, the Carboniferous outliers within the region have thin attenuated successions, with numerous non-

Table 5 Stratigraphical classification of Carboniferous rocks in the south of Scotland.

98002964-T0601

LITHOSTRATIGRAPHY		CHRONOSTRATIGRAPHY				
		European			International	
N	S	Substage	Stage	Subsystem	Stage	Subsystem
WARWICKSHIRE GROUP			Autunian	SILESIAN*	Gzhelian	PENNSYLVANIAN
			Stephanian		Kasimovian	
Unconformity		Asturian (Westphalian D)	Westphalian		Moscovian	
SCOTTISH COAL MEASURES GROUP	PENNINE COAL MEASURES GROUP	Bolsovian (Westphalian C)				
		Duckmantian (Westphalian B)			Bashkirian	
		Langsettian (Westphalian A)				
CLACKMANNAN GROUP	YOREDALE GROUP	Yeadonian	Namurian			
		Marsdenian				
		Kinderscoutian				
		Alportian				
		Chokierian				
		Arnsbergian			Serpukhovian	MISSISSIPPIAN
		Pendleian				
STRATHCLYDE GROUP		Brigantian	Visean	DINANTIAN*	Visean	
		Asbian				
	BORDER GROUP	Holkerian				
		Arundian				
		Chadian				
INVERCLYDE GROUP	INVERCLYDE GROUP	Courceyan	Tournaisian		Tournaisian	
Southern Upland outliers: Oldhamstocks, Sanquhar, Thornhill, etc	Scottish Borders: Berwickshire, Cheviot, Solway					

* Although the Carboniferous Subcommission of the International Commission on Stratigraphy has recommended that the terms 'Dinantian' and 'Silesian' should no longer be used, they are such fundamental units in the description of British Carboniferous rocks that they are likely to be encountered throughout the currency of this guide

sequences and unconformities. However, along the southern and eastern margins of the massif deposition was more continuous and successions can generally be correlated with those in the contemporaneous basins developing in the Northumberland–Solway Basin, where relatively thick and complete marine and deltaic sequences were laid down.

STRATIGRAPHICAL PRINCIPLES

The Carboniferous rocks in the south of Scotland have a long history of geological study and a variety of stratigraphical schemes, based on a range of characteristics, has been applied to them. This has led to a confusing plethora of local stratigraphical names and uncertainty over the correlation of the sequences from one area to another. Historically, three main divisions of the Carboniferous rocks of Britain were recognised based on the character of the rock sequence, the Carboniferous Limestone, the Millstone Grit and the Coal Measures. These lithostratigraphical (rock) divisions later became broadly equated with the chronostratigraphical (time) divisions of the Dinantian (divided into the Tournaisian and Visean), Namurian and Westphalian; the last two of these divisions, together with younger Carboniferous divisions poorly represented in Britain, made up the Silesian (Table 5). International convention has now replaced the Dinantian and Silesian with, respectively, the Mississippian and Pennsylvanian subsystems, though the boundaries do not coincide precisely (Table 5). The more detailed subdivision into chronostratigraphical stages and substages (Table 5) is largely based on biostratigraphy, which relies on the identification of key fossils and fossil assemblages, and the recognition of how these features change through successions of strata. The Westphalian Stage has traditionally been divided into four stages, originally lettered A to D but now formalised as substages: respectively Langsettian, Duckmantian, Bolsovian and Asturian.

Plate 34 *(opposite)* *Selection of Carboniferous fossils from the south of Scotland region.*

Alga

a Saccamminopsis fusulinaformis *(McCoy 1849). BGS GSE 9131. Lower Limestone Formation, East Lothian (P693040).*

Gastropod

b Staparollus (Euomphalus) *aff.* acutus *(J Sowerby 1818). BGS GSE 9151. Alston Formation, Dumfries & Galloway (P693053).*

Corals

c Caninophyllum archiaci monense *Lewis 1929. BGS GSE 6195. Tyne Limestone Formation, Dumfries & Galloway (P693052).*

d Lithostrotion clavicatum *Thomson 1883. BGS GSE 6103. Tyne Limestone Formation, Dumfries & Galloway (P693044).*

Marine bivalves

e Actinopteria persulcata *(McCoy 1851). BGS GSE 11790. Tyne Limestone Formation, Dumfries & Galloway (P693047).*

f Wilkingia maxima *(Portlock 1843). BGS GSE 11939. Tyne Limestone Formation, Dumfries & Galloway (P693055).*

g Posidonia becheri *(McCoy 1851). BGS GSE 11779. Alston Formation, Dumfries & Galloway (P693042).*

Brachiopods

h Gigantoproductus *cf.* maximus *group? (McCoy 1844). BGS GSE 14799. Tyne Limestone Formation, Dumfries & Galloway (P693048).*

i Punctospirifer redesdalensis *North 1920 (and crinoid fragment) BGS GSE 14807. Tyne Limestone Formation, Dumfries & Galloway (P693046).*

Crinoid

j Woodocrinus *cf.* macrodactylus *de Koninck 1858. BGS GSE 14727. Alston Formation, Dumfries & Galloway (P693045).*

Nonmarine bivalve

k Anthracosia *sp. (intermediate between* A. aquiina *(J de C Sowerby 1840) and* A. ovum *(Trueman and Weir 1951). BGS GSE 6915. Pennine Coal Measures Group, Northumberland (P693054).*

a
2 cm
b
2 cm
c
1 cm
d
2 cm
e
1 cm
f
2 cm
g
2 cm
h
3 cm
i
1 cm
j
2 cm
k
1 cm

Biostratigraphy

Biozones are stratigraphical intervals characterised by the fossil biotas that they contain, and are defined by criteria such as the successive first appearances of distinctive fossil species, the total stratigraphical range of a species, the overlapping ranges of two or more species, or by particularly abundant occurrences of a distinctive species. Biozonation based on the occurrence of corals and brachiopods has been of historical importance in the classification of platform carbonates, mostly of Visean age, though the faunas are now considered to have been strongly facies controlled. In other Tournaisian and Visean lithofacies, and through the Namurian, the relatively diverse molluscan fauna was utilised. It is now also common to utilise microfauna such as conodonts and foraminifera, which facilitate correlation at international level. Particularly useful for the lower Carboniferous are the corals, foraminifera and conodonts, whilst for the upper Visean, Namurian and Westphalian the best resolution is provided by ammonoids. Nonmarine bivalves, which are important to the zonction of late Namurian and Westphalian sequences, tend to occur in association with fish material and ostracods; at a number of stratigraphical levels bivalve shells have built up to form distinctive 'mussel bands'. Accumulations of estheriids (small crustaceans that occupied brackish waters) can also be stratigraphically useful. A selection of Carboniferous macrofossils is illustrated in Plate 34.

The ammonoids (goniatites) occur within thin marine bands that developed during discrete marine incursions at periods of high global sea level. The marine bands are present over wide areas and contain characteristic fossil assemblages so that, together with palynology (see below), they have become the primary means of correlation within and between coalfields in both the Pennine and Midland Valley basins, and thence as far as eastern Europe. They have also allowed preliminary correlation of Westphalian sedimentary rocks onshore with those in the southern North Sea. Marine bands represent geographically widespread, short-lived, marine flooding events and are fundamental to a proper understanding of the succession since they reflect the most significant changes in base level (relative sea level). Three widespread marine bands — Subcrenatum, Vanderbeckei and Aegiranum — are used to define the substages of the Westphalian forming, respectively, the bases of the Langsettian, Duckmantian and Bolsovian (Table 5). However, a plethora of local names has been applied and different marine bands have been taken as the boundaries of the Coal Measures lithostratigraphical formations in the Pennine and Scottish sequences. Correlation is provided in Table 6 with local examples taken from the Sanquhar and Canonbie coalfields in the south of Scotland. Only the Vanderbeckei marine band, separating the Lower and Middle Coal Measures formations, has the same stratigraphical significance in both the Pennine and Scottish successions.

The remains of coal-swamp vegetation (which included relatives of present-day tree ferns, horsetails and club mosses) are common fossils in the Coal Measures strata and following significant studies in the first half of the 20th Century, a set of plant biozones was established for the Westphalian. More recently palynology, which in Carboniferous rocks is focused on plant spores and pollen (miospores), has become a standard biostratigraphical technique. Its great advantage is that only a small rock sample is needed to provide many microfossils (miospores). As a result, it has proved invaluable in the correlation of borehole sequences offshore and enables comparison between them and the onshore basins. It has been much applied to interpretations of Westphalian strata from beneath the North Sea. Miospores are present in both marine and terrestrial environments and so are particularly useful for correlations of marine successions with nonmarine rocks lacking stratigraphically useful macrofossils, and for correlation between separate nonmarine successions.

Sequence stratigraphy

Seismic profiling has become the primary technique for analysing basin evolution, and from it has developed the discipline of sequence stratigraphy, which affords an alternative approach to the description, interpretation and correlation of the Carboniferous successions. Sequences are defined as stratigraphical units bounded by regional unconformities or disconformities

Table 6 *Stratigraphical classification of Westphalian strata in the south of Scotland. Marine band names refer to the immediately subjacent line.*

STAGE	SUBSTAGE	NONMARINE BIVALVE ZONE	MIOSPORE ZONE	STANDARD MARINE BAND NAMES	MARINE BANDS: Canonbie	MARINE BANDS: Sanquhar	LITHOSTRATIGRAPHY: Canonbie	LITHOSTRATIGRAPHY: Sanquhar
Westphalian	Asturian (Westphalian D)	*tenuis*	*Thymospora obscura*				WARWICKSHIRE GROUP	
	Bolsovian (Westphalian C)	*phillipsi*	*Torispora securis*	Cambriense	Riddings		PENNINE COAL MEASURES GROUP: Pennine Upper Coal Measures Formation	SCOTTISH COAL MEASURES GROUP: Scottish Upper Coal Measures Formation
		Upper *similis-pulchra*		Shafton	Rowanburnfoot		Pennine Middle Coal Measures Formation	
			Vestispora magna	Edmondia	Viaduct	Lagrae		
				Aegiranum	Skelton	Skipsey's		
	Duckmantian (Westphalian B)	Lower *similis-pulchra*		Sutton	Knottyholm	Eastside		Scottish Middle Coal Measures Formation
			Dictyotriletes bireticulatus	Haughton	Sandwich	Bankhead		
				Clowne				
				Maltby				
		modiolaris	*Schulzopora rara*	Vanderbeckei	Queenslie	Queenslie		
	Langsettian (Westphalian A)	*communis*	*Radiizonates aligerens*	Kilburn			Pennine Lower Coal Measures Formation	Scottish Lower Coal Measures Formation
				Burton Joyce				
				Langley	?Templeman's			
		lenisulcata	*Triquitrites sinan – Cirratriradites saturni*	Amaliae				
					?	Tait's		
				Meadow Farm		Correlation uncertain		
				Parkhouse				
				Listeri				
				Honley				
				Springwood				
				Holbrook				
				Subcrenatum				

98002964-T0602

related to cyclical changes of sea level. Sequence stratigraphical models then distinguish between sedimentary units (or systems tracts) produced during different parts of the cycle of variation in relative sea level. So, for example, transgressive systems tracts formed when sea level was rising, high-stand systems tracts formed when sea level was high, and low-stand systems tracts formed when sea level was low. Sequences are on different scales, and are arranged in a hierarchy from 1st order (of longest duration) to 5th order. Sequence stratigraphy emphasises the allocyclic controls on sedimentary successions (tectonism, climate and eustacy) and can provide an improved understanding of depositional history and sedimentary system development. In places, especially in concealed sequences with limited borehole control, it can enable a higher resolution correlation than can be obtained by existing lithological and palaeontological methods. However, and despite the advantages, the recognition and correlation of surfaces bounding systems tracts in borehole core or at outcrop can be difficult, and biostratigraphy commonly provides important controls and constraints when integrated with sequence stratigraphy interpretations. Given the fossiliferous nature of most Carboniferous rocks, biozonation is likely to remain an independent tool for stratigraphical subdivision and the establishment of a chronological framework for the succession. A useful equivalance between the two schemes is provided by the marine bands — long established and well tested in stratigraphical correlation — which equate to maximum flooding surfaces in sequence stratigraphy terminology.

Lithostratigraphy

In common with Carboniferous successions elsewhere in Britain, those of southern Scotland contain a range of economic mineral products that have long attracted attention. Coal, limestone and sandstone have been worked locally, particularly so in the Canonbie and Sanquhar coalfields, and to help identify the beds they were commonly given local names. For example, in the Langholm and Canonbie districts, the Carboniferous stratigraphical nomenclature applied in the 19th and early 20th century divided the rocks into the Whita Sandstone, the Cementstone Group, the Fell Sandstones, a Marine Limestone Group, the Millstone Grit, two coal groups and the Red Sandstones of Canonbie. Volcanic rocks of Birrenswark were also recognised at the base of the succession and evidence of other short-lived volcanic episodes was noted above the Fell Sandstones. Subsequent work in the 1950s and 1960s divided the Dinantian sequences into a tripartite (Lower, Middle and Upper) Border Group and a bipartite (Lower and Upper) Liddesdale Group. These units were primarily distinguished on biostratigraphical grounds with unit tops and bases defined by marker horizons such as marine bands.

In an attempt to unify the various localised stratigraphical schemes the British Geological Survey has recently rationalised the nomenclature of British Carboniferous lithostratigraphy, grouping together eight broad types of sedimentary lithofacies associations (Table 7). The regional group names used in this account derive from that rationalisation; their lithofacies associations and principal formational components are summarised in Figure 35. The lithostratigraphy defined for the Midland Valley of Scotland can be extended southwards to accommodate the Carboniferous outliers within the Southern Uplands massif: in upward succession this comprises the Inverclyde, Strathclyde, Clackmannan and Scottish Coal Measures groups. The Inverclyde Group can be usefully extended southwards as the lowermost unit of the Carboniferous sequence in the Northumberland and Solway basins. There however, above the Inverclyde Group, the northern England stratigraphy applies with, in upward succession, the Border, Yoredale, Pennine Coal Measures and Warwickshire groups (Table 5 and Figure 35).

REGIONAL STRUCTURE

The Southern Uplands and the Northumberland–Solway Basin form the northern components of a regional arrangement of tectonic blocks and basins that extends southwards, as detailed in the companion volume — *Northern England* — in the British Regional Geology series. Only the northern edge of the Northumberland–Solway Basin extends across the border into the south of Scotland. In general, the Tournaisian lithofacies characteristic of the Midland Valley succession — the Inverclyde Group — extends southwards across the Scottish Border region into the northern margin of the Northumberland–Solway Basin. During the Visean, the Cheviot Block separated the basin from the Tweed Basin in Berwickshire (Figure 34) but ceased to be an effective barrier by the beginning of the Namurian.

Early to mid Carboniferous extension

Lithospheric extension under the influence of broadly north–south tension affected much of Laurussia from late in the Devonian; it was particularly active until the end of the Visean but continued intermittently through much of Carboniferous time. The interaction of the north–south extensional regime with the pre-existing Caledonian fault pattern led to reactivation of

Table 7 *Summary of the lithofacies characteristics of the Carboniferous successions in the south of Scotland (after Waters et al., 2007. BGS Research Report RR/07/01).*

Facies	Subfacies	Lithologies	Depositional environments
Continental and peritidal	Continental fluvial clastic ('cornstone')	Purple-red conglomerate, sandstone and red mudstone, with nodules and thin beds of concretionary carbonate (calcrete)	Alluvial fan, fluvial channel and floodplain overbank; semi-arid climate
	Peritidal marine and evaporite ('cementstone')	Grey mudstone, siltstone and sandstone, characterised by nodules and beds of ferroan dolostone ('cementstone') and evaporites (mainly gypsum and anhydrite)	Alluvial plain and marginal marine flats subject to periodic desiccation and fluctuating salinity; semi-arid climate
Heterolithic clastic and nonmarine carbonate		Interbedded grey sandstone, siltstone, mudstone and locally oil shale. Thin subordinate beds of lacustrine limestone and dolostone, seatearth, coal and sideritic ironstone	Fluviatile, deltaic, lacustrine and rarely marine, commonly alternating in thin cycles
Open marine platform and ramp carbonate		Strata representing these two facies are not present in the south of Scotland	
Hemipelagic			
Mixed shelf carbonate and deltaic ('Yoredale')		Typically upwards coarsening cycles of limestone, mudstone, sandstone, seatearth or ganister and coal	Marine with episodic progradation of lobate deltas
Fluviodeltaic ('Millstone Grit')	Shallow-water, sheet-like delta	Upwards-coarsening cycles of black mudstone, grey siltstone, fine- to very coarse-grained feldspathic sandstone; coal and seatearth common, but thin	Cyclic delta progradation with common black shale marine bands resulting from delta abandonment and marine transgression. Common disconformities
	Delta-top	Condensed, predominantly upward-fining cycles of structureless clayrock to sandstone. Thick high-alumina seatclay, fireclay and bauxitic clay are common. Beds of limestone, ironstone, cannel and coal rare	Millstone Grit cyclic sequences are thicker and fewer than Coal Measure ones
Fluviodeltaic ('Coal Measures')		Upward-fining and upward-coarsening cycles of grey to black mudstone, grey siltstone, fine- to medium-grained sandstone, with frequent seatearth or ganister and relatively thick coals	Wetland forest and soils, floodplain, river and delta distributary channel, prograding deltas and shallow lakes
Alluvial ('Barren Measures')	'Red-bed'	Red, brown or purple-grey mudstone, siltstone and sandstone. Minor grey mudstone, thin coal, lacustrine ('Spirorbis') limestone and pedogenic limestone (calcrete)	Fluvial channel. Red beds oxidised at, or close to, time of deposition; semi-arid climate

98002964-T0603

Caledonian faults as the marginal structures of transtensional basins. One such major basin complex opened in the Midland Valley of Scotland, with a variable, mostly dextral, strike-slip component. Thence the transtensional regime was propagated across the Southern Uplands massif where several small basins opened with their margins controlled both by the major north-east-trending strike faults and by the north-west- and north-north-east-trending Caledonian cross faults. To the south, the Solway and Northumberland basins opened in response to north–south extension across the Iapetus Suture Zone, the pre-existing orientation of which determined the broadly east-north-east trend of the basin boundary faults. Interpretation of cross-basin geophysical profiles indicates that the faults defining the southern margin of the basin (Maryport–Stublick–Ninety Fathom) were rooted into the suture zone, where extension and growth faulting was facilitated by the pre-existing intracrustal detachment.

Throughout much of its length the northern margin of the Northumberland–Solway basin is also formed by a system of *en echelon*, synsedimentary dislocations downthrowing predominantly to the south-east; they include the North Solway, Waterbeck, Gilnockie and Alwinton

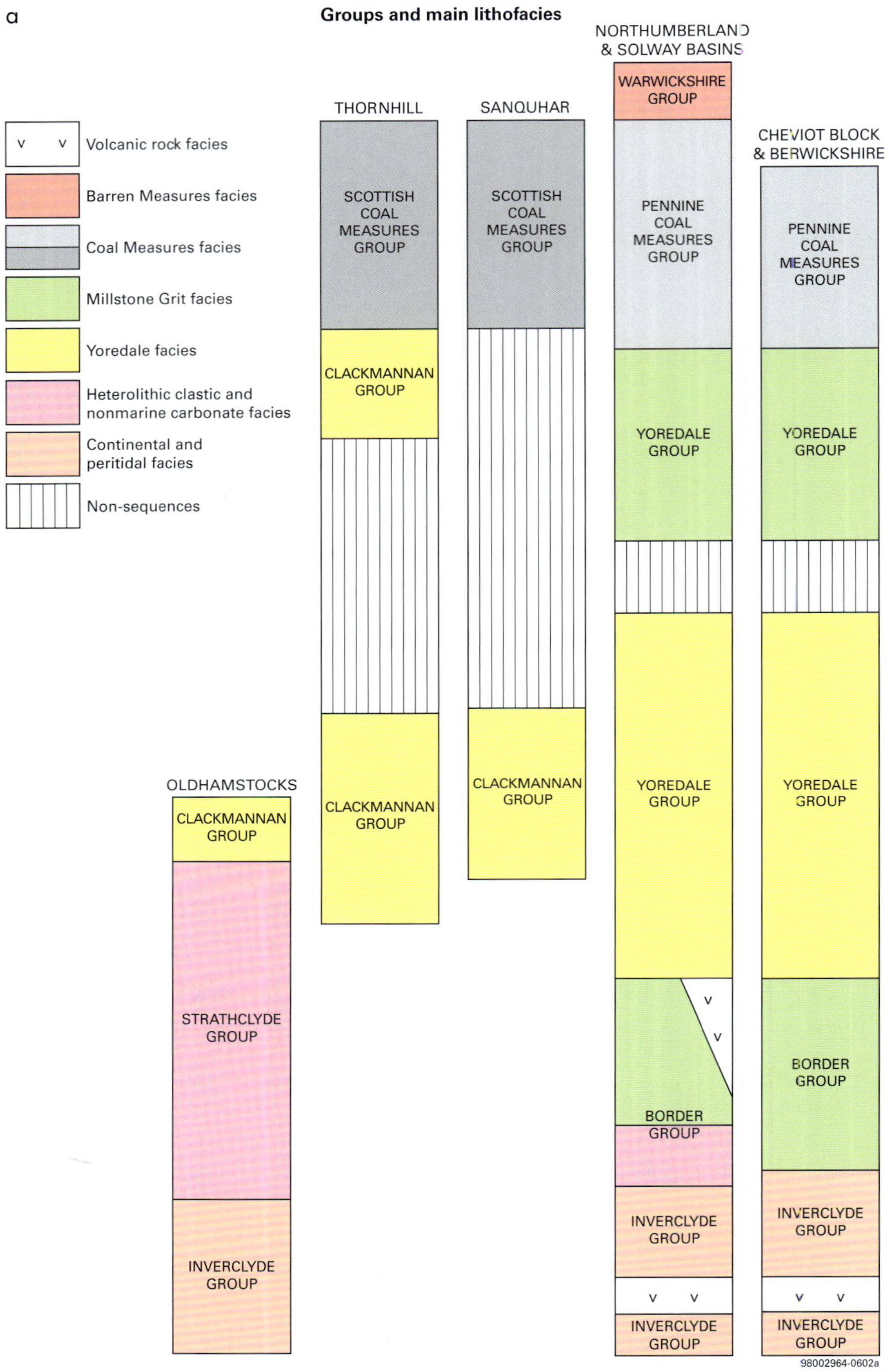

Figure 35a *Summary chart of lithofacies for the Carboniferous successions in the south of Scotland (after Waters et al., 2007. BGS Research Report RR/07/01).*

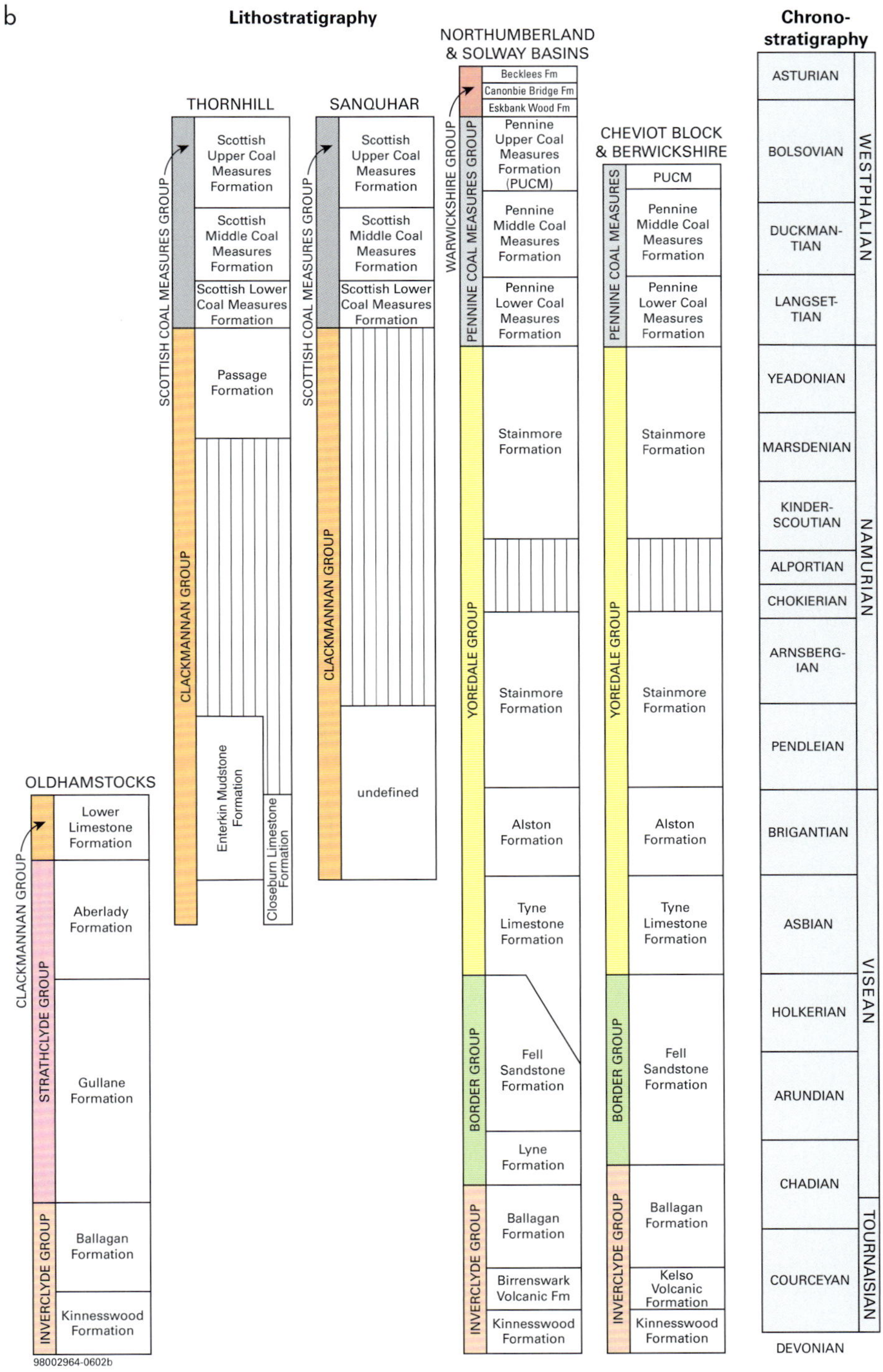

Figure 35b Summary chart of stratigraphy for the Carboniferous successions in the south of Scotland (after Waters et al., 2007. BGS Research Report RR/07/01).

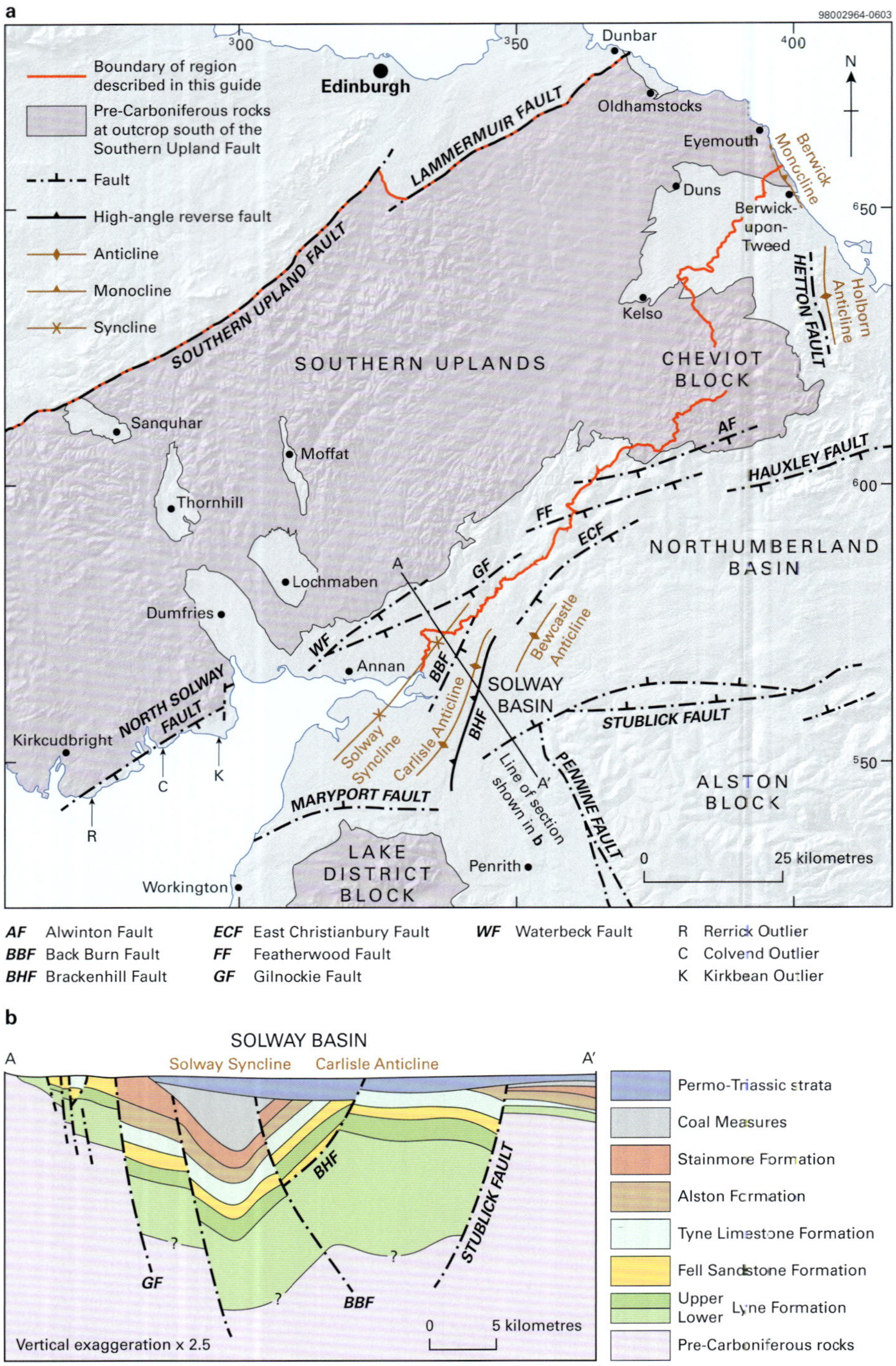

Figure 36 *a. Map showing the main structures associated with Carboniferous deposition and subsequent late Carboniferous to early Permian deformation. b. Line of section A–A′.*

faults (Figure 36). North-westerly trending cross faults acted to transfer downthrow between different fault blocks. The extensional tectonics active at the basin margin early in the Carboniferous facilitated the eruption of lavas and emplacement of a variety of minor igneous intrusions. So, for example, the deposition of the Kinnesswood Formation alluvial-plain sediments was followed by eruption of the lavas of the coeval Birrenswark Volcanic and Kelso Volcanic formations. These volcanic formations are considered below in their lithostratigraphical context, but the broader manifestations of Carboniferous magmatism are considered separately in the final section of this chapter.

Syndepositional faulting

The northern margin of the Solway Basin saw rapid extensional subsidence during the Tournaisian deposition of the Inverclyde Group, with normal fault movement continuing through the Visean to influence deposition of the Border Group. An example of a major north-easterly fault with such a syndepositional history is the Waterbeck Fault, north of Annan (Figure 36). It throws Yoredale Group strata down to the south-east and cuts out parts of the Border Group formations to the north. There, and in general within the Border and Yoredale groups, abrupt transitions are observed from relatively flat lying beds to zones of tightly folded strata. This pattern is particularly well observed in the Fell Sandstone Formation as exposed in the River Annan north of Brydekirk [NY 186 705]. Elsewhere in the region, an analagous structural pattern is observed in the Tournaisian and Visean strata of Berwickshire, where zones of steeply dipping and folded strata are believed to have formed during periods of synsedimentary deformation.

Close to the north-east to south-west trending North Solway Fault, an inherited Caledonoid structure, there is much evidence within the Tournaisian to Visean, basin-margin, coarse clastic sedimentary rocks of the Rerrick and Colvend Outliers of periodic synsedimentary, dip-slip, normal movement. Synsedimentary fault deformation of hanging-wall strata increases towards the North Solway Fault, whilst cyclicity within alluvial fan deposits of the Rerrick Outlier may be attributed to variation in rate of fault-controlled subsidence. Further away from the basin margin, strata of the Kirkbean Outlier exhibit characteristics of a more stable environment of

Plate 35 *Thirlstane Sandstone Beds, Powillimount Sandstone Member, showing liquefaction structures most probably induced by seismic activity shortly after deposition of the original sediment; viewed from a position close to the Thirlstane natural arch [NX 992 568] (P001161).*

marginal to shallow marine deposition, but rates of subsidence may nevertheless have been controlled tectonically. Evidence of synsedimentary seismic activity is particularly convincing within the coastal outcrop of the Thirlstane Sandstone Beds (Powillimount Sandstone Member, Fell Sandstone Formation) (Plate 35).

Late Carboniferous (Variscan) compression

Basin shortening and inversion took place in late Westphalian times as a peripheral effect of the Variscan Orogeny farther south; many of the basin margin faults were reactivated with a reverse sense of displacement. By this time, the thermal subsidence that had controlled the Namurian and Westphalian patterns of sedimentation had all but ceased whilst, to the south, the Gondwanan continent had collided with Laurussia to create the supercontinent of Pangaea (Figure 3d). The Laurussia–Gondwana collision generated the Variscan Orogeny with a mountain fold belt produced as its culmination. The approximate northern limit of the fold belt (the Variscan Front) extends from the mainland of Europe across southern England and Wales and into Ireland. The south of Scotland lay in a foreland setting well to the north of the deformation front. Although tectonic activity in the foreland region was much weaker than in the fold belt to the south, some folds were formed, both independently and as adjuncts to fault reactivation. An anticline affecting the Penton Limestone (Alston Formation), as exposed in the Liddel Water near Langholm, is illustrated in Plate 36.

Seismic profiles across the Solway Basin show Carboniferous strata folded into the north-north-east-trending Solway Syncline and, to the south-east, the complementary Carlisle Anticline (Figure 36). To the north-west of the syncline, and along the northern margin of the Solway Basin, are north-east orientated and plunging anticlinal structures, well seen in the Kirkbean Outlier. Farther north-east, along the northern margin of the Northumberland Basin, smaller scale, generally north-east-trending structures predominate, but the Cheviot Block produces much local complication. Around the eastern side of the block the structural trend swings toward a north–south

Plate 36 *Anticline in the Penton Limestone (Alston Formation) exposed in the Liddel Water at Penton Linns [NY 432 774] (P221694).*

orientation, as shown in north-east Northumberland by the Holborn Anticline and its associated Hetton Fault (Figure 36). North of the border, in eastern Berwickshire, the north-north-west-trending Berwick Monocline faces east, with the steep limb coinciding with a high angle reverse fault that juxtaposes Carboniferous rocks against Silurian and Devonian strata. The complexity of these relationships is inferred to arise from the reverse reactivation of original syndepositional faults.

COAL-BEARING SEQUENCES

The economic importance of the coal-bearing parts of the Carboniferous succession, particularly those of Westphalian age, has led to a plethora of research. Some general characteristics are worth considering here in their regional context since they apply equally to the different successions described in subsequent sections of this chapter.

The Westphalian coal-bearing strata now seen spanning the border between north-west England and south-west Scotland, originally accumulated along the northern margin of the Northumberland–Solway Basin, itself the northern sector of the Pennine Basin, a depositional province that was continuous with the north-west European paralic belt, from which it became separated by later Variscan folding. The basin was bounded on its northern side by the Southern Uplands and associated small scale landmasses such as the Cheviot Block (Figure 34). Farther to the north, a separately subsiding basin incorporated the Midland Valley of Scotland. The dominant transtensional tectonic regime opened up several smaller basins within the Southern Uplands massif that show stratigraphical links with either the Midland Valley succession to the north or the Pennine Basin succession to the south.

Depositional environment

During much of the Carboniferous Period, southern Scotland occupied an equatorial position and experienced a humid, tropical climate characterised by high precipitation rates. This combination of factors provided ideal conditions for the development of a high water table, poorly drained palaeosols, peat swamps and the eventual formation of coal seams. The coal-bearing sequences mostly accumulated in a fluviolacustrine environment, a low-lying and largely waterlogged plain that was subjected both to intervals of emergence and to intermittent marine transgressions. Across north-west England and the south-west Scottish border area, the earliest Westphalian strata were deposited in a gently subsiding, lower delta plain environment under the dominant influence of mainly fluvial delta systems. Southward progradation of this depositional system through time resulted in middle Westphalian strata being deposited in a more proximal, upper delta plain environment, dominated by river distributary channels and lacustrine deltas, which was particularly conducive to the development of coal swamps. Later in the Westphalian, a waning fluvial influence caused re-establishment of lower delta plain conditions. Accordingly, the maximum coal development is seen in the middle part of the succession, the upper Langsettian and lower Duckmantian.

The Coal Measures strata contain an abundant and varied fossil fauna that includes both nonmarine and marine species. Nonmarine invertebrates include worms, gastropods, bivalves, eurypterids, crustaceans and fish, whilst the marine faunas include brachiopods, goniatites, foraminifera and conodonts (Plate 34). The mussel bands formed by nonmarine bivalves (Plate 37) are of particular importance for biostratigraphical purposes and bivalves are the basis of a widely used biozonal scheme (Table 6). Nonmarine fossils are normally concentrated in the few metres of argillaceous strata that form the basal zone of the various cyclothems, but many cyclothems have no preserved fauna or contain only a small range of undiagnostic invertebrate fossils. Marine fossils are naturally restricted to the marine bands.

Remains of the coal swamp vegetation are common fossils in the Coal Measures. Masses of compressed plant material make up the coal seams, roots are found in situ in seatearths, drifted leaf fronds and plant stems occur in mudstone and siltstone beds, and chaotic 'log-jams' of broken tree trunks and branches are a feature of some of the thicker sandstones. The Westphalian flora of the floodplains was dominated by pteridosperms with some ferns, sphenopsids (Plate 38a) and lycopods, and that of the peat-forming swamps by lycopods.

Plate 37 *Section of a mussel band with species of the nonmarine bivalve* Carbonicola *(P693035). The specimen is from the persistent mussel band that occurs at about 2–3 m beneath the Kirkconnel Splint Coal, Scottish Lower Coal Measures Formation, in the Sanquhar Basin.*

Today, the lycopod group is only represented by low-growing plants, but during the Westphalian some lycopods were tree sized with *Stigmaria* root systems, the most familiar being *Lepidodendron* and *Sigillaria* with their distinctive bark patterns of rhomboidal scales (Plate 38b). *Calamites*, a giant relative of the present-day horsetail *Equisetum* was also common and grew around lakes and on point bars, whilst a range of pteridosperms grew on the levées alongside meandering rivers.

Clastic lithologies and cyclothems

The Coal Measures are built-up from a repetition of cyclothems in each of which the main lithologies commonly follow one another in an ascending order of mudstone at the base (overlying the coal at the top of the underlying cycle), siltstone, sandstone, seatearth, coal. The nature and origin of the cyclothems have been much discussed, and it appears that no single explanation will suffice. Such cyclicity is a natural reflection of the interplay of sedimentary processes, and the only external mechanism needed to produce them is continuous subsidence. However, it is generally accepted that the periodic changes in sea level leading to marine flooding events are related to global glacial events, in this case the late Carboniferous glaciation of southern Gondwana which at that time lay over the South Pole. In contrast to circumstances earlier in the Carboniferous, contemporaneous fault activity is not considered to have been a major influence on sedimentary patterns, and only local fault-controlled effects have been recorded.

The inter-coal sequences were deposited during the gradual infilling of shallow interdistributary bays and lakes by shallow-water delta complexes. They are interbedded with a number of prominent sandstone bodies that were deposited by the low-sinuosity, distributary channels feeding crevasse-splay systems (Figure 37). These bodies can be stacked, one above another, to produce substantial thicknesses of sandstone locally.

Mudstones are generally well bedded and grey, but darker and more fissile when carbonaceous. Siderite commonly occurs in thin layers or as flattened nodules. Many of the mudstones contain a shelly fauna and this is usually indicative of a brackish-water depositional environment; marine mudstones are rare. Siltstones are also grey, commonly laminated and with a range of bioturbation structures. They grade imperceptibly into both mudstone and sandstone, the latter being mostly fine grained and quartzofeldspathic.

Seatearths, as preserved palaeosols, may be developed from any of the clastic lithologies and are gradational into the underlying facies. In contrast, the contact with any overlying coal seam is sharp. The seatearths contain abundant carbonaceous plant material (most commonly

a

b

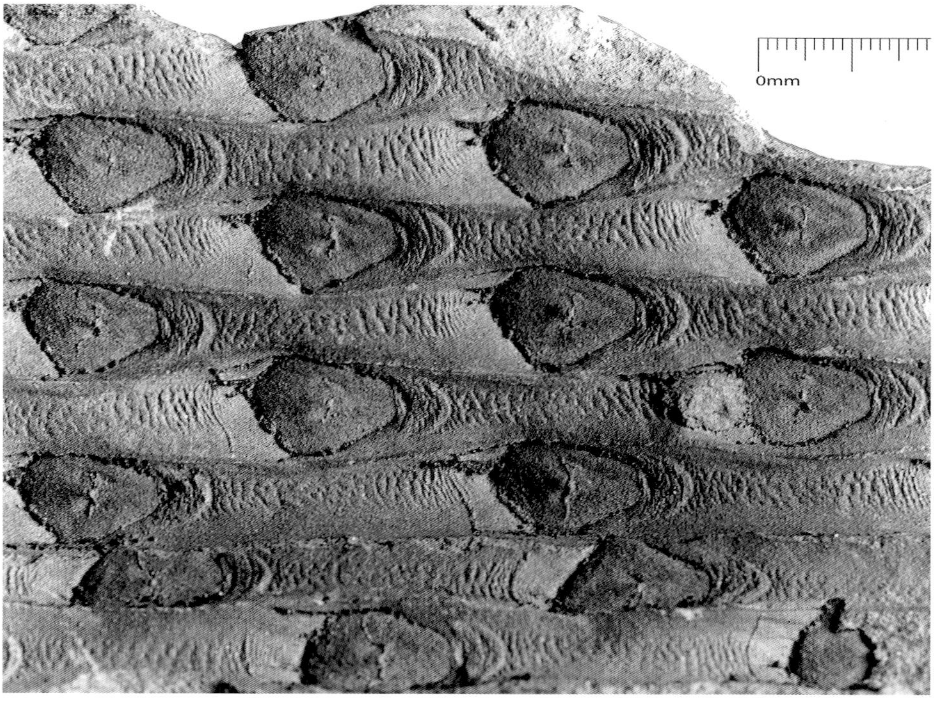

Plate 38 *Plant fossils from the Scottish Coal Measures Group, Sanquhar outlier.*

a Sphenopteris nummularia *Gutbier 1835, Polbower Burn, Kirkconnel (P688593).*
b Sigillaria *sp., The Gullet, Sanquhar (P688051).*

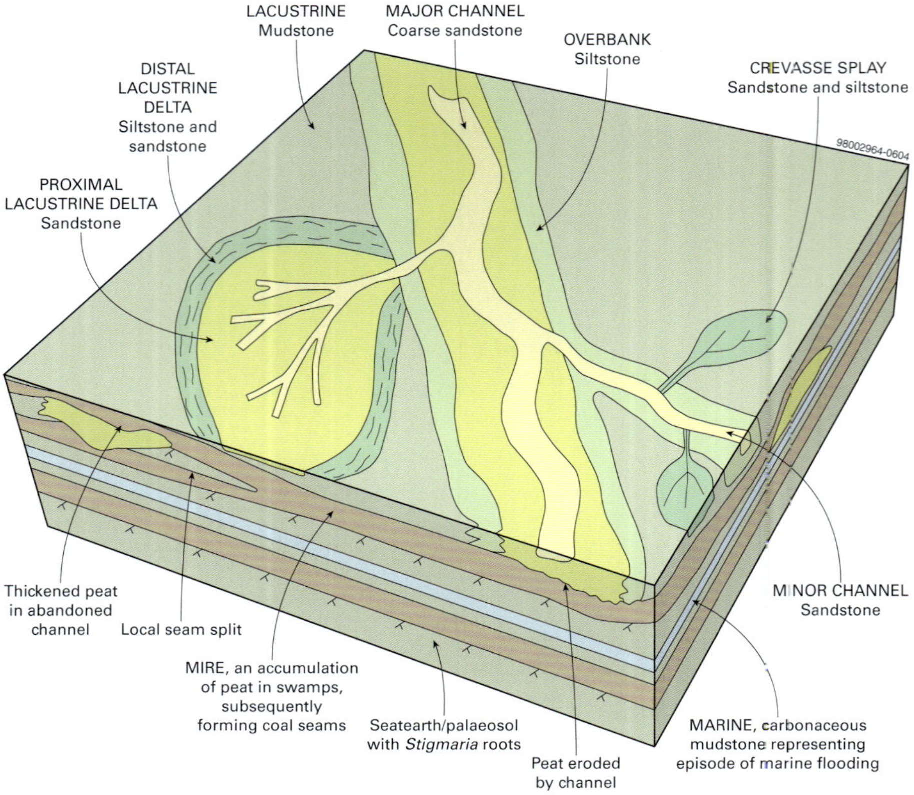

Figure 37 *Schematic illustration of the facies variation within the upper delta plain depositional environment of the Westphalian Scottish and Pennine Coal Measures groups (after Guion et al., 1995).*

Stigmaria rootlets), and bright coaly stringers with disseminated pyrite. The rootlets increase in abundance upwards through the seatearth and disrupt any original lamination that might have been present. The seatearths typically break along irregular fracture surfaces that may be either covered with slickensides, or curved and polished.

Coal

Coal swamps formed as mires developed across low-lying alluvial plains and abandoned lacustrine delta systems (Figure 37). Individual abandoned delta systems are thought to have been up to 10 km wide and 20 km long. The associated peat-forming mires developed over prolonged periods of time and were not necessarily contemporaneous from one delta system to another. Lateral continuity and synchroneity would have been particularly unlikely along the more active, marginal parts of the basin, as preserved in southern Scotland, where variations in coal seam thickness and localised seam splits would be expected.

Peat-forming mires thrive under waterlogged conditions of rising base level and are able to maintain themselves for thousands of years in water depths of up to about 1 m, probably the optimum water depth for swamp growth. A lowered water table leads to oxidation and destruction of organic matter as the swamp environment gives way to better-drained conditions; no coals form, only overthickened palaeosols. An accelerating rise in water level allows the mire to be buried by clastic sediment. Across parts of southern Scotland, peat-forming environments ranged from low-lying and brackish mires to seasonally flooded forest swamps. The resulting coal seams range up to about 2 m in thickness and so a peat:coal compaction ratio

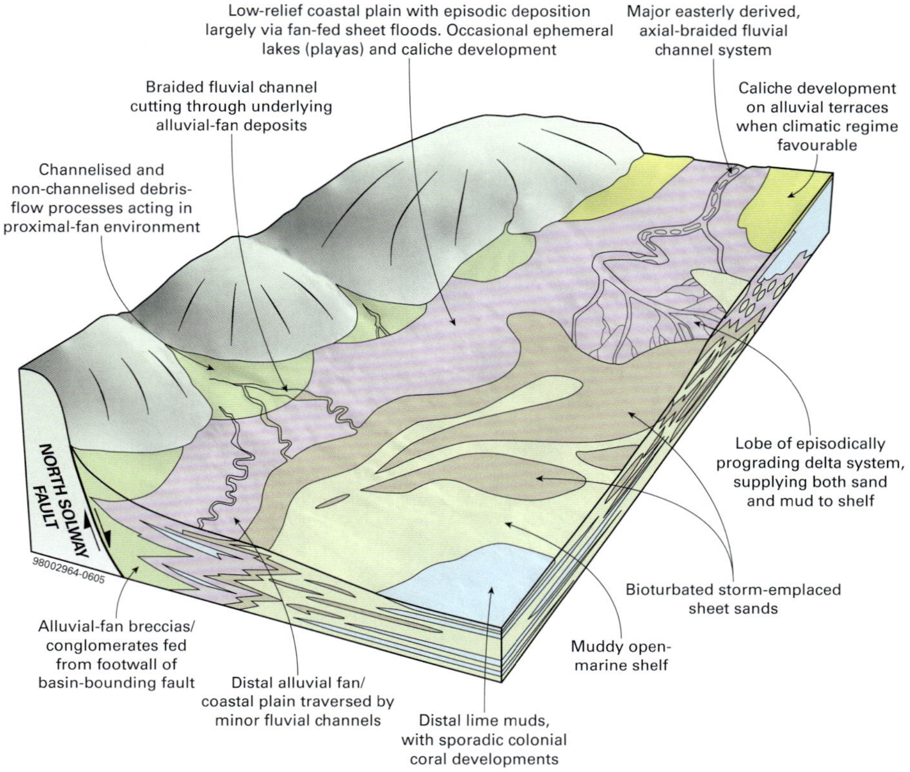

Figure 38 *The depositional environments for the Tournaisian and early Visean successions of the Solway Basin, a schematic landscape sketch after Maguire et al., 1996.*

in the order of 10:1 would indicate that peats were originally up to 20 m thick; autocompaction during peat growth would reduce that decompacted thickness.

TWEED AND NORTHUMBERLAND–SOLWAY BASINS

In the Scottish Borders, along the northern margin of the Northumberland–Solway Basin and underlying the Tweed Basin, the oldest Carboniferous rocks, of Tournaisian age, belong to the **Inverclyde Group**. They include the basaltic lavas of the **Kelso Volcanic Formation** ('Kelso Traps') and the **Birrenswark Volcanic Formation** which overlie red-bed strata assigned to the **Kinnesswood Formation**. The latter rests either conformably on red-bed strata of Devonian age (Stratheden Group) or with marked angular unconformity upon an irregular surface eroded across the steeply dipping, Lower Palaeozoic turbidite sequence. Stratigraphically above the lavas lies the **Ballagan Formation**, which forms the upper part of the Inverclyde Group. This earliest Carboniferous sedimentary succession of the Tweed and Solway basins is distinct from the rest of the Carboniferous succession in being lithologically similar to coeval successions in the Midland Valley of Scotland. Accordingly, the same lithostratigraphical formational terminology is applied (Table 5 and Figure 35b).

Higher parts of the Carboniferous succession are not present in the Scottish sector of the Tweed Basin, with the exception of a structurally isolated coastal outcrop north of Berwick. Elsewhere, along the north-west margin of the Nothumberland–Solway Basin, there is a marked departure from the Midland Valley lithofacies, reflected in the application of different lithostratigraphical terminology. There (and across northern England) the Inverclyde Group is succeed by the **Border Group**, which includes the **Lyne Formation**, partly coeval with the

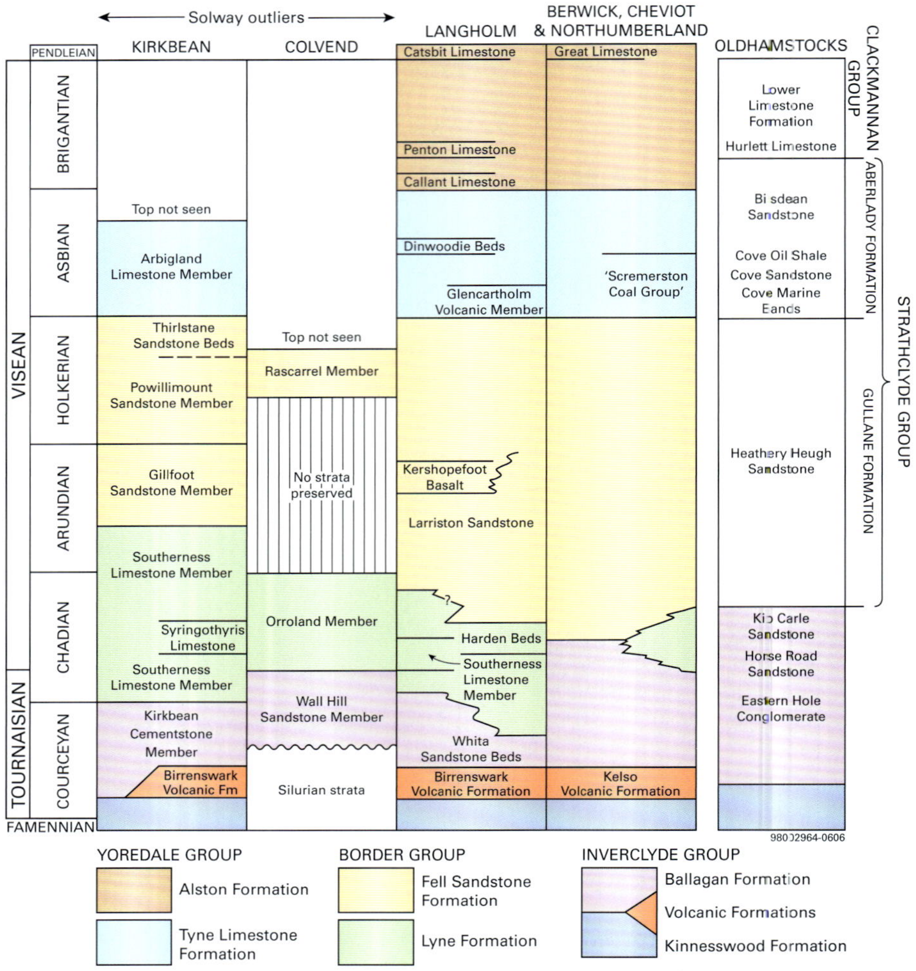

Figure 39 *Correlation of the Tournaisian and Visean ('Dinantian') successions in the south of Scotland.*

Ballagan Formation and largely restricted to the Solway area, and the **Fell Sandstone Formation**.

The Border Group is largely of early Visean age but marine lithofacies are rare within it and there are few stratigraphically useful fossils. Consequently there are difficulties in making a detailed correlation of the succession across the region, particularly toward the bottom of the sequence. For much of this early Carboniferous interval the basins formed narrow gulf-like extensions of the open sea, widening to the south-west, and their sedimentary rock successions reflect the interplay of fluviodelatic and paralic depositional systems (Figure 38). The northern emergent margins of the Northumberland and Solway basins were sources of clastic sediment during the early period of deposition, but for much of Dinantian time axial drainage systems were dominant, building from the north and east towards a shallow sea in the west. However, marginal clastic deposition was relatively persistent in the Solway Basin adjacent to the active North Solway Fault system. The variations in local lithostratigraphy for this 'Dinantian' interval are summarised and correlated in Figure 39.

The Border Group is succeeded by the **Yoredale Group**, a Visean to Namurian succession characterised by repeated upward-coarsening sedimentary cycles from limestone through mudstone to sandstone, capped by seatearth and coal. The 'Yoredale' cycles range

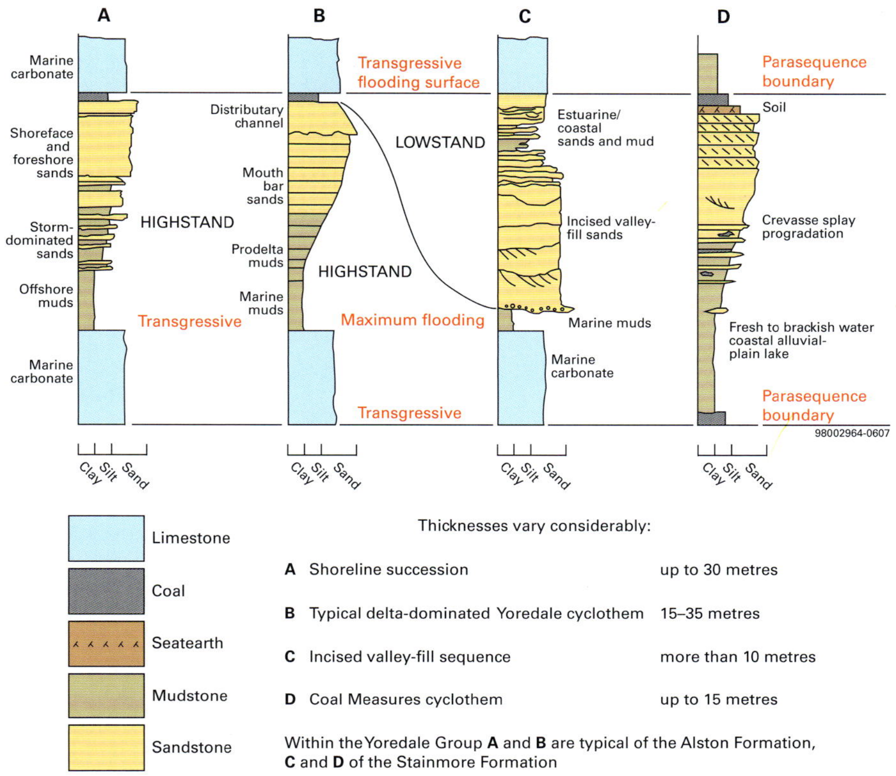

Figure 40 *Illustrative logs and interpretations for some high-frequency clastic sequences within the Yoredale and Scottish Coal Measures groups in the south of Scotland (after Tucker et al., 2003).*

in thickness up to about 35 m with an average of about 20 m; some are incomplete, and the proportions of the different lithologies vary in response to subtle changes in depositional environment (Figure 40). The Yoredale Group is divided, in upward sequence, into the **Tyne Limestone**, **Alston** and **Stainmore formations** based largely on the relative abundance of the different rock types within cycles. The base of the group (and of the Tyne Limestone Formation) is of early Asbian age on the northern margin of the Northumberland–Solway Basin, but becomes diachronously younger southwards. The Alston Formation is mostly of Brigantian age but its uppermost strata are Pendleian so that the formation straddles the Visean–Namurian chronostratigraphical boundary. The Stainmore Formation (which is only sparsely represented in southern Scotland) extends upwards to the top of the Namurian.

The Yoredale Group is succeeded in the Langholm–Canonbie area of the Scottish Borders by Westphalian fluviodeltaic strata of the **Pennine Coal Measures Group**. The group includes the **Pennine Lower**, **Pennine Middle** and **Pennine Upper Coal Measures formations** and comprises repeated mudstone–sandstone–coal cyclothems that individually range up to about 15 m in thickness (Figure 40). Late in the Westphalian, the Pennine Coal Measures Group was succeeded in the Solway Basin by the red-bed succession of the **Warwickshire Group**.

Inverclyde Group

Kinnesswood Formation

Along the southern margin of the Southern Uplands massif the Upper Devonian Stratheden Group passes conformably up into the Inverclyde Group. The base of the latter is taken at the first appearance of calcrete palaeosol and so is likely to be diachronous; in many places it remains ill defined. The oldest strata of the Inverclyde Group belong to the **Kinnesswood**

Plate 39 Ballagan Formation sandstone exposed in the north bank of the River Tweed near Lennel [NT 858 411] (P667218).

Formation of Late Devonian (Famennian) to early Tournaisian (Courceyan) age. The best biostratigraphical resolution is provided by a Courceyan miospore microflora from the Tweed Basin near Eyemouth, but fish macrofaunas from the Eyemouth and Langholm areas could be either Famennian or Courceyan in age.

The Kinnesswood Formation comprises red and yellow-brown sandstones and siltstones with locally developed conglomerate and thin beds of palaeosol containing concretionary carbonate nodules (calcrete or 'cornstone'); thicker beds of calcrete appear near the top of the formation. The clastic rocks were deposited by ephemeral streams that drained the Southern Uplands landmass under semi-arid, seasonally wet, weathering conditions. In the Tweed Basin, the Cheviot volcanic massif was the source of the conglomerates, whilst farther west their provenance lay in the Caledonian igneous and Lower Palaeozoic sedimentary rocks of Galloway.

Birrenswark and Kelso volcanic formations

The initially rapid, extensional fault-controlled subsidence of the Northumberland–Solway Basin was accompanied by extrusion of basaltic lavas along its northern margin. In the west, around Langholm, the **Birrenswark Volcanic Formation** comprises thin flows of microphyric, feldspar-rich basalt totalling 100 m in thickness at Burnswark but reducing to some 15 m at Kirkbean. An olivine basalt flow at Watch Hill [NY 435 908], north-east Langholm, has given a K-Ar whole-rock age of 361 ± 12 Ma. Several small volcanic vents containing agglomerate and/or intrusive material occur in the vicinity of Langholm, and whilst no direct

Plate 40 *Section through the Ballagan Formation exposed in the Tarras Water near Langholm [NY 388 818] (P579357).*

genetic link has been established, it is likely that they acted as feeders for the lavas of the Birrenswark Volcanic Formation, or possibly for the slightly younger Glencartholm Volcanic Member of the Tyne Limestone Formation (see below). Sedimentary debris in the agglomerate-filled vents includes Lower Palaeozoic lithologies and dolostone and chert of probable Late Devonian age; the principal igneous component is olivine-basalt.

Farther east, in Roxburghshire, the **Kelso Volcanic Formation** comprises at least six flows, and possibly as many as twelve, of alkaline olivine basalt together with subordinate tuff beds. The formation is about 120 m thick. A scattering of small basaltic intrusions is spatially associated with the lavas and one, from Mellerstain Hill [NT 641 397] has given a K-Ar whole-rock age of 361 ± 7 Ma. Like the date from the Birrenswark lavas, this suggests volcanism at around the Devonian–Carboniferous boundary. The outcrop of the Kelso Volcanic Formation extends southward across the English border near Carham [NT 799 384], whilst a very similar volcanic assemblage, the **Cottonshope Volcanic Formation**, crops out just to the south of the border on the south-west flank of the Cheviot Hills.

Ballagan Formation

The lavas of the Kelso and Birrenswark volcanic formations interfinger with, and are overlain conformably by the **Ballagan Formation**; where it does not overlie the lavas the Ballagan Formation conformably follows the underlying Kinnesswood Formation. The Ballagan Formation is of late Tournaisian to early Visean (Courceyan–Chadian) age and crops out in the Tweed Basin, the Langholm area, and on the Solway Coast. In the Tweed Basin, underlying the Merse of Berwickshire, the Ballagan Formation comprises about 430 m of mudstone with thin interbeds of argillaceous dolostone ('cementstone') and sandstone (Plate 39), with sporadic lenticular bodies of cross-bedded, channel-fill sandstone up to 30 m thick. Locally, the sequence succeeding the Kelso Volcanic Formation also contains concretionary calcrete, the most notable example being the Carham Limestone Bed ('Carham Stone'), a cherty magnesian limestone up to 7.5 m thick, thought to have accumulated by chemical precipitation from waters enriched in lime by showers of volcanic dust.

In several parts of the Tweed Basin the Ballagan Formation contains miospores and a fossil fauna that includes environmentally tolerant bivalves and ostracods, and at Foulden Burn in Berwickshire [NT 921 553] has yielded twelve or more species of fresh- to brackish-water fish from the mudstone of the Foulden Fish Bed. Farther west, at the margin of the Solway Basin around Langholm, the Ballagan Formation contains plant fossils and a sparse faunal assemblage that includes gastropods, ostracods, modiolid bivalves and fish remains. The fossils occur in discrete associations of alluvial, fluvial and fluviodeltaic sedimentary rocks, intercalated with lacustrine, coastal plain and sporadic marine deposits (Figure 38).

In its outcrop from the Langholm area, westward along the Solway coast, the Ballagan Formation is in the order of 100 to 150 m thick (thinning westward) and is made up of thinly interbedded sandstone, mudstone and lagoonal dolostone ('cementstone') with some anhydrite (Plate 40). It is a terrestrial/fluvial to peritidal assemblage known in its Solway coastal outlier as the **Kirkbean Cementstone Member**. It was deposited in shallow channels within mudflats in a low-lying, coastal environment but with only a limited and intermittent marine influence (Figure 38). The Kirkbean Cementstone Member contains plant fragments and a quasi- to nonmarine fauna of bivalves, gastropods and ostracods, although fully marine bivalves are present locally. Elsewhere, in Annandale around Dalton [NY 104 740], boreholes prove the Kirkbean Cementstone Member to be at least 100 m thick and to contain abundant fibrous gypsum (satin spar) bands up to 4 cm thick. The 'cementstones' are underlain by dark red, fine-grained to conglomeratic sandstone, the **Annandale Sandstone Beds**. The anhydrite (or its precursor, gypsum) formed either subaqueously in shallow water or subaerially in coastal sabkhas, but following burial to depths greater than a few hundred metres, $CaSO_4$ is usually preserved as anhydrite. Both gypsum and anhydrite are readily removed in the surface weathering zone and so are rarely seen at outcrop.

Another 'cementstone' sequence is seen in the Langholm–Newcastleton area, where it is up to 200 m thick and passes downwards into fine-grained, massive, fluviatile and deltaic sandstones, derived from the north, which are collectively referred to the **Whita Sandstone Beds**. These 'cementstones' were formerly regarded as part of a 'Lower Border Group' (Figure 41) and demonstrate the uncertainties that bedevil correlation between the various cementstone–sandstone sequences. The 'cementstones' above the Whita Sandstone could equally well be regarded as part of the Ballagan Formation or the Lyne Formation (Border Group — see below), whilst the Whita Sandstone is assigned to the Ballagan Formation.

The most westerly outcrop of the Ballagan Formation forms part of the Rerrick Outlier on the north coast of the Solway Firth. There, the **Wall Hill Sandstone Member** rests with angular unconformity on Wenlock strata of the Raeberry Castle Formation. It has a maximum thickness of about 360 m and comprises upward-fining sequences of conglomeratic sandstones,

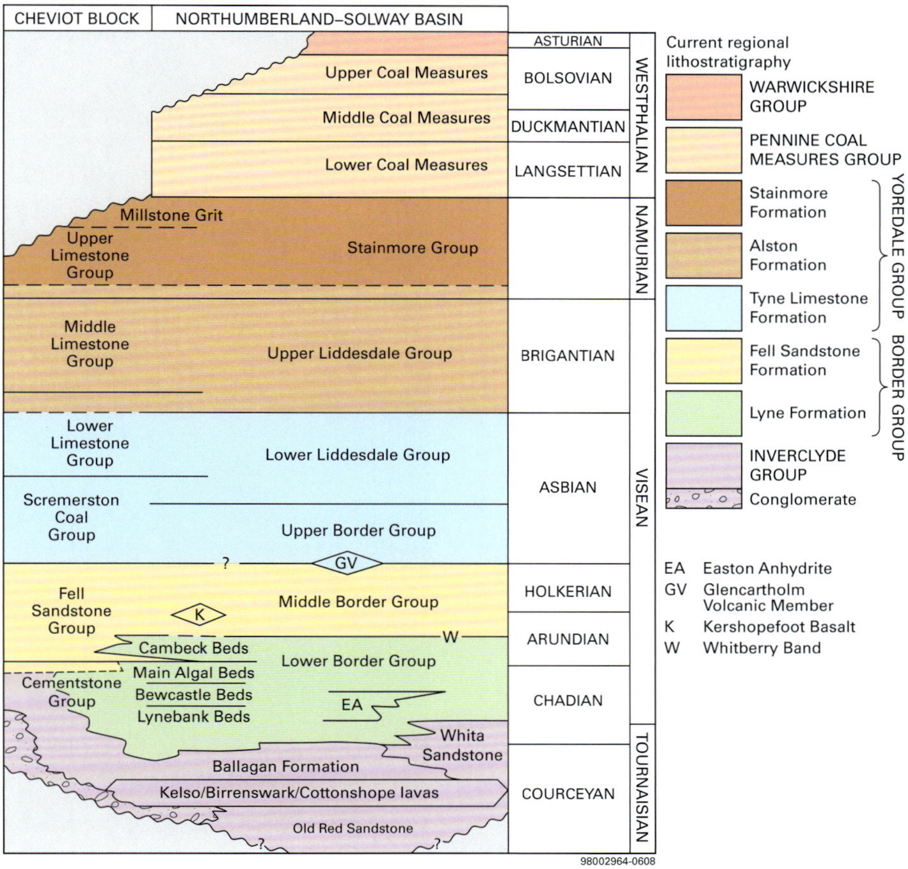

Figure 41 *Correlation chart for the traditional district-based Carboniferous lithostratigraphies (named on the figure) and the regional lithostratigraphy for the Northumberland–Solway Basin adopted in this account (identified by colour).*

siltstones and mudstones. The fine-grained rocks were laid down on a floodplain by a braided fluvial system of low to moderate sinuosity, with the coarse-grained rocks deposited in channels. Palaeocurrent data indicate a predominately north and north-westerly source from the Southern Uplands terrane, and the conglomerates include clasts derived from the roof of the Criffel–Dalbeattie pluton. Local but intense sediment fluidisation developed at a number of localities, perhaps influenced by synsedimentary activity on the nearby North Solway Fault.

Border Group

The Border Group has a more-or-less continuous outcrop along the northern margin of the Northumberland–Solway Basin but in the east this lies entirely on the English side of the border (Figure 2). The outcrop spans the border to the north of Keilder and thence extends westward to Newcastleton, Ecclefechan and the Southerness area of the Kirkbean outlier.

Lyne Formation

In general, the Ballagan Formation becomes more marine in character, both upward and laterally towards the south-east, as it passes conformably and diachronously up into the Lyne Formation of the Border Group. The Lyne Formation is late Tournaisian–Arundian in age and in the main part of its outcrop, which lies south of the English border, it comprises cyclical sequences of fine-grained subarkosic sandstone, siltstone, mudstone, thin beds of limestone

Plate 41 *Lowest algal bed in the Southerness Limestone Formation exposed on the shore at Southerness Point [NY 972 541] (P220451).*

or dolostone, and anhydrite: the Lynebank, Bewcastle, Main Algal, Cambeck, and Easton Anhydrite members are recognised (Figure 41). At outcrop, the Lyne Formation is up to about 1000 m thick but the Easton Anhydrite Member has been proved in boreholes to comprise anhydrite abundantly interbedded through about 1300 m of clastic strata at about the level of the Lynebank and Bewcastle members. Deposition took place in a fluctuating network of peritidal, deltaic and fluvial environments subject to occasional marine incursions, with the anhydrite probably accumulating under sabkha conditions.

Much of the Scottish succession is assigned to the **Southerness Limestone Member**, about 135 m of thinly interbedded calcareous mudstone, siltstone and limestone. The member is best seen in the Kirkbean Outlier where it is richly fossiliferous, with a varied marine fauna of Chadian to Arundian age, and contains two distinctive algal stromatolite bands each about 1 m thick (Plate 41). The stromatolites resemble dome types from algal beds of the Lyne Formation in the Northumberland Basin. Individual domes are up to 30 cm in diameter with a relief of 10–15 cm, and are set in a calcareous mudstone matrix. They probably formed in a low intertidal to shallow subtidal depositional environment since there are signs of limited reworking. The algal bands are developed in two of three sedimentary cycles identified in the upper part of the member as follows:

1\. Flaggy micaceous sandstone with ripple marks and plant remains
5\. Interbedded limestone and mudstones
4\. Nodular algal band (in two of the three cycles)
3\. Argillaceous limestone interbedded with calcareous mudstone
2\. Fine grained, rippled sandstone
1\. Flaggy, micaceous sandstone with ripple marks and plant remains

Each cycle commences with sandstones deposited in a littoral environment. These are succeeded by calcareous beds, formed under shallow subtidal conditions, but with some sandy limestones containing oolites as an indication that the sediments were at times affected by wave action. The cyclicity may be attributed to variation in subsidence rate and terriginous sediment input, both perhaps related to movement on the basin boundary faults.

Around Newcastleton and Ecclefechan, the member contains an Arundian fauna of brachiopods, bivalves, foraminifera and ostracods that suggests a stratigraphical level equivalent to the Cambeck Member. Also present, both there and in the Kirkbean Outlier, are limestone beds (respectively the 'Harden Beds' and Syringothyris Limestone Bed) of possible equivalence to the Whitberry Marine Band that farther south marks the base of the Fell Sandstone Formation. However, stratigraphically above the Syringothyris Limestone Bed at Southerness are the two distinctive algal stromatolite bands of 'Lyne Formation type'. If these stratigraphical correlations are correct, then the Southerness Limestone Member ranges diachronously higher than the top of the Lyne Formation elsewhere in the Solway and Northumberland basins.

In the Rerrick Outlier, the **Orroland Member** of the Lyne Formation comprises about 280 m of red beds formed under subaerial conditions on the margin of the Solway Basin against the syntectonically active North Solway Fault (Figure 38). The fining-upward cycles of conglomerate, sandstone, siltstone and calcrete–palaeosol are sporadically interrupted by marine beds of greyish red, fossiliferous limestone, which is locally oolitic. A mostly detrital fauna of corals, brachiopods and bivalves suggests correlation with the Syringothyris Limestone in the Southerness succession.

Fell Sandstone Formation

The **Fell Sandstone Formation** has an arcuate outcrop around the flanks of the Cheviot massif from Burnmouth in south-east Scotland through Northumberland and northern Cumbria, crossing back into Scotland in the Newcastleton area. Thence the outcrop continues westward along the northern shore of the Solway Firth, where the Gillfoot and Powillimount sandstone members are identified in the Kirkbean Outlier, and the Rascarrel Member in the Rerrick Outlier. The formation is mostly of Chadian–Holkerian age, though the highest strata may range up into the early Asbian. The lower boundary in Northumberland is taken at the Whitberry Marine Band, but this fails westward as the lithostratigraphical boundary becomes diachronously younger. It is Chadian in the Cheviot district where the Fell Sandstone directly overlies the Ballagan Formation, and Chadian to Holkerian in the north Solway Basin where it overlies the Lyne Formation.

Subarkosic sandstone is the dominant lithology and in its thickest development of about 350 m in central Northumberland, the formation consists almost entirely of sandstone. Northwards, the proportion of sandstone to mudstone varies considerably and in the vicinity of Berwick-upon-Tweed the Fell Sandstone Formation may contain up to 40 per cent of finer-grained lithologies. It was laid down by a fluvial depositional system that advanced from an uplifted source area to the north and east within the Grampian and Fenno–Scandinavian structural blocks. A continuous steady uplift of the source area provided a constant supply of submature clastic sediment, with complementary subsidence of the basin maintaining the fluviodeltaic depositional environment. Deposition was probably effected by several braided river systems that were constrained to the axial region of the Northumberland Basin by intrabasinal faulting. Temperate humid conditions prevailed during deposition. The succession is largely unfossiliferous, although ostracods and the large bivalve *Archanodon jukesi* (Bailey) have been recorded together with some plant fossils. Along the axis of the Northumberland Basin the fluvial sandstone passes westward and diachronously into a succession of fluviodeltaic and shallow marine deposits.

The earliest fluvial deposits accumulated in the north-east. There, to the south of Burnmouth [NT 956 610] on the Berwickshire coast, some 200 m of pale-coloured, cross- and convolute-bedded sandstone contain bands of quartz pebbles and lenses of red sandy mudstone. The sandstone is overlain by some 75 m of interbedded mudstone, sandstone and seatearth with a few thin coals (worked at Lamberton, just north of the border, in the 19th century) and argillaceous dolostone. Near the top of this sequence lies the 1.5 m thick 'Lamberton

Limestone' which is a likely correlative of the Cove Marine Bands, part of the Aberlady Formation (Strathclyde Group), farther north in the Oldhamstocks Basin (see below).

To the south-west, in the vicinity of Langholm, a coarsening-upwards siliciclastic sequence culminates in a fluvial sandstone unit up to 250 m thick, known locally as the 'Larriston Sandstone', overlain by a more mixed, thinly bedded sandstone–mudstone succession. The 'Larriston Sandstone' was derived from the north-east as channel-fill deposits within a braided river system, which by the mid Arundian was of increasing regional importance and fed into a delta system farther to the south-west. There, around Ecclefechan, the sequence comprises sandstone (some conglomeratic, with quartz pebbles), mudstone and sporadic seatearth with roots, the latter suggesting intermittent emergence of the delta surface. In the Newcastleton outcrop there is evidence for a syndepositional, extensional tectonic regime from the presence, in mid sequence, of about 30 m of porphyritic olivine basalt lava: the Kershopefoot Basalt (Figure 41).

In the Kirkbean Outlier, on the north Solway coast, the 120–150 m-thick **Gillfoot Sandstone Member** comprises quartzose sandstone, conglomerate and red mudstone, with a few thin beds of sandy limestone containing detrital fossils. Conglomerates form about 20 per cent of the succession and have a calcareous matrix. They contain intraformational fragments in addition to pebbles of Silurian wacke-type sandstone and Devonian dyke rocks. The member is sparsely fossiliferous and an Arundian age is likely. The lithologies indicate a marginal coastal depositional setting with the conglomerates deposited by sheetfloods flowing over low-lying supratidal areas.

Succeeding the Gillfoot Sandstone are about 160 m of strata, of Arundian to Holkerian age, that make up the **Powillimount Sandstone Member**. Lithologies include calcareous and

Plate 42 *Cross-bedded sandstone of the Thirlstane Sandstone Beds, Powillimount Sandstone Member, exposed in the Thirlstane natural arch [NX 992 567], section height is approximately 1.5 m (P220455).*

Plate 43 *Breccio-conglomerate of the Rascarrel Member (Fell Sandstone Formation) drapes the North Solway Fault at Castlehill Point, near Rockcliffe [NX 855 525] (P220436).*

quartzose sandstones, sandy limestone and mudstone, with sporadic thin coals and associated rooted seatearth. Sandstone beds range in thickness from 0.3 to 3 m and are laterally extensive. They are well sorted and commonly exhibit ripple cross lamination, contain abundant carbonaceous plant remains and are extensively bioturbated. Limestones range from arenaceous to argillaceous types and contain detrital fossil remains, oolites and rolled algal nodules (or oncolites). The characteristic lithologies, and the evidence of oncolites and detrital fragments, point to deposition in a shallow marine environment subject to wave action. Conversely, the presence of thin developments of coal and associated seatearths indicates periodic shallowing of lagoonal waters and the development of highly vegetated coastal plains and low lying supratidal flats.

It is likely that the sediments of the Powillimount Sandstone Member were deposited in a tidal lagoon protected from the effects of severe storms by some form of offshore sand barrier. That barrier may now be represented by the prominent and thickly bedded **Thirlstane Sandstone Beds**, 25 m of quartzose sandstone, which form the top of the member. At the base of the Thirlstane Sandstone, the contact with the underlying strata is irregular and intraformational fragments and wood remains are present in the lowest beds. The rocks are characterised by large-scale cross-bedding (Plate 42) and an increasing number of penecontemporaneous liquefaction deformation structures from south to north along the strike of the outcrop (Plate 35). The magnitude and frequency of the liquefaction structures may perhaps be attributed to seismic activity on the nearby and syndepositionally active North Solway Fault.

Farther west, in the isolated faulted blocks of the Rerrick and Colvend outliers, the Fell Sandstone Formation is represented by its **Rascarrel Member**. At Colvend, east of Castlehill Point, the spectacular basal Carboniferous unconformity is represented by a veneer of carbonate cemented breccia, overlain by conglomerate, clinging to cliffs cut in Silurian strata and a microdiorite dyke (Plate 43). The breccia clasts comprise Lower Palaeozoic sandstone and igneous rock derived from the hinterland to the north. The breccia and conglomerate dip seawards at high angles as a result of syndepositional faulting on the line of the unconformity, and the conglomeratic sandstone grades upwards into mudstone; thin interbeds of sandy limestone contain algal fragments and a sparse marine fauna of Arundian to Holkerian aspect. At both Rerrick and Colvend, at least 350 m of much-faulted strata are present, comprising arkosic and conglomeratic fluvial sandstone, siltstone, calcareous mudstone and a few coals, making this potentially the thickest development of the Fell Sandstone Formation, a situation strongly influenced by the proximity of the syndepositionally active North Solway Fault.

Yoredale Group

The Yoredale Group does not crop out in the Scottish sector of the Tweed Basin *sensu stricto*, but does appear on the North Sea coast to the north of Berwick, separated from the Tweed Basin succession by the Eyemouth outcrop of Silurian strata. The following account therefore refers mostly to the northern margin of the Northumberland–Solway Basin.

Tyne Limestone Formation

The lowest division of the Yoredale Group is the **Tyne Limestone Formation**, a sequence of highly variable, marine to deltaic, 'Yoredale'-type upward-coarsening cycles in which a thin but extensive bed of marine limestone is overlain by mudstone and sandstone commonly topped by a seatearth and a thin coal seam (Figure 40); there is much lithostratigraphical variation both vertically and laterally. The formation is largely Asbian in age and locally exceeds 2000 m in thickness. It has a wide outcrop across Northumberland and northern Cumbria and extends across the border into Scotland from south-west Liddesdale to Ecclefechan and the north Solway coast.

The base of the Tyne Limestone Formation is locally marked, between Langholm and Canonbie, by the volcanic rocks of the **Glencartholm Volcanic Member**, about 150 m of tuffs and basaltic lavas. Sedimentary rocks associated with the tuffs contain a marine fauna of brachiopods and bivalves, but are notable for rich arthropod and fish faunas, the former including crustaceans, eurypterids and scorpions. The fossils indicate a level near the base of the Asbian Stage. Above the Glencartholm Volcanic Member, the Tyne Limestone Formation comprises 'Yoredale' cyclical sequences consisting predominantly of marine mudstone and limestone with subordinate sandstone and a few seatearths and thin coal seams. Many of the beds are fossiliferous with a fauna noted for extensive coral colonies and layers of large brachiopods. More arenaceous strata follow, succeeded in turn by 30 m of marine mudstone, the **Dinwoodie Beds**, which are notable for their rich and varied fauna of corals, brachiopods and bryozoans, perhaps the most diverse fauna of the entire Carboniferous sequence.

In the lower part of the Tyne Limestone Formation there is a lateral eastward transition into a sequence of lacustrine–deltaic cycles of limestone, mudstone, sandstone and thick coal seams (up to 2 m) traditionally known in Northumberland and Berwickshire as the **'Scremerston Coal Group'**. In Berwickshire, this division is about 300 m thick (though much thicker in Northumberland) and is succeeded by a return of the marine–deltaic lithofacies. In this, typically, the sedimentary cycles comprise a limestone overlain by mudstone, which commonly shows evidence for storm-driven reworking of the sediment, followed by shallow-marine sandstone. Some of the cycles are topped by a terrestrial development of calcrete, seatearth and coal, perhaps indicating a semi-arid but seasonally wet environment.

The Tyne Limestone Formation strata that crop out along the north Solway coast in the Kirkbean Outlier are assigned to the **Arbigland Limestone Member**. Strata include thickly bedded, bioturbated, calcareous sandstones with coalified plant casts, thin sandy limestones, locally with ooliths and algal debris, dark grey carbonaceous mudstones and thin coal partings. The limestones and mudstones have an abundant and diverse fauna that includes brachiopods, bivalves, gastropods, crinoids, bryozoa, orthocones and corals, the latter including colonies of *Lithostrotion*. An Asbian age is likely and the sedimentary features are consistent with deposition in a restricted, lagoonal environment in which there was only limited reworking of sediment. A more active depositional regime is shown by the stratigraphically highest part of the member, where thickly bedded, bioturbated, ripple cross-laminated sandstone is interbedded with calcareous mudstone and some argillaceous limestone. These strata have been much affected by slump folding and are steeply inclined and locally overturned. Shallow, sandstone filled scours and washouts are common.

Alston Formation

The **Alston Formation** is the lowest division of the Yoredale Group present on the south side of the Northumberland–Solway Basin, where it succeeds the platform limestone of the Great Scar Limestone Group. The base is less well defined on the north side of the basin, where Brigantian 'Yoredale' cycles conformably succeed those of Asbian age that are assigned to the Tyne Limestone Formation. The top of the Alston Formation is defined by the **Great Limestone Member** and its correlatives, which are of Pendleian (earliest Namurian) age. The Alston Formation has a relatively restricted outcrop on the northern side of the Northumberland and Solway basins, but is present between Langholm and Canonbie and near Ecclefechan; from these localities it dips beneath the Canonbie Coalfield (Westphalian — see below) where its concealed strata have been proved in boreholes.

Between Langholm and Canonbie the limestone–mudstone–sandstone, 'Yoredale' cyclicity is well developed. Nine or ten cycles are developed over an interval of about 120 m, with the highest of the fossiliferous limestones, the Catsbit Limestone, being a correlative of the Pendleian Great Limestone Member; it therefore marks the top of the Alston Formation. A lower limestone, the Penton Limestone, is exposed in the hinge zone of the anticline illustrated in Plate 36. All of the limestones contain a relatively rich fauna with corals and brachiopods (Plate 34), and the overlying mudstones yield many brachiopods and molluscs. Assemblages of foraminifera occur in places.

Stainmore Formation

Across parts of northern England the **Stainmore Formation** comprises a largely deltaic, cyclical succession of sandstone, mudstone and poor-quality coal: a few limestone beds occur, mostly in the lower part of the formation. Some of the sandstones occur as large, channel-fill deposits, fining upwards and with erosional bases cross-cutting the underlying strata. Although the general term 'Millstone Grit' has been used for this Namurian sucession, it has long been recognised that the lithological assemblage in the Northumberland–Solway Basin is distinct from the thick development of coarse 'gritstone' and marine mudstone of the Millstone Grit Group's type area in the south Pennines. Rather, across northern England and into the Canonbie area of southern Scotland, 'Yoredale' cyclicity dominates, albeit limestone beds are relatively scarce. This northern Namurian sequence is now defined as the Stainmore Formation. Its base overlies the Pendleian Great Limestone Member (and correlatives); the top of the formation is defined at the base of the Subcrenatum Marine Band, which marks the base of the Westphalian Pennine Coal Measures Group.

Strata of the Stainmore Formation in the Canonbie area are about 400 m thick but, in a further lateral variation in lithofacies, limestone and calcareous mudstone generally take the place of the dark goniatite-bearing mudstone that forms the basis for the traditional biostratigraphical zonation farther south. At Canonbie, goniatites are rare and the macrofossils, though abundant at many horizons, are not particularly diagnostic of stratigraphical position. Rather, the fauna is an impoverished continuation of that seen in the Alston Formation. Towards the top of the formation at Canonbie, the 'Yoredale' cycles pass up into a sequence of sandstones, seatearths and thin coals, collectively called the **Penton Coals,** which have been worked locally. This division also contains several interbedded units of mudstone with *Lingula*, whilst a few thin beds of limestone and calcareous mudstone have marine faunas.

Pennine Coal Measures Group

Westphalian Coal Measures strata crop out in the Canonbie Coalfield, a limited area of such strata that lies mainly to the north of the border in south-west Scotland, but has a concealed extention southwards into Cumbria, linking with the Cumbrian Coalfield (Figure 42), as described in the companion volume for *Northern England*. The Canonbie Coalfield is situated at the north-

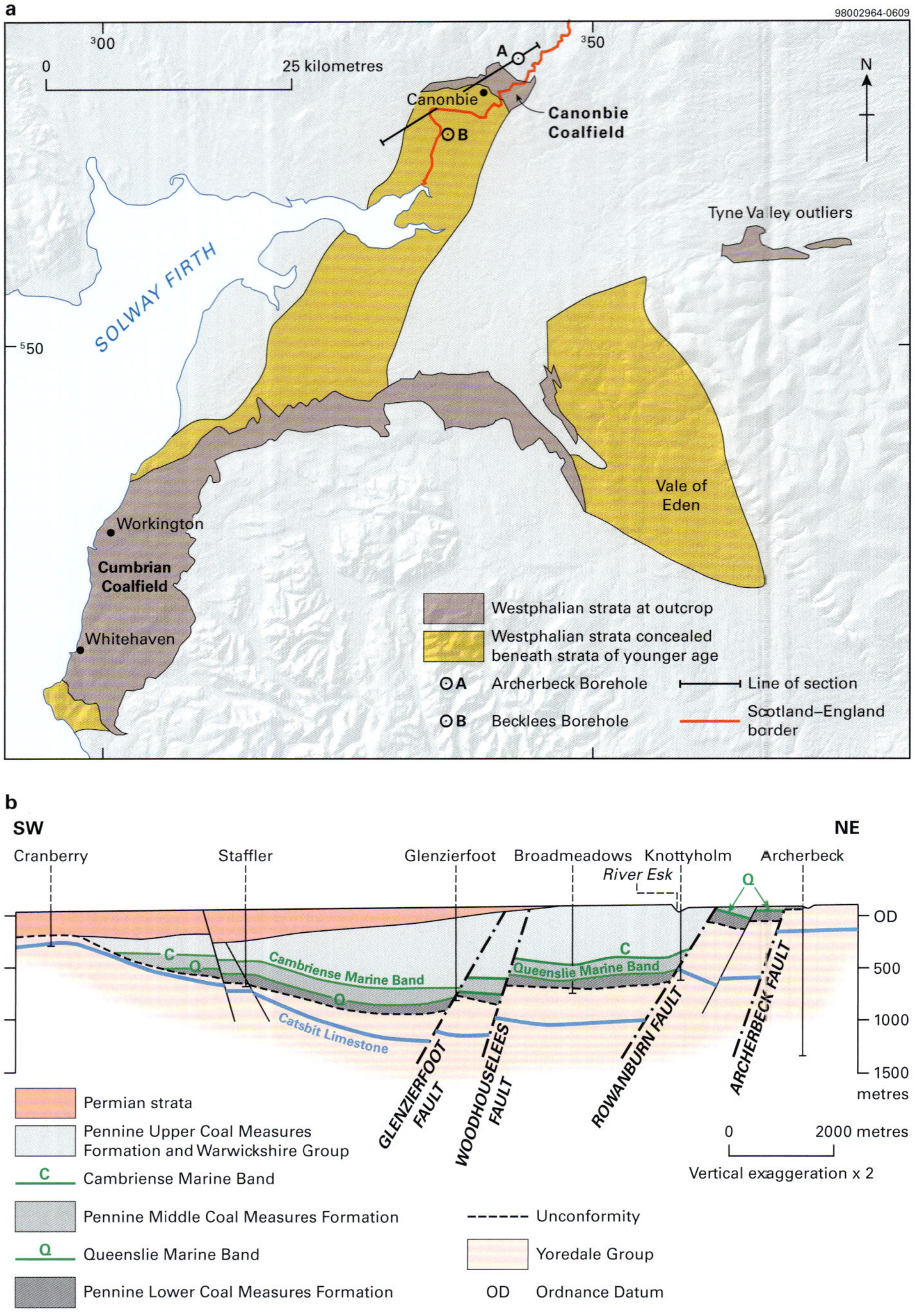

Figure 42 Regional setting of the Canonbie Coalfield with the deep structure of the basin illustrated by a representative cross-section.

west margin of the north-east to south-west-trending Solway Syncline, beneath which some coal seams may persist to reappear in Cumbria. Although their continuation has not been proved by deep boreholes, it is supported by seismic surveys. Application of both of these techniques in the 1980s indicated that the concealed coalfield is larger and has more economic potential than was previously envisaged. The exposed Canonbie coalfield is relatively small and was exploited on a limited scale in the 19th and early 20th centuries; deep mining ceased in the 1920s.

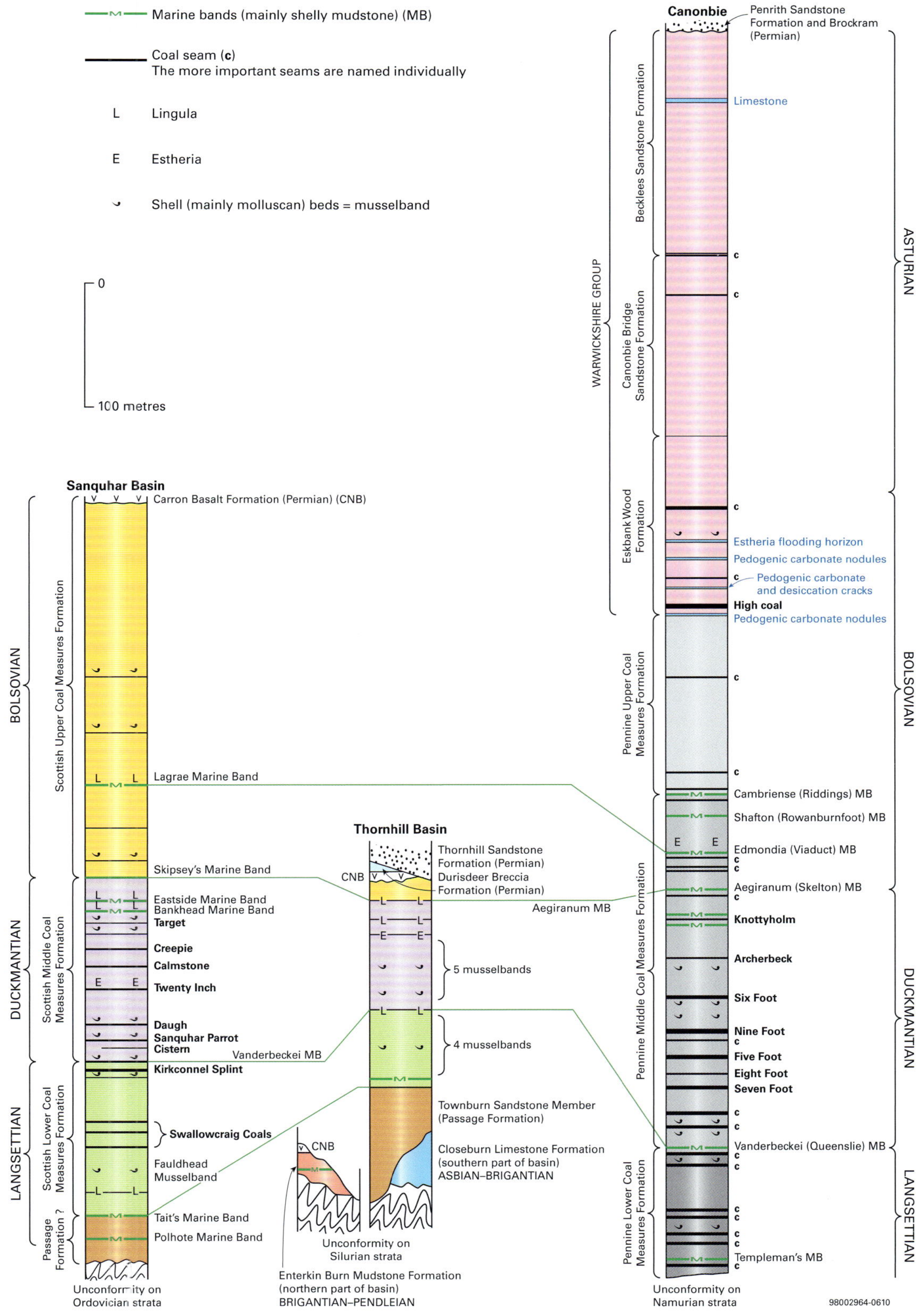

Figure 43 *Representative sections and correlations for the Scottish Coal Measures Group, the Pennine Coal Measures Group and the Warwickshire Group in the south of Scotland.*

Deposition in much of the Solway Basin continued unbroken from the Namurian into the Westphalian, but along the northern margin of the basin there is an unconformity at the base of the Pennine Coal Measures Group, such that strata high in the Pennine Lower Coal Measures Formation rest unconformably on Yoredale Group rocks. There is also evidence for at least one unconformity in the overlying, late Westphalian 'red beds' that in the Canonbie Coalfield form part of the Warwickshire Group (see below).

The Pennine Coal Measures Group is predominantly nonmarine and is characterised by coarsening-upward cycles, commonly up to 15 m thick, composed of mudstone, siltstone and sandstone capped by a seatearth and coal, though coal forms only a minor part of the total sequence (Figure 40); clay ironstone occurs within some mudstones. The Solway Basin and its Westphalian strata were subject to some uplift and deformation during the late Carboniferous to early Permian Variscan tectonic episode, with development of open folds (Plate 36) and much faulting, whilst regionally the strata of the Solway Basin were compressed into the major Solway Syncline (Figure 36). Uplifted areas were soon eroded and Westphalian beds are now preserved only in the downwarped, coalfield areas. Subsequent sedimentation produced an unconformable cover of Permo-Triassic rocks over the Westphalian beds, which are commonly reddened to a depth of more than 100 m beneath the sub-Permian unconformity. Whilst erosion has since removed parts of the cover, much of the Canonbie Coalfield remains concealed beneath Permo-Triassic strata (Figure 42).

In the Scottish part of the Canonbie Coalfield, the Pennine Lower Coal Measures Formation is about 110 m thick, the Middle Coal Measures about 250 m and Upper Coal Measures about 150 m (Figures 42 and 43): the sequence thickens into the hinge zone of the Solway Syncline. Coals in the Pennine Lower Coal Measures Formation are commonly unnamed and difficult to correlate, with borehole information showing them to be generally less than 0.8 m in thickness. The main seams are from the Duckmantian part of the Pennine Middle Coal Measures Formation, with only a few thin coals present in its uppermost, Bolsovian section. The few coals in the Pennine Upper Coal Measures are mostly located close to the base of the formation but an exception is the aptly named High Coal, which occurs about 150 m above the base of the Upper Coal Measures. It forms a convenient marker for the base of the overlying Warwickshire Group.

Pennine Lower Coal Measures Formation

At Canonbie, the Pennine Lower Coal Measures (Langsettian) rest unconformably on Namurian strata of the Yoredale Group. In common with the situation seen in other coalfields of the region, the Subcrenatum Marine Band, which defines the base of the Langsettian, is not present. Instead, correlation rests on the likely equivalence of the stratigraphically higher Langley Marine Band of west Cumbria and Templeman's Marine Band, identified at Canonbie near the bottom of the Becklees Borehole [NY 3517 7158]. Mudstone–siltstone–seatearth cyclothems are most common, but substantial sandstone bodies occur as likely channel fills, and several thin coals are present. A distinctive mudstone 'mussel band', with nonmarine bivalves and ostracods lies near the top of the formation some 4–6 m below the Vanderbeckei (Queenslie) Marine Band.

Pennine Middle Coal Measures Formation

The Pennine Middle Coal Measures (Duckmantian to lower Bolsovian) at Canonbie include strata between the Vanderbeckei (Queenslie) Marine Band and the Cambriense Marine Band, both of which can be identified. The Vanderbeckei Marine Band contains megaspores, the inarticulate brachiopod *Lingula*, and fish remains; slightly higher is a 'mussel band' in about 7.5 m of mudstone with ironstone bands that also contains nonmarine bivalves and fish material. Mudstone–sandstone–seatearth cyclothems tend to be well developed on a scale of 3 to 30 m, with a few nonmarine mussel bands in the mudstone. Several coals are also well developed so that the lower half of the Pennine Middle Coal Measures has been the most economically important part of the Canonbie succession. A 'mussel band' just below the Archerbeck Coal has a particularly varied fauna of at least 20 species of non-marine bivalves. In the upper part of the Pennine Middle Coal Measures, above the Archerbeck Coal, marine bands become more common and there is a commensurate reduction in coal development.

The Aegiranum (Skelton) Marine Band, marking the base of the Bolsovian Substage is developed in mudstone with foraminifera, inarticulate brachiopods, fish remains and conodont elements. Above this, two further marine bands appear in the succession before the Cambriense (Riddings) Marine Band marks the onset of the Pennine Upper Coal Measures. It appears as a bioturbated mudstone with ironstone nodules and contains foraminifera, inarticulate brachiopods, fish fragments, and conodont elements.

Pennine Upper Coal Measures Formation

The Pennine Upper Coal Measures Formation (upper Bolsovian) at Canonbie largely comprises mudstone–siltstone cyclothems with few substantial sandstones and only sporadic coals. Those that are present are relatively thin and located close to the base of the formation. The exception is the High Coal, which occurs about 150 m above the base and marks the onset of the overlying Warwickshire Group.

Warwickshire Group

In the Canonbie Coalfield, the Pennine Coal Measures Group is overlain by a red-bed succession that, together with other, similar upper Carboniferous sequences elsewhere in northern and central England, make up the Warwickshire Group (Table 6). A key characteristic is reddening that occurred soon after sediment deposition, but which is not uniformly developed throughout; as a result, red beds alternate with unreddened, grey-green intervals. This phenomenon introduces local uncertainty as to the exact age of the Warwickshire Group strata and their stratigraphical relationship to apparently coeval, but unreddened rocks in the Pennine Coal Measures Group.

The Canonbie outcrop of the Warwickshire Group consists of interbedded mudstones, siltstones and sandstone similar to those of the underlying Pennine Coal Measures but mostly reddened; coal is rare. The group is late Bolsovian to Asturian in age. Its base at Canonbie is conformable, grading up from the grey mudstones of the Pennine Upper Coal Measures Formation into the overlying red beds. The sharp, unconformable base of the Permian succession defines the top of the group. At outcrop, about 290 m of Warwickshire Group red beds are exposed along the banks of the River Esk, but the maximum proved thickness is the 530 m seen in the Becklees Borehole, close to the central axis of the Solway Syncline. Seismic reflection data indicate that elsewhere in the centre of the syncline the group could be up to about 700 m thick.

Three formations have been recognised within the group, each with distinctive geophysical log signatures that allow them to be readily correlated in the subsurface. The lowermost, **Eskbank Wood Formation** (late Bolsovian to Asturian), ranges in thickness from 145 m to 175 m. Its base is gradational (and probably diachronous) across alternations of grey and primary red-bed strata but can conveniently be taken at the first appearance of pedogenic carbonate nodules just below the High Coal, which forms a prominent marker horizon (Figure 43). The formation comprises red mudstone and some sandstone, calcrete palaeosols, thin beds of *Spirorbis* limestone and *Estheria*-bearing mudstone. In the lower part there are a few thin coals and beds of grey mudstone, some of which contain nonmarine bivalves. The overlying **Canonbie Bridge Sandstone Formation** (Asturian) ranges in thickness from 131 m to 154 m. The base of the formation is sharp, marked by the incoming of thick units of medium- and coarse-grained, cross-bedded channel sandstone. A noticeable feature of these sandstones is their greenish grey colour, which can be related to the presence of abundant lithic grains. There are sporadic interbeds of mudstone and calcrete. The **Becklees Sandstone Formation** (Asturian) is the highest unit recognised from the Warwickshire Group of Canonbie and is overlain unconformably by Permian strata. Its full thickness is not known, but about 200 m is proved in the Becklees Borehole. Fine-grained sandstone with a distinct orange-brown colour is the dominant lithology, with some thin beds of mudstone and calcrete. Borehole cuttings suggest limestones and thin coals may also be present, though rare.

Warwickshire Group sedimentation in the Canonbie area largely took place on an alluvial plain drained by braided river systems, and was characterised by an early, primary oxidation of the strata. Palaeocurrent data from channels in the Canonbie Bridge Sandstone show that the rivers flowed towards the north. Overbank and floodplain mud was deposited between the channels, where soils were able to form during intervals of low aggradation. The *'Spirorbis'* limestones of the Eskbank Wood Formation are most probably of lacustrine origin.

OLDHAMSTOCKS BASIN, DUNBAR

South-east of Dunbar, as far as Cockburnspath, the Midland Valley lithostratigraphy (Table 5) can be recognised in a coastal outlier of Carboniferous rocks. The coastal section exposes a succession of strata, the **Inverclyde**, **Strathclyde** and **Clackmannan groups**, of Courceyan to Brigantian age (Figure 39), which accumulated at the south-western margin of the Oldhamstocks Basin, an area of independent subsidence peripheral to the south-east margin of the much larger Midland Valley Basin (Figures 34 and 36). The Dunbar–Cockburnspath outlier is bounded to the south-west by the Innerwick Fault, but inland exposure is very limited. To the south, the outlier is separated from the Tweed Basin by the Lammermuir Hills and Coldingham Moor outcrops of Lower Palaeozoic strata, which probably formed a contemporary, terrestrial barrier between the two areas of subsidence. Many ot the Strathclyde and Clackmannan groups' limestones have been worked locally for agricultural lime, with some kilns still preserved. Clackmannan Group limestones are now worked on a large scale near Dunbar for cement production.

Inverclyde Group

At the base of the group, the **Kinnesswood Formation**, here probably of Tournaisian (Courceyan) age, comprises red and green sandstones and pedogenic limestones ('cornstones'). Fragmentary fish fossils have been found at several levels, and of particular importance is the occurrence at Hawk's Heugh (NT 790 714) of *Remigolepis* (the only British example of this fish), *Holoptychius* and *Bothriolepis*. Generally, this combination would be taken to indicate a Famennian (latest Devonian) age, only possibly ranging up into the lower part of the Tournaisian. However, from Hawk's Heugh, a miospore assemblage confirms a Courceyan (earliest Tournaisian) age. Though in general the formation conformably overlies Upper Devonian strata of the Greenheugh Sandstone Formation (Stratheden Group), in the coastal section at Hawk's Heugh the boundary is faulted. Close to Oldhamstocks, probable Kinnesswood Formation strata are associated with a thin succession of red volcanic breccia and basaltic lava that may correlate with the **Kelso Volcanic Formation**, described earlier in this account from its outcrop farther south where it is thought to be of Courceyan age and to succeed the Kinnesswood Formation. As discussed in Chapter 5, there is considerable uncertainty over the precise position of the Devonian–Carboniferous boundary within the lithostratigraphy.

The sandstones of the Kinnesswood Formation pass up into the **Ballagan Formation**, characterised by the appearance of grey mudstones and ferroan dolostones ('cementstones'). The palynology of the Ballagan Formation shows it to be Courceyan to Chadian in age, and it has a macrofossil fauna that includes quasimarine bivalves and ostracods. The lower part of the Ballagan Formation comprises about 60 m of silty, ripple cross-laminated sandstone, grey mudstone, and dolostone. These are alluvial floodplain deposits and culminate in the distinctive Eastern Hole Conglomerate, with its red, coarse-grained sandstone matrix and fossil content of fish scales and spines, and the overlying *Sanguinolites* band with coalified wood *(Lepidodendron)* fragments, the eponymous bivalve and rare ostracods and fish scales. Higher in the sequence are thin beds of calcareous sandstone, siltstone and dolostone, the sandstone containing bivalves, ostracods and fish fragments. About 56 m of thicker sandstone beds, the Horse Road Sandstone, succeed the highest dolostone and are predominantly cross-bedded with common slump structures. Some of the sandstone beds contain rip-up clasts of mudstone, and some feature distinctive, spherical, calcareous concretions up to 1.5 m in diameter.

The Horse Road Sandstone is unfossiliferous and is interpreted as the infill of a major fluvial channel. The succeeding sequence, about 90 m thick, sees the recurrence of fine-grained sandstone, dolostone and mudstone of floodplain origin. Finally, at the top of the Ballagan Formation, thick sandstone beds reappear in a sequence that includes the Kip Carle Sandstone at Cockburnspath, which is 21 m thick and has been interpreted as a fluvial channel-fill deposit.

Strathclyde Group

In the Dunbar–Cockburnspath outlier, the Strathclyde Group is divided into the Gullane and Aberlady formations (Figures 35 and 39). About 125 m of **Gullane Formation** strata are present. The lower part of the formation comprises mostly mudstone interbedded with thin coals (<0.3 m), succeeded by 56 m of cross-bedded, channel-fill sandstone, the Heathery Heugh Sandstone, which has a median parting of soft mudstone, 3.4 m thick. The upper part of the formation is made up of argillaceous sandstone and soft carbonaceous mudstone. Miospores of early Visean age are present in places, as well as plant fragments and nonmarine bivalves; poorly developed 'marine bands' contain the brachiopod *Lingula* and various molluscs.

The **Aberlady Formation** is about 140 m thick at Cove Harbour and includes red-orange weathering sandstone (including the prominent Cove Harbour and Bilsdean sandstones), bioclastic limestone with a varied shelly fauna, and the two Cove Marine bands, the lower of which is a 3.5 m-thick sandstone, and the upper a somewhat thinner, blue-grey silty mudstone. The marine bands contain a shelly fauna (including goniatites) and miospores of Asbian age. The Cove Marine Bands have been correlated with the Macgregor Marine Bands of the Lothians and East Fife, which represent the first fully marine incursions to affect east central Scotland in Carboniferous times. There, the lowermost of the multiple Macgregor Marine

Plate 44 *Palaeokarst surface of the Blackbyre Limestone (Aberlady Formation) as exposed on the shore at Catcraig [NT 715 774]. The seatearth-filled hollows are thought to mark the growth position of trees (P006000).*

Bands is taken as the formal base of the Aberlady Formation so, if the correlation is secure, the lower of the Cove Marine Bands would mark that level in the Oldhamstocks Basin. Above the upper Cove Marine Band lies the Cove Harbour Sandstone, about 25 m thick, and then a sequence of grey silty mudstone with nodular siderite horizons that is capped by the probably lacustrine Cove Oil Shale containing algal traces and pyritised plant stems. Together with the succeeding Bilsdean Sandstone, the sequence is interpreted as mostly crevasse splay and crevasse channel floodplain deposits, with the thicker, cross-bedded, red-stained sandstones originating as fluvial channel fills.

The Longcraig to Barns Ness section of the Aberlady Formation is composed of about 19 m of interbedded coals, mudstones and limestones, the latter including the Hollybush and Blackbyre limestones, formerly known respectively as the Lower and Middle Longcraig limestones. The formation was deposited in fluviodeltaic and marine transgressive environments, and most of the limestones contain a rich and varied fauna that includes corals, brachiopods, bivalves and crinoids. The Blackbyre Limestone is nodular and contains abundant *Siphonodendron* corals, brachiopods and bryozoan debris. It overlies bioturbated sandstone and is overlain by a sequence of mudstones, seatclays with *Stigmaria*, and thin coals. The upper surface of the Blackbyre Limestone, best seen at Catcraig [NT 715 774] is remarkable in that it is pitted with seatearth-filled hollows that are thought to mark the sites of trees growing on what was probably a palaeokarst surface (Plate 44). The top of the Aberlady Formation is taken at the base of the Hurlet Limestone, formerly known as the Upper Longcraig Limestone.

Clackmannan Group

The Hurlet Limestone marks the base of the Clackmannan Group and of its lowest division, the **Lower Limestone Formation**. The latter is characterised by strongly cyclical, upward-coarsening sequences of limestone, mudstone, siltstone and sandstone, which may be capped by seatearth and coal. The limestones are laterally extensive and have a rich and diverse fauna of marine fossils that includes foraminifera, corals, bryozoa, brachiopods, gastropods, bivalves, nautiloids, crinoids and echinoids. The formation was deposited in progradational lobate deltas during marine transgressive–regressive cycles.

The Lower Limestone Formation is Visean (Brigantian) in age on the basis of its shelly fauna and miospore flora. The Hurlet Limestone notably includes *Koninckophyllum* corals whilst the higher Craigenhill (formerly the Lower Skateraw) Limestone is characterised by the brachiopod *Gigantoproductus*. Farther up the sequence the Blackhall (formerly the Middle Skateraw) Limestone is crinoidal and contains a 25 cm-thick band packed with foraminifera, whilst the Neilson Shell Bed immediately above it contains brachiopods, trilobites and crinoids. The succeeding Chapel Point and Barns Ness limestones (local names applied to two of the more widely identified Hosie limestones) are only sparsely fossiliferous but bellerophontid gastropods, coiled nautiloids and crinoids are present locally, whilst trace fossils are abundant.

CENTRAL OUTLIERS (WHITECLEUCH, SANQUHAR AND THORNHILL)

Carboniferous (and Permian — see Chapter 7) rocks occupy a series of outliers mostly located in what were palaeovalleys in Upper Palaeozoic times, with only limited connection to the contemporaneous Midland Valley and Solway basins. During the early Carboniferous, these valleys were probably areas of erosion rather than deposition, since the oldest Carboniferous rocks present belong either to the Strathclyde Group or to the Clackmannan Group.

All three inliers are fault controlled to some extent. The **Whitecleuch** outlier is essentially an isolated fragment of the Douglas Coalfield of the Midland Valley of Scotland, preserved by downfaulting in a small graben within Ordovician rocks immediately south of the Southern Upland Fault. Its stratigraphy therefore closely matches that of its parent, the Douglas Basin. A similar situation is seen nearby at New Cumnock, where part of the Ayrshire Coalfield succession oversteps the Southern Upland Fault. The **Sanquhar** outlier is formed by a half-graben structure orientated north-west to south-east, with the bounding Sanquhar Fault on the north-east margin downthrowing Carboniferous against Ordovician rocks. For the **Thornhill** outlier, a plexus of north–south orientated faults associated with the major Carronbridge Fault controls the eastern side of a complex half-graben. At both Sanquhar and Thornhill, some smaller subsidiary outliers with limited stratigraphy occur to the east of the main outliers, suggesting that deposition intermittently overstepped the basin boundary faults.

Strathclyde Group

The Strathclyde Group has very resticted outcrops at the north-west and south-east margins of the Whitecleuch outlier, and to the south-east of New Cumnock. In both cases the lowermost beds rest directly on Ordovician strata. In the Whitecleugh outlier, a thin succession of sandstone and some mudstone is assigned to the **Lawmuir Formation**. Much of the sandstone is variably mottled green and red, but some is white and quartzose and prominent in exposure. The New Cumnock outcrop consists of thinly interbedded sandstone and mudstone. No strata of the Strathclyde Group are known from either the Sanquhar or Thornhill outliers.

Clackmannan Group

Sanquhar outlier

At the south-west margin of the Sanquhar outlier, strata assigned to the Clackmannan Group form a condensed sequence up to 30 m thick of coarse-grained and carbonaceous sandstone and grey siltstone. They rest with abrupt angular unconformity on Ordovician strata, and include the **Polhote Marine Band** mudstone, of Pendleian age. This mudstone, with a composite fauna of plant scraps, brachiopods, bivalves and gastropods, records a rare marine incursion into the outlier and allows valuable, if limited, correlation with the Clackmannan Group in the Midland Valley of Scotland.

Around the south-eastern margin of the main outlier, and in several of the adjacent minor outliers at Bogs Burn, Auchentaggert, and Muirhead, sandstones, siltstones and mudstones referred to the Clackmannan Group form isolated lenses preserved between the supra-Ordovician unconformity and the unconformity at the local base of the Scottish Coal Measures Group. The mudstones and calcareous siltstones contain plant remains and a marine fauna that includes bryozoa, brachiopods, gastropods, bivalves, and nautiloids. The fauna is most probably late Visean in age, with the presence of the brachiopod *Latiproductus latissimus*? indicative of the Brigantian substage.

Thornhill outlier

Within the Thornhill outlier, the **Closeburn Limestone Formation** of Visean (upper Asbian to Brigantian) age crops out around the southern margin and consists of about 25 m of dolomitic limestone, sandstone, siltstone and mudstone. The limestone was of sufficient thickness and quality to have been mined at Closeburn and Barjarg and burned for agricultural lime in the late 18th century, with well-preserved limekilns present at Croalchapel on the east bank of the River Nith. It has a fauna of corals, bryozoa, brachiopods, gastropods, bivalves, large orthocone nautiloids and fish. The partly coeval **Enterkin Mudstone Formation** forms the base of the Carboniferous succession in the north of the outlier and consists of up to 25 m of mudstone, limestone, siltstone and sandstone that crop out discontinuously in small areas on its western and northern margins. The mudstone and limestone include at least two marine beds with a fauna of bryozoa, brachiopods, gastropods, bivalves and nautiloids that suggest a Brigantian (to perhaps early Pendleian) age.

The **Townburn Sandstone Member** of the **Passage Formation** comprises pebbly sandstone (with plant stems) up to 40 m thick and rests unconformably on the Enterkin Mudstone, the Closeburn Limestone, or where these are overstepped, Lower Palaeozoic rocks. Whilst the member lacks age-diagnostic fossils, it is the youngest unit in the Clackmannan Group of the Thornhill Basin, and is most probably a Namurian and/or Westphalian deposit.

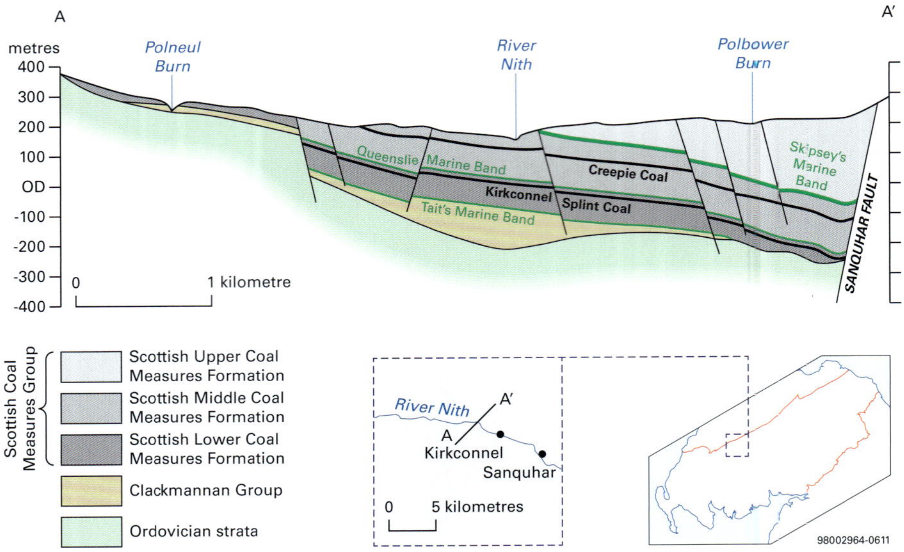

Figure 44 *Representative cross-section of the Sanquhar coalfield to illustrate the half-graben structure of the basin.*

Whitecleuch and New Cumnock

In the Whitecleuch outlier, the **Lower Limestone Formation**, the lowest division of the Clackmannan Group (Figure 35), crops out in several fault-bounded blocks where it is represented by a transgressive cyclic sequence of marine limestone and mudstone, passing up into siltstone, sandstone and sporadic seatearth and coal. It is succeeded across a low-angle disconformity by the pebbly sandstone and siltstone of the **Passage Formation**.

The small lenticular area of Carboniferous rocks which overstep the Southern Upland Fault immediately south of New Cumnock contains representatives of the Lower Limestone, Limestone Coal, Upper Limestone and Passage formations of the Midland Valley of Scotland, which rest with gross angular unconformity on Ordovician sandstones of the Marchburn Formation. These Clackmannan Group rocks form a southern extension of the Carboniferous succession seen in the Ayrshire Coalfield and are best considered, in context, in the companion volume for the *Midland Valley of Scotland*, British Regional Geology series.

Scottish Coal Measures Group

Coal Measures strata, sandstone–mudstone–seatearth–coal cyclothems (Figure 40), occupy most of the Sanquhar outlier and form a substantial part of the Thornhill outlier. Much smaller outcrops are present within the Whitecleugh outlier, and at the base of the mostly Permian succession occupying the Stranraer Basin.

Strata of the **Scottish Lower Coal Measures Formation** crop out in the north-west of the Whitecleugh outlier. They conformably succeed the Passage Formation but the top of the sequence is faulted out. Less than 100 m of sandstone and mudstone are present, with a few interbedded seatearths and thin coals, but there is a lack of distinctive horizons and only general correlation with the Douglas Coalfield is possible.

On the west (less-faulted) side of the Loch Ryan half-graben, north-west from Stranraer, a thin Westphalian sequence rests with marked angular unconformity on steeply inclined Lower Palaeozoic sandstone, and is covered unconformably by Permian breccias. The Westphalian sequence, the **Leswalt Formation**, comprises about 30 m of grey, red, and mottled yellow-brown sandstones interbedded with purple-grey mudstone and rare seatearth. The micaceous sandstone beds, and more particularly the thin interbedded silty mudstones, contain a Westphalian plant assemblage, most probably correlating with either the Scottish Lower or Middle Coal Measures.

Sanquhar Scottish Coal Measures Group strata form the bulk of the Carboniferous succession in the Sanquhar outlier (Figures 43 and 44). On the western margin of the basin, there are about 120 m of the **Scottish Lower Coal Measures Formation**, but this division thins eastwards due to overlap onto an irregular topography of subvertical Ordovician sandstone beds. The lower part of the succession is mainly arenaceous, with rare thin coal seams, while the upper part includes a group of up to eight **Swallowcraig Coals** and the important **Kirkconnel Splint Coal**, the most extensively worked seam in the coalfield and normally between 0.75 and 1.5 m thick.

Towards the base of the Sanquhar succession lies **Tait's Marine Band**, a possible correlative of the Lowstone Marine Band that elsewhere is taken to define the base of the Scottish Coal Measures Group. Tait's Marine Band has a composite fauna including miospores, brachiopods, gastropods, bivalves, nautiloids and crinoids, but passes abruptly northwards into a '*Lingula* band' mudstone with a more resticted fauna. Higher in the succession there are several mudstone beds containing nonmarine bivalves (mussel bands). The **Fauldhead Mussel band**, below the Swallowcraig Coals, has a fauna of nonmarine bivalves, ostracods and fish remains. The mussel band below the Kirkconnel Splint Coal, contains nonmarine bivalves (Plate 37) and fish remains whilst another mussel band is locally developed in the roof of that seam.

The **Scottish Middle Coal Measures Formation** includes strata between the base of the **Queenslie (Vanderbeckei) Marine Band** and the base of **Skipsey's (Aegiranum) Marine Band** and has a maximum thickness of about 135 m at Sanquhar (Figure 43). Apart from several named and extensively worked coal seams, the succession consists mainly of sandstone, siltstone, mudstone and seatearth, with ironstone ribs in places. As with the underlying Scottish Lower Coal Measures, there is a general thinning of the succession towards the eastern part of the Sanquhar Basin, consistent with its palaeo-connection into the thicker and more extensive Carboniferous succession to the north-west, in the Midland Valley.

The Queenslie Marine Band has a fauna of foraminifera, sponge spicules and bivalves. There are several mussel bands (Plate 37) higher in the succession, including one just above the Queenslie Marine Band, one in the roof of the **Sanquhar Parrot Coal**, and one in the roof of the **Daugh Coal**; the fauna in the roof of the Sanquhar Parrot Coal is relatively the richest of the three. There are several higher mussel bands: one about 12 m above the '*Estheria* band' that overlies the **Twenty Inch Coal**; one above an '*Estheria* band' below the **Target Coal**; the **Bankhead (Haughton) Marine Band**, which contains fish in addition to *Lingula mytilloides*; and the **Eastside (Sutton?) Marine Band** with a more varied assemblage of foraminifera, sponge spicules, inarticulate brachiopods and trace fossils.

The **Scottish Upper Coal Measures Formation** at Sanquhar comprises about 300 m of mainly mudstones, siltstones and sandstones with minor thin coals. It includes strata between the base of **Skipsey's Marine Band** and the regional unconformity at the base of the Permian succession (Figure 43); at Sanquhar (and Thornhill, see below) the unconformity is partially overlain by weathered, olivine basalts of the lower Permian Carron Basalt Formation. The lower part of the Scottish Upper Coal Measures Formation is grey in colour but reddened strata appear towards the top, as seen in the late Westphalian sequences farther south.

Skipsey's Marine Band is a carbonaceous siltstone (or locally an impure limestone) with a varied benthonic fauna that includes foraminifera, brachiopods, bivalves, nautiloids, ammonoids, ostracods, crinoids, conodonts and fish. Higher in the succession, the **Lagrae (Edmondia) Marine Band** contains foraminifera, inarticulate brachiopods, bivalves and fish remains. Several mussel bands are present: one a little above Skipsey's Marine Band and two more much higher in the succession, respectively about 113 m and about 195 m above Skipsey's Marine Band.

Thornhill In the Thornhill Basin, some 140 m of strata from the Scottish Coal Measures Group have been proved in the Crichope Linn Borehole [NX 9093 9511], a highly attenuated succession compared to that at Sanquhar. The main lithologies are sandstones, mudstones and seatearths, with the intervals where coals would be expected commonly represented by thin iron-rich red mudstones, sometimes with relict coaly matter. Perhaps because of the limited thickness of Carboniferous strata, the reddening seen only in the Scottish Upper Coal Measures Formation at Sanquhar extends throughout the entire Scottish Coal Measures Group at Thornhill.

Lithologies in the **Scottish Lower Coal Measures Formation** (Langsettian) at Thornhill are mainly medium-grained, grey-white sandstone, purple and red siltstone and mudstone, and mottled seatearth. The strata are almost 50 m thick in the Crichope Linn Borehole. Although there are no coals present, five levels can be tentatively correlated by their nonmarine bivalves with mussel bands in the Sanquhar succession (Figure 43).

The **Scottish Middle Coal Measures Formation** (Duckmantian) in the Crichope Linn Borehole consists of 75 m of cyclical red mudstone, siltstone, sandstone and seatearth between the Vanderbeckei Marine Band at the base of the formation and the base of the Aegiranum Marine Band at the base of the Scottish Upper Coal Measures Formation. Other boreholes at Closeburn [NX 903 945] and Carronbank [NS 882 013] confirm the presence of six marine and nonmarine fossil horizons in the Thornhill outlier, which can be correlated with assemblages at Sanquhar to the north and the Solway Basin to the south (Figure 43).

The **Scottish Upper Coal Measures Formation** (Bolsovian) is only about 15 m thick in the Crichope Linn Borehole where it consists of red or purplish red mudstone and siltstone with ironstone nodules and plant detritus, as well as thin sandstone and seatearth. The Aegiranum Marine Band at the base of the Formation has been identified at several localities at Thornhill (including the Crichope Linn Borehole) as a foraminifera-bearing, mudstone mussel band. It can be correlated with the Skelton Marine Band of the Canonbie succession and Skipsey's Marine Band in the Sanquhar outlier (Figure 43 and Table 6). At Thornhill, strata above this level and up to the base of the Permian Carron Basalt Formation have proved unfossiliferous.

MAGMATISM

The Birrenswark and Kelso volcanic formations (Inverclyde Group) are the extrusive manifestations of Late Devonian to early Carboniferous magmatism, which also gave rise to a range of minor intrusions across the south of Scotland. The presence higher in the succession of the Kershopefoot Basalt (Fell Sandstone Formation) and the Glencartholm Volcanic Member (Tyne Limestone Formation) demonstrates that magmatism continued well into the Visean. Although there are many small intrusions and agglomerate-filled vents scattered across the Scottish Borders between Langholm and Duns, and though most are regarded as of Carboniferous age, it is not possible to identify any of them as a specific centre from which the lavas and tuffs were erupted. Nevertheless, a general association seems highly likely. Several of the minor intrusions cut Border Group strata and these can therefore be tentatively related to the younger volcanic units.

Most of the regional swarm of early Carboniferous minor intrusions — vents, plugs, dykes and sills — are, in compositional terms, alkali basalts, with some of the larger and coarser-grained intrusions, which cooled relatively slowly, being texturally doleritic. This assemblage is accompanied by a range of more evolved trachytic intrusions, some of which are formed by silica-undersaturated phonolites whilst others consist of more silica-rich rhyolitic or microgranitic rocks. The relatively silica-rich trachytic magmas were more viscous than the basaltic types and their intrusion commonly resulted in sills and laccoliths rather than in extrusive rocks. The trachytic rocks are also particularly tough and resistant to erosion and so tend to give rise to prominent landscape features. The best-known of the evolved, silica-undersaturated intrusions now forms the Eildon Hills (Plate 5) [NT 550 325], near Melrose, but other examples in that general vicinity occur at Skelfhill Pen (phonolite), Millstone Edge (phonolitic trachyte), Linhope Burn (phonolitic trachyte), Tudhope Hill (phonolitic trachyte) and Pikethaw Hill (phonolite). Farther east, a cluster of trachytic and rhyolitic sills and laccoliths intruded into

Plate 45 *Vertical dyke of Carboniferous olivine-dolerite intruded into Devonian (Lower Old Red Sandstone) Reston Group conglomerate at Fairy Castle [NT 702 699]. The conglomerate has been preferentially eroded (P000941).*

Upper Devonian strata of the Stratheden Group crops out to the west of Duns around Dirrington Great Law [NT 698 549].

The Eildon Hills intrusion (Plate 5) is a composite laccolith made up from several thick sheets of locally columnar jointed, trachytic riebeckite-microgranite and phonolite. It cuts

Silurian strata and an outlier of the Upper Devonian Stratheden Group. A radiometric date of about 352 Ma suggests that intrusion occurred quite early in Carboniferous times. A scattering of dykes and small irregular intrusions of similar composition to the main Eildon Hills body is seen within a few kilometres of it, and suggests that the original intrusive mass was much larger than the preserved remnant. To the north-west of the Eildons, a large volcanic vent (Chiefswood, approximately 2 km^2) of rhyolitic tuff is thought to be sightly younger than the laccolith intrusion, whilst the latter is itself intruded by a small basaltic vent. The Eildon Hills intrusion straddles one of the major strike faults in the Lower Palaeozoic Southern Uplands terrane and it is tempting to think that intrusion was localised by that pre-existing structure.

Throughout the region, there are minor intrusions, mostly mafic, which are of uncertain age and association but which are assigned a general Carboniferous age. An example is the sparse swarm of broadly north-west-trending olivine-dolerite dykes that cut the Lower Devonian Great Conglomerate Formation in Lauderdale (Plate 45). Whilst a Carboniferous age for these dykes seems most likely, a Mid Devonian age has also been proposed and cannot be discounted. Analcime-bearing dolerite intrusions thought to be of early Carboniferous age occur around Duns, where they have been quarried for aggregate.

Late Carboniferous to early Permian mafic minor intrusions are widely distributed throughout the Midland Valley of Scotland, with major sill complexes developed both there and in northern England. However, in the Southern Uplands minor intrusions of this age are rare and are restricted to scattered alkali-dolerite and quartz-dolerite sills and dykes; the quartz-dolerite dykes are late Carboniferous in age (about 300 Ma) but the alkali-dolerites may be older. Together, the dolerite dykes form a sparse, east–west-trending swarm that is most extensive in the north-eastern part of the Southern Uplands, though possibly the best exposed examples are to be seen in the River Esk and the Liddel Water north of Canonbie.

Rather more unusual igneous rocks of late Carboniferous to early Permian age are 'camptonite' or 'monchiquite' dykes, lamprophyric rocks respectively with plagioclase or feldspar-free with analcite. Although they are rare, they are fairly widespread, with two small intrusions of 'essexite', a type of nepheline-gabbro, near Wanlockhead and Abington. In the Sanquhar Basin, the Westphalian Coal Measures are cut by a number of highly altered analcime dolerite sills and a few north-west-trending 'monchiquitic' or 'camptonitic dolerite' dykes. The dykes are generally decomposed and have fed sheets intruded along the coal seams, the sheets being converted to 'white trap' in the process and rendering the coal unworkable locally. Both the dykes and sills are thought to be related to thin (<50 m) lower Permian lava sequences locally present at Sanquhar and in the nearby Thornhill Basin: the lower Permian lavas are described in Chapter 7. The north-west-trending, upper Carboniferous or lower Permian dyke-like 'Essexite' intrusion exposed in Craighead Quarry near Abington (formerly known as the Crawfordjohn Essexite) is now classified as a nepheline-gabbro. This distinctive rock is characterised by the presence of large phenocrysts of euhedral Ti-rich augite set in a finer-grained groundmass of plagioclase, clinopyroxene, olivine, nepheline and analcime. It has become well known as the raw material for high-quality curling stones, rivalling the better-known Palaeogene microgranite lithology from Ailsa Craig, the small island off Girvan in the Firth of Clyde.

seven

Permian and Triassic

Late in Carboniferous times, a major continental collision far to the south of southern Scotland produced the Pangaea supercontinent, and in the process drove the Variscan Orogeny. As a peripheral effect, the Carboniferous basins that had formed across southern Scotland and northern England were inverted, their strata uplifted, folded and faulted. The Permian Period commenced (at about 299 Ma) with the erosion of these strata, at a time when Britain lay, as a part of Pangaea, in tropical latitudes about 10 degrees north of the Equator (Figure 3e). Pangaea continued to drift slowly northward and by Triassic times (from about 251 Ma) the south of Scotland region had moved to about 30 degrees north. The palaeogeography is summarised in Figure 45. Desert conditions prevailed throughout this interval, creating depositional environments that included sandy dunefields, alluvial plains, ephemeral lakes and mudflats. There was localised volcanic activity early in the Permian, evidence for which is now best preserved in the Thornhill Basin.

Permian and Triassic continental clastic sedimentary rocks were formerly referred to the 'New Red Sandstone'. Early descriptions of the 'New Red Sandstone' of the Dumfries Basin date back to the 1850s, with the desert red beds long known for their vertebrate trackway trace fossils. However the strata of southern Scotland lack biostratigraphically diagnostic fossils and so are difficult to date with any precision. Further, because the Permo-Triassic sedimentary basins in which they were deposited are tectonically and geographically isolated (Figure 46), a separate lithostratigraphical nomenclature has been established for each basin. For the exclusively Scottish basins these local schemes are now all included within the Stewartry Group. The large Carlisle Basin spans the border between southern Scotland and north-west England, and for this sequence the lithostratigraphy developed for the extensive English outcrop is extended across the border into the relatively small Scottish outcrop along the northern margin of the basin. Only general correlation is possible between the wholly Scottish and the trans-border successions (Figure 47).

SOUTHERN UPLANDS BASINS: STEWARTY GROUP

Rocks of the Stewartry Group, mostly red sandstones and breccio-conglomerates, are confined to a series of fault-bounded, north-west- or north-trending basins that cut across the Southern Uplands massif: Stranraer, Ballantrae, Thornhill, Dumfries, Lochmaben, Moffat and the Snar Valley (Figure 46). The basin boundary faults followed reactivated Caledonian trends in response to broadly east–west extension, as mainly half-graben structures developed on either side of a compound horst underpinned by the Galloway granitic plutons. For the most part, the boundary faults do not appear to have an earlier history of reactivation and in places cut folded Dinantian strata and displace the North Solway Fault, one of the most active of the early Carboniferous syndepositional faults. It would seem probable that there was rotation of the regional stress field through Carboniferous times and into the Permian Period.

Stranraer Basin

The Stranraer Basin is a half-graben bounded to the east by the Loch Ryan Fault. Thick, massive breccio-conglomerate of the Permian **Loch Ryan Formation** is exposed along the Loch Ryan coast to the north-west of Stranraer. Geophysical evidence shows that the basin closes northwards and is asymmetric, with up to 1200 m of strata at its eastern margin. The formation rests unconformably on Carboniferous or Lower Palaeozoic rocks, the latter providing the locally sourced, angular and subangular clasts of wacke sandstone that dominate the breccia (Plate 46). A northerly provenance has been deduced for the deposit, which is interpreted to be the product of either debris flow or sheet flood in alluvial fans. In broad agreement with

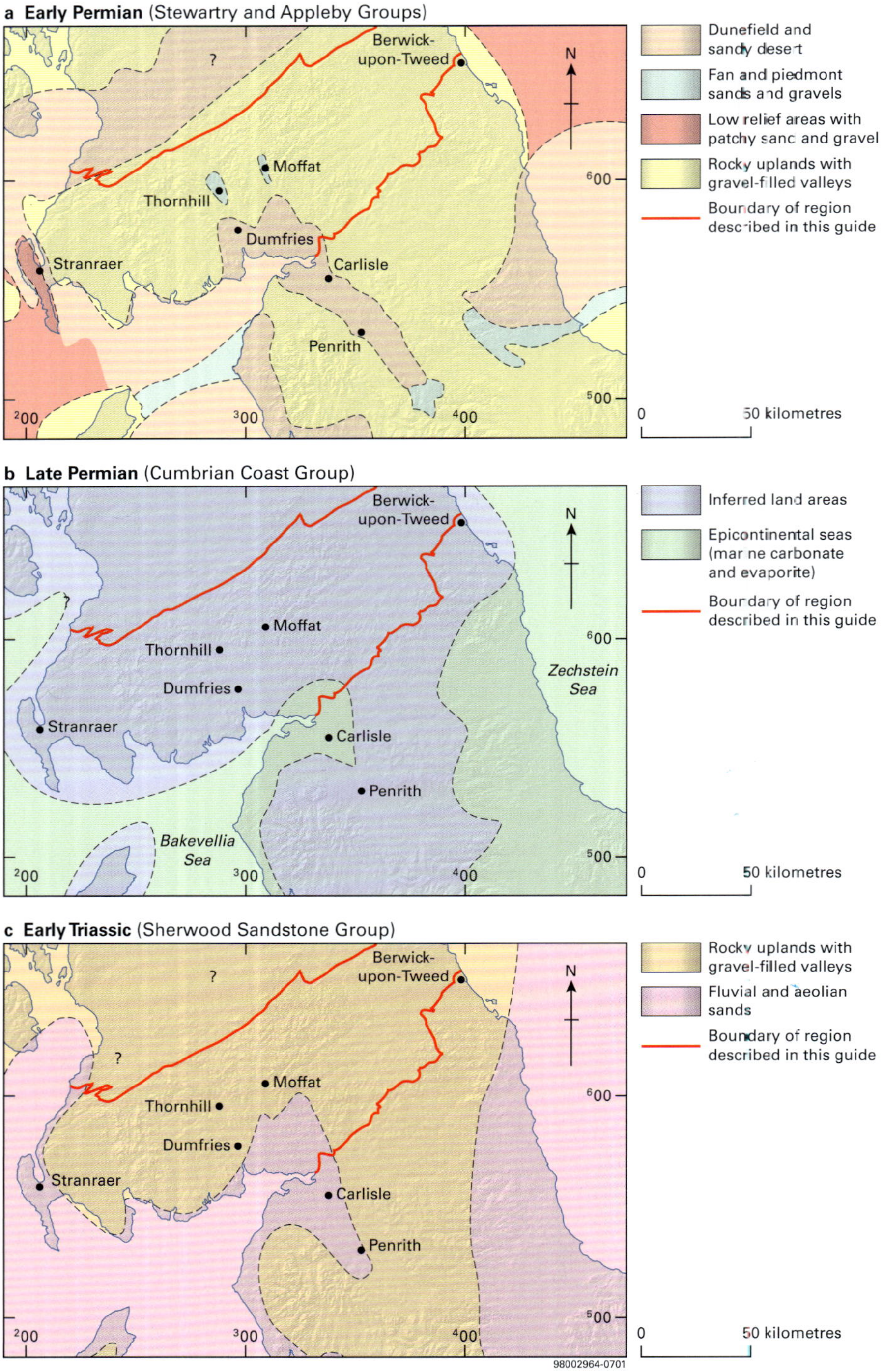

Figure 45 Distribution of depositional environments across the south of Scotland and northern England during Permian and Triassic times (after Cope et al., 1992. Atlas of Palaeogeography and Lithofacies, Geological Society of London Memoir No. 13).

an overall southwards pattern of sedimentary distribution is the southwards increase in the proportion of red sandstone interbeds in the succession. There is also a suggestion that the highest strata preserved become younger southwards, with possible Triassic red sandstone

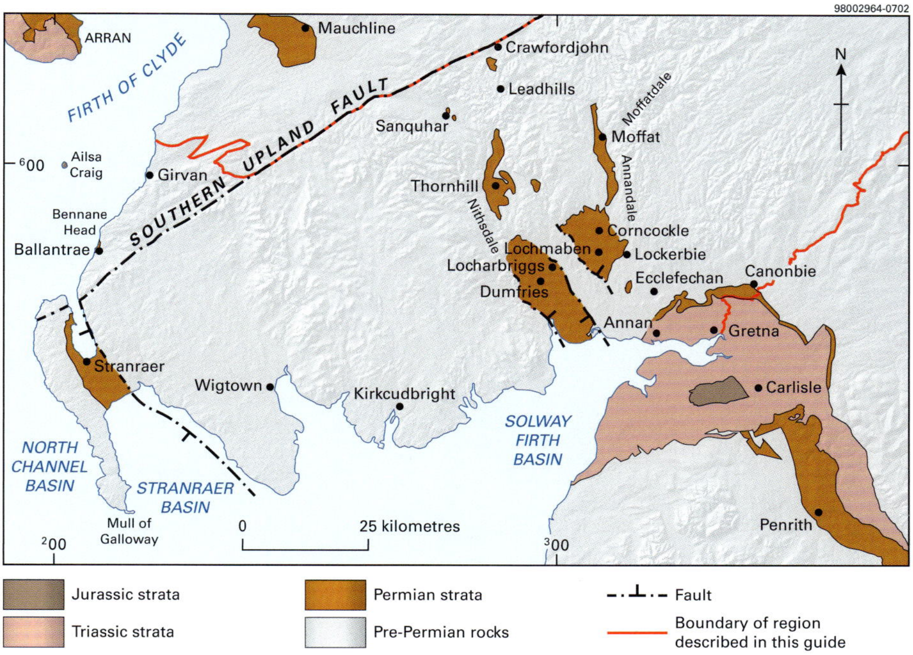

Figure 46 *Outcrops of Permian and Triassic strata in the south of Scotland and north-west England.*

(correlated with the Sherwood Sandstone Group of the Carlisle Basin — see below) recorded in a borehole in Luce Bay.

Ballantrae Basin

In Ayrshire between Ballantrae and Bennane Lea [NX 092 859] some 750 m of lower Permian red beds form coastal exposures at the margin of the mostly offshore Ballantrae Basin. At the base of the succession, the **Park End Breccia Formation** comprises about 250 m of red breccio-conglomerate with rare silty sandstone lenses. Over 90 per cent of the clasts are of wacke sandstone with the remainder being composed of mudstone, chert, basalt and various granitic and gabbroic coarse-grained igneous rocks. The conglomerates are probably braided stream deposits. Within the formation is a conformable layer of doleritic rock, 1 m thick, which may be either a lava flow or a sill. It shows a much higher degree of alteration than nearby Palaeogene dykes and is thought to be a manifestation of the early Permian magmatism seen elsewhere in the region.

Conformably overlying the Park End Breccia is the **Corseclays Sandstone Formation**, which comprises about 500 m of red, micaceous, fine-grained sandstone and siltstone. The strata represent channel and overbank deposits of ephemeral streams. Along most of the inland, eastern margin of its outcrop, the formation is faulted against the Ordovician Ballantrae Complex, but in the north, at Bennane Lea, the highest part of the Permian succession laps onto the Ordovician rocks. There, clasts of spilitic basalt and serpentinite are contained in a red sandstone matrix, and define the **Bennane Lea Breccia Member**.

Sanquhar Basin

The youngest rocks seen in the Sanquhar Basin are olivine basalt lavas that form a small outlier in its south-eastern extremity where they unconformably overlie strata of the Scottish Middle Coal Measures Formation. The lavas, no more than 40 m thick, are assigned to the lower Permian **Carron Basalt Formation**, which has a much more extensive outcrop to the

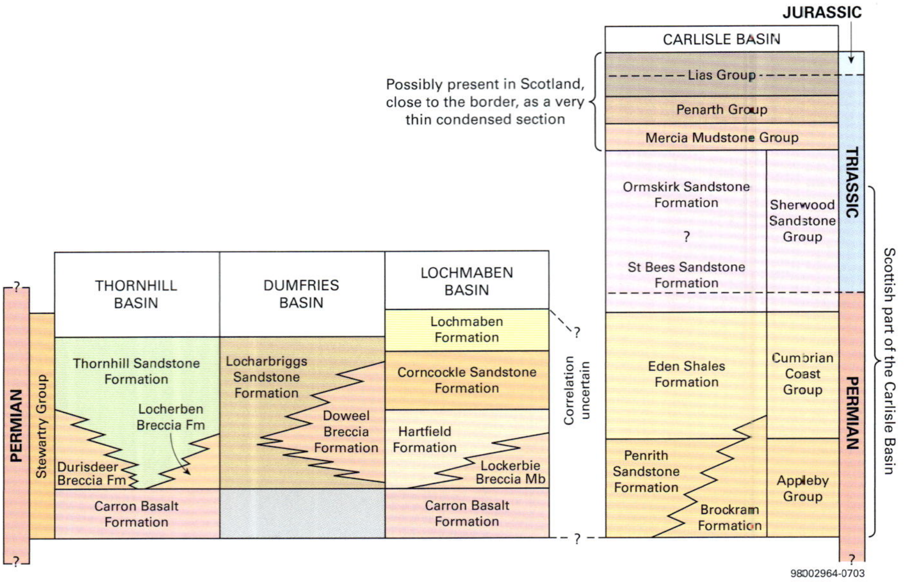

Figure 47 *Correlation chart for the Permian and Triassic successions of the south of Scotland and north-west England.*

south in the Thornhill Basin (see below). The basaltic lavas are also of similar composition, and are probably equivalent in age, to those seen farther north-west within the Mauchline Volcanic Formation of the Mauchline Basin in the Midland Valley. There, sedimentary rocks interbedded with the basalts have yielded a flora that is currently regarded as early Permian.

Thornhill Basin

The Thornhill Basin was initiated as a complex graben structure controlled by the reactivation of faults from the broadly north–south trending Caledonian set. However, for the most part the faults are overstepped by the basinal strata so that the currently exposed margins of the outlier commonly show Permian beds resting unconformably on Ordovician or Silurian strata. At the base of the Permian sequence, in the northern part of the basin, thin olivine basalt lava flows with interbedded conglomerate and sandstone make up the Carron Basalt Formation and are overlain unconformably by fluvial breccio-conglomerate and sandstone of the Durisdeer Breccia Formation (Figures 43 and 47). The Durisdeer Breccia interdigitates with and passes up into aeolian sandstones of the Thornhill Sandstone Formation. In the deepest part of the basin, west of the Carronbridge Fault, geophysical modelling indicates the maximum combined thickness of Permian and underlying Carboniferous strata is about 400 m. In a small subsidiary outlier (c. 3 km^2) at Locherben, to the east of the main basin, a local breccia development towards the base of the Thornhill Sandstone Formation is assigned to the Locherben Breccia Formation.

In the Thornhill Basin, the olivine basalts of the **Carron Basalt Formation** are typically much weathered and comprise a few thin lavas, no more than 20 m in total thickness. At the type section in the Carron Water [NS 885 017 to 888 022], in the northern part of the basin, thin lava flows are interbedded with streamflood breccias and sandstones and either overlie Coal Measures strata or overstep them to rest directly on Lower Palaeozoic wacke sandstone with marked unconformity. Angular clasts within the breccias are of basalt and wacke sandstone. A thin amygdaloidal lava is also exposed in the subsidiary Permian to Carboniferous outlier of Locherben, at the base of the Locherben Breccia Formation. About 6 km to the south-south-east of Locherben, remnants of lava overlie a red volcanic breccia in three

Plate 46 *Breccia from the Loch Ryan Formation near Soleburn Bridge [NX 035 645] on the western shore of the loch. (P772045).*

small outliers close to the confluence of Windyhill Burn and the Water of Ae [NX 975 905]. These remnants, together with the presence of the Carron Basalt Formation in another small outlier around Lakehead Farm [NY 001 880] close to the north-west margin of the Lochmaben Basin (Figure 46), and of locally derived basalt-dominated breccias elsewhere around the margin of that basin (see below), demonstrate the considerable original extent of early Permian volcanicity.

In all of its outcrops, the lava of the Carron Basalt Formation typically comprises a lower aphanitic layer and an upper amygdaloidal layer. The tops of the flows are fissured, infilled and overlain by medium to fine-grained laminated sandstone and fine-grained breccia. In hand specimen, the rocks exhibit a red speckling due to the alteration of the abundant olivine crystals.

In the northern part of the Thornhill Basin, the **Durisdeer Breccia Formation**, a sequence of sandy breccias, fluviatile sandstones and siltstones, overlies the Carron Basalt Formation (Figure 47). The type section in the Hapland Burn [NS 888 023 to 889 025] reveals interbedding of tabular, laminated sandstones and sandy breccia in which wind-facetted clasts, up to 20 cm in diameter, are mostly of locally derived basalt; also present, but rare, are clasts of wacke-type sandstone derived from Lower Palaeozoic strata. The clasts are contained in a silty-sand matrix with up to 10 per cent 'millet seed', aeolian quartz and basalt sand grains, and the formation passes upwards into the aeolian Thornhill Sandstone Formation (see below).

The general depositional environment of the Durisdeer Breccia Formation was the lower part of a wadi fan complex in which ephemeral streams flowed outwards from the edges of fans across a desert floor. The deposits range from the coarse breccias, formed in streamflood channels cut into a basalt pediment, to the fine-grained, planar and cross-laminated sandstones and siltstones of desert-floor lakes and braided streams. At East Morton [NS 885 008], a disturbed sequence of thinly laminated sandstone, siltstone and mudstone is interpreted as an inland sabkha or lake deposit disrupted by the solution of evaporite minerals which may once have cemented the sediments.

In the small outlier of Locherben to the east of the main Thornhill Basin, breccio-conglomerate with a predominance of wacke sandstone clasts, derived from a local source, are assigned to the **Locherben Breccia Formation**. This unit contains wacke-type sandstone clasts up to 10 cm diameter with sporadic larger clasts of purple amygdaloidal basalt, but generally the breccia lacks basaltic detritus. The breccia matrix is composed of subangular to well-rounded grains of quartz sand with up to 20 per cent of 'millet seed' character. The Locherben Breccia Formation built up as a series of sheetflood deposits from wadi fans. The formation interdigitates with and passes up into aeolian dune sandstone of the Thornhill Sandstone Formation.

The **Thornhill Sandstone Formation** crops out across a large part of the Thornhill Basin and comprises up to 200 m of red, cross-bedded, aeolian dune sandstones. The sandstones were formerly well exposed in quarries at Gatelawbridge and Closeburn which supplied building stone. Cross-bedding in the sandstones shows a consistent north-easterly wind direction, in common with the other major Permian basins. The type section, a 30 m thick sequence in the Crichope Burn at Crichope Linn [NX 911 955] (Plate 47) shows characteristically well sorted, fine to coarse grained, laminated ('pin-striped') quartz arenite in tabular to wedge-shaped, cross-stratified sets. Individual quartz grains are well rounded and are accompanied by a small proportion of feldspar and basalt grains, which are also rounded. The basal 1.5 m of the formation is not exposed in the burn but in the nearby BGS Crichope Linn Borehole [NX 909 955], aeolian sandstones overlie a basal 1.4 m thick, coarse-grained, fluviatile sandstone with rounded pebbles of quartz and flakes of red mudstone, which rests with angular unconformity on Coal Measures strata.

The aeolian sandstones are readily distinguished from fluviatile sandstones by the absence of mica, their generally good sorting, the excellent sorting of individual laminae of contrasting

Plate 47 *Thornhill Sandstone Formation exposed in Crichope Linn [NX 911 955], near Thornhill (P258072).*

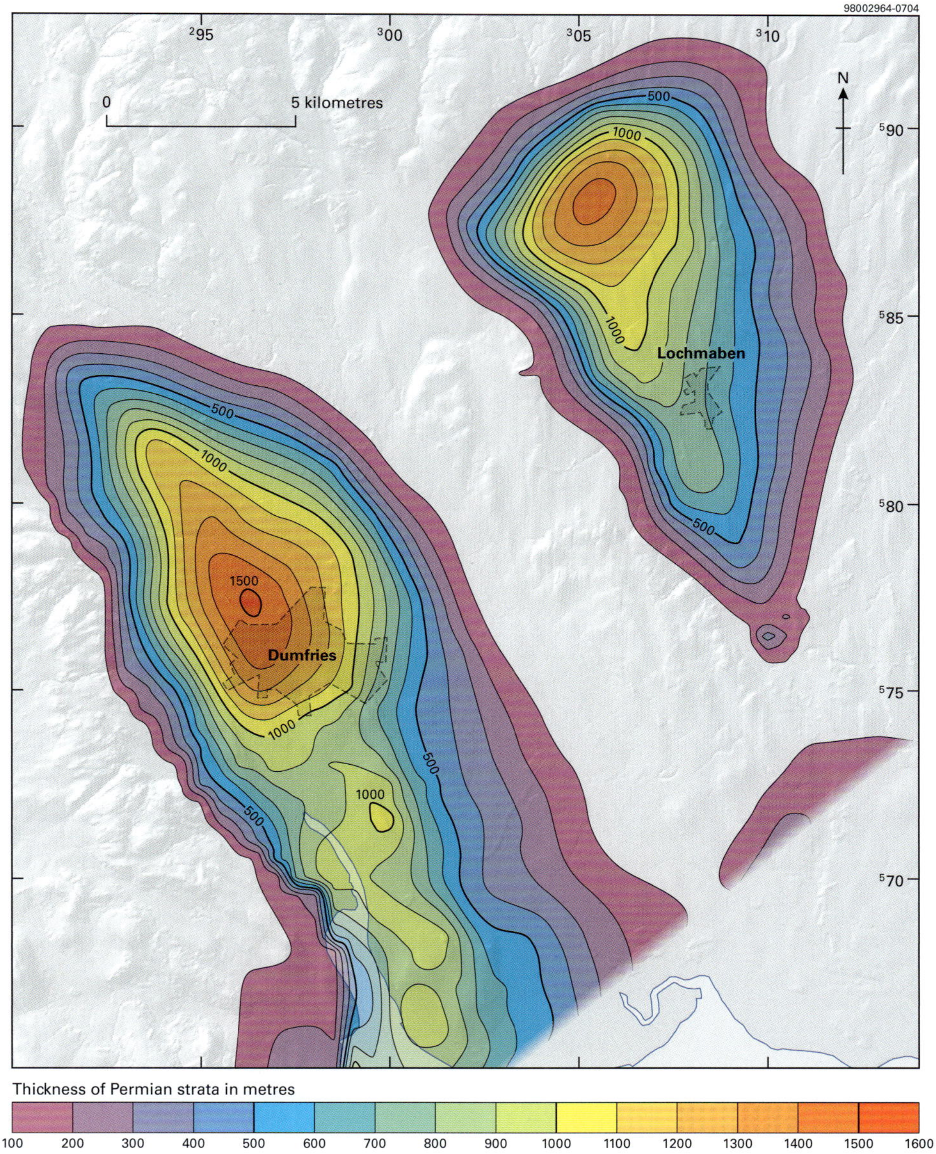

Figure 48 *Estimates of the thickness of the sedimentary fill in the Dumfries and Lochmaben basins based on 3D gravity modelling and assuming an average density for the infill of 2.32 Mg/m*^{3}*. Contours are at 0.1 km intervals.*

grain size, the presence of thick cross-bedded units and the scarcity of silt or clay layers. Typically two types of aeolian sequences are present. Dune deposits consist principally of medium- to fine-grained sandstones in cross-bedded units bounded by tabular and trough-shaped surfaces. Interdune accretion deposits consist of parallel-laminated coarse-grained sandstones alternating with finer-grained laminae; they have a bimodal grain-size distribution.

Dumfries Basin

The sequence of breccio-conglomerate and red sandstone that fills the Dumfries Basin is estimated from geophysical modelling to attain, in the centre of the basin, a maximum thickness of about 1.5 km (Figure 48). The basin is a broadly symmetrical, graben structure

a

b

c

Plate 48 *Doweel Breccia Formation: a) and b), as exposed in Castledykes Park (an old quarry) [NX 976 746], Dumfries, where channel-fill breccia overlies fluvial red sandstone. a) General view of cliff face (P220789). b) Detail of lithologies (P220788). c) Detail of angular clasts of wacke-sandstone as seen in borehole core from Cargen [NX 963 720], to the south of Dumfries (P001430).*

extending north-west to south-east, though the marginal faults are mostly overstepped by the basinal strata. The Doweel Breccia Formation crops out in the south and west of the basin and the Locharbriggs Sandstone Formation crops out in the north and east. Although in places the Locharbriggs Sandstone Formation is seen to underlie the Doweel Breccia Formation, the boundary is strongly diachronous and the two are interpreted to interfinger in the subsurface.

The lower part of the **Doweel Breccia Formation** comprises an interbedded sequence of breccia, pebbly fluviatile sandstone and some aeolian sandstone. Locally, breccias within this interbedded sequence crop out as prominent ridges, whilst an exposure within Castledykes Park, Dumfries [NX 976 746], includes a cross-section of a stream-flood channel containing a gravel breccia with a sandy matrix (Plate 48a and b). The erosive channel base cuts down into underlying aeolian cross-bedded sandstone and is, in turn, overlain by bedded sandstone and pebbly sandstone interpreted as sheet-flood deposits. Many of the Silurian sandstone clasts in the Doweel Breccia are sharply angular (Plate 48c).

The upper part of the Doweel Breccia Formation is lithologically less varied than the lower part. In an old railway cutting [NX 9383 7388] near the eponymous Doweel Farm, a section typical of the formation's upper part comprises very thick beds of breccio-conglomerate with thin interbeds of red laminated sandstone. In the cutting, angular clasts of locally derived, Lower Palaeozoic sandstone, mudstone and granodiorite, up to about 25 cm across, are matrix supported within fine-grained red sandstone. The sandstone beds, which contain some well-rounded aeolian grains, are discontinuous and pass laterally into clast-supported coarse breccio-conglomerates. The matrix-supported rudaceous rocks are interpreted as debris flow deposits that accumulated in a basin-margin, alluvial-fan setting with deposition of the overlying sandstones during the waning stages of flow. The clast-supported breccio-conglomerates probably originated through the reworking, by winnowing, of the sediment surface prior to deposition of the overlying debris flow.

Vertical and lateral variation in clast composition within the breccias allows them to be distinguished as the products of debris flows and alluvial fans fed from the west by two separate wadi valleys. This division is made possible by the local derivation of the clasts. In the south-west of the basin they are mostly of granodiorite and metasedimentary rocks derived from the Criffel Pluton and its thermal aureole. In the north-west they are predominantly of Lower Palaeozoic sedimentary and 'Caledonian' dyke rocks.

Progradation of alluvial fan deposits into the basin centre is likely to have been associated with source rejuvenation by movement of the faults that define the western margin of the basin. The relatively small maximum clast size and apparent absence of a down-fan decrease in maximum grain size, are considered to reflect reworking of already accumulated breccias after source uplift. Interbedding of alluvial fan deposits with aeolian sandstones demonstrates the advance of wind-blown deposits over the fan surface and the persistence of arid 'desert' conditions. However, an increase in both the proportion of debris flow deposits and fine-grained matrix sediment toward the top of the sequence, together with an absence there of reworked aeolian sand grains, may indicate a change towards more humid conditions.

The **Locharbriggs Sandstone Formation** crops out along the north-east side of the Dumfries Basin though natural exposure is scarce. The formation unconformably overlies Lower Palaeozoic strata and is interbedded with or underlies the Doweel Breccia Formation. Although the maximum thickness of the formation is not proven, 63 m was penetrated by water boreholes in the vicinity of Holywood [NX 960 816]. The type section of the formation is at Locharbriggs North Quarry [NX 990 810] (Plate 49). Here, and in two adjacent quarries, 25 m of red, cross-bedded, fine-grained sandstone were worked for building stone. At the beginning of the 21st century, the North Quarry was disused and partly water filled, with a new, active quarry developed a little to the south. The cross-bedding sets are mostly wedge shaped or tabular and between 0.5 and 2.0 m thick. The aeolian character of the sandstone is inferred, in part, from the rounded to subrounded and frosted nature of the component grains and

Plate 49 *Locharbriggs Sandstone Formation as exposed in Locharbriggs Quarry [NX 990 810] near Dumfries in 1984 (P001579). The rock face shows dune bedding, with the units wedging out to the left.*

is supported by the record of reptilian trackways. First, second and third order bounding surfaces may be attributed to the development of large draas (large-scale, composite dune bedforms), migration of dunes over a draa, and small-scale erosional discontinuities associated with changing airflow patterns, respectively (Figure 49).

Lochmaben Basin

From geophysical evidence, the Lochmaben Basin contains some 1300 m of sedimentary strata (Figure 48). It is a graben structure, though the faulted south-western margin is now most apparent, the north-east marginal faults having been overstepped by the basinal strata. To the north-west of the main basin, in the small outlier (c. 1 km^2) around Lakehead, olivine basalt lava and lava breccia of the **Carron Basalt Formation** (see section on Thornhill, above) rest unconformably on Silurian strata and are succeeded by the red clastic rocks of the **Hartfield Formation**. These overstep the lavas in the Lakehead outlier to rest directly on the Silurian rocks with marked unconformity, and carry that relationship into the main Lochmaben Basin where the formation crops out on the basin's western, northern and eastern margins. The Hartfield Formation comprises up to 170 m of red, laminated and cross-laminated micaceous sandstone interbedded with pebbly sandstone and lenses of breccio-conglomerate. The sandstone forms thick beds, some brick red with dune cross-bedding, and is relatively fine, with many frosted and rounded, aeolian quartz grains. The clasts in the pebbly sandstone and breccio-conglomerate range up to about 3 cm across and are mostly tabular. They are matrix-supported in red, silty sandstone but are aligned parallel or near parallel to bedding and so define a crude lamination. Lithologies commonly present in the clast assemblage

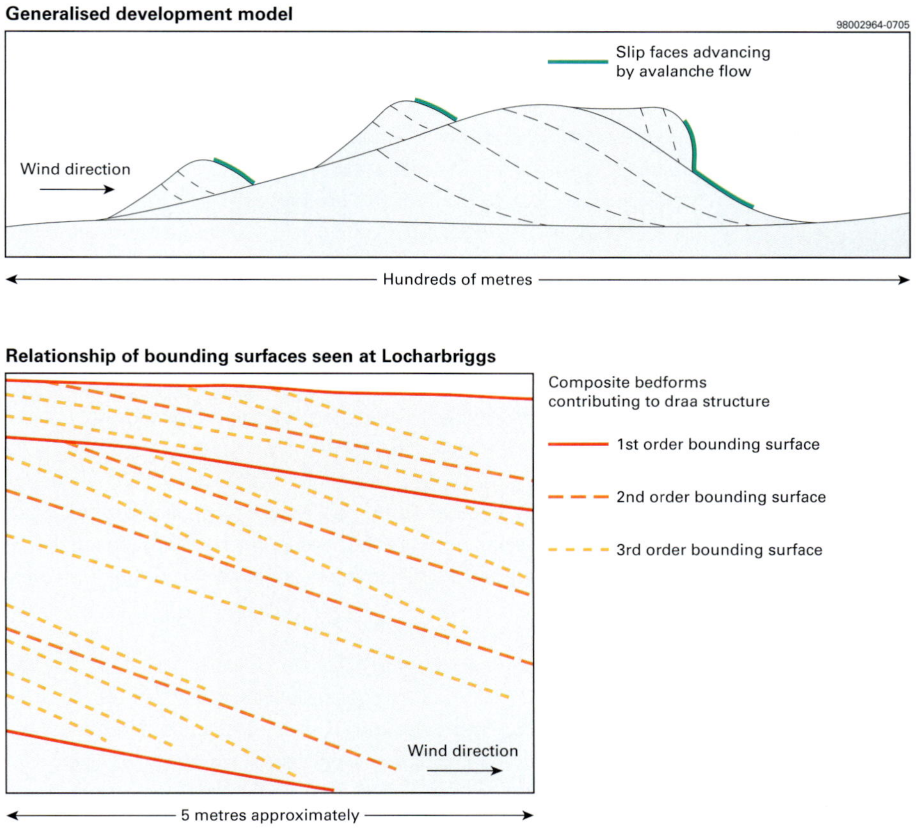

Figure 49 *Cross-section through an idealised large dune structure illustrating the formation of bounding surfaces (not to scale) and development of the large-scale composite bedforms that comprise a draa (adapted from Brookfield, 1979).*

are red or grey mudrock, grey micaceous sandstone, cream-coloured quartzite and, locally, amygdaloidal basalt. On the eastern margin of the basin near Lockerbie, the lowermost part of the Hartfield Formation is formed by the **Lockerbie Breccia Member**, up to 15 m of red-brown breccio-conglomerate with thin interbeds of red sandstone. The distinctive assemblage of lithic clasts in this member includes reddened limestone, wacke sandstone, mafic volcanic rocks, siltstone, mudstone and rounded quartz pebbles.

The **Corncockle Sandstone Formation** incorporates the majority of the Lochmaben Basin sequence. It is about 900 m thick and is made up of fine- to medium-grained, well-sorted, red quartzose sandstone with large-scale aeolian cross-bedding. Some of the best sections are seen in Corncockle Quarry [NY 085 870] which has long supplied good building stone. Exposed bedding planes in the quarry formerly revealed well-preserved reptilian trackway trace fossils assigned to the ichnogenus *Chelichnus*, examples of which were also recorded from Locharbriggs Quarry near Dumfries (see above). Trackways of this ichnogenus are well known from the middle Permian, Cornberger Sandstone (Germany) and the lower Permian, Coconino Sandstone (Arizona, USA).

At Lochmaben, in the centre of the eponymous basin, the Corncockle Sandstone is overlain by the **Lochmaben Formation**, 50–60 m of red, irregularly bedded siltstone and fine-grained, locally cross-bedded sandstone. The Lochmaben Formation also includes beds of coarse-grained sandstone with well-rounded and frosted, aeolian grains contained in a fine matrix, and some beds of breccio-conglomerate; there are slump structures, small gypsiferous nodules, and near-horizontal gypsum-cemented veins.

Moffat Basin

The Moffat Basin is a narrow elongate trough containing about 200 m of breccio-conglomerate and red sandstone assigned to the **Hartfield Formation**. The strata are partly unconformable on a highly irregular Lower Palaeozoic surface and partly faulted against these older rocks along reactivated, north-north-west-trending Caledonian faults. Basal breccias, interpreted as scree deposits, contain clasts that are mostly derived from the underlying Lower Palaeozoic sandstone, but a few clasts contain Carboniferous fossils and so suggest that the Permian strata may in part have been deposited on Carboniferous sedimentary rocks. The basal breccias are directly overlain by interbedded fine-grained sandstones and breccio-conglomerate (with clasts of wacke sandstone and mudstone) that were deposited by sinuous braided streams. This fluvial sequence is then overlain by aeolian sandstone, finely laminated, fine grained and cross-bedded, with interbeds of evenly laminated coarse sandstone.

Snar Valley outlier

Up to 700 m of moderately well-sorted breccio-conglomerate, containing mostly subangular clasts of wacke-type sandstone, mudstone and chert, are present in the small outlier (c.3 km^2) cut by the Snar Water between Leadhills and Crawfordjohn. There is no overt fault control on the outlier and the Permian strata appear to occupy a palaeotopographical depression in the underlying Ordovician turbidite sequence. The Permian beds make up the **Glendouran Breccia Formation** and are thought to have originated as debris-flow and sheet-flood deposits.

CARLISLE BASIN

Along the southern margin of the Southern Uplands massif, the Permian to Triassic strata in southern Scotland form the marginal sequence of the Carlisle Basin, most of which lies to the south of the border (Figure 46) and is described in the companion volume for *Northern England*. The basin has an extensive onshore outcrop across northern Cumbria, where the top of the sequence is Early Jurassic in age, and extends offshore under the innermost part of the Solway Firth. Thence, towards the south-west, it is linked over a low ridge to the somewhat larger, offshore Solway Firth Basin that also contains Permian, Triassic and Lower Jurassic rocks.

The lowermost Permian strata of the Carlisle Basin, though of probable mid Permian (Guadalupian) age, rest with strong unconformity on Carboniferous rocks. The latter were folded and uplifted during Variscan basin inversion in late Carboniferous to early Permian times and rapidly eroded to a relatively smooth peneplain prior to Permian sedimentation. In places, for example where Tournaisian or Visean beds lie beneath the sub-Permian unconformity, several thousand metres of strata have been removed. Conversely, in the down-folded areas such as the Solway syncline, late Westphalian beds lie beneath the unconformity and the thickness of eroded strata is probably no more than a few hundred metres. There is ubiquitous reddening of the subjacent Carboniferous strata for several metres beneath the unconformity.

The Scottish outcrop of Carlisle Basin strata extends across the Canonbie–Gretna–Annan area. Exposure is very limited and it is only relatively recently that boreholes and geophysical surveys have enabled a proper assessment of the sequence. Around the Canonbie Coalfield, early workers had observed basal breccias and 'brick red' sandstones above the 'Barren Red Coal Measures' (now, in part, the Pennine Upper Coal Measures Formation and Warwickshire Group). Locally, the breccias were shown to be overlain by an argillaceous unit, the 'Robgill Marls' or 'St Bees Shales', which passed up into an upper sandstone division, variously named and divided as part of the 'Annan Series'. Data from recently drilled boreholes and geophysical surveys have radically improved the interpretation of the sequence, age, lithology and stratigraphy of these Permian to Triassic strata, and allowed their correlation with the established Carlisle Basin succession. Late Permian evaporites (gypsum–anhydrite) and microfloras have also recently been recognised and described for the first time from onshore Scotland.

Three groups of strata are now recognised in the Scottish outcrop of the Carlisle Basin sequence (Figure 47). The locally derived, basal sedimentary breccias and sandstones are referred, respectively to the Brockram and Penrith Sandstone Formation of the Permian Appleby Group. Above the Appleby Group, the argillaceous and evaporite-bearing strata form the Eden Shales Formation of the Cumbrian Coast Group. The Eden Shales grade upwards into the St Bees Sandstone Formation, which is the lowermost division of the Sherwood Sandstone Group. The lowest beds of the St Bees Sandstone are late Permian in age, but most of the Sherwood Sandstone Group is Triassic.

Appleby Group

Along the northern margin of the Carlisle Basin, the basal few metres of the Permian sequence comprise sedimentary breccia and sandstone of local origin, which rest unconformably on Carboniferous or older strata. Most of these strata can be referred to as 'Brockram', a general term for the coarse clastic, basal Permian breccias of north-west England. These are alluvial fan and flash-flood deposits that interfinger with red aeolian sandstones, the characteristic lithology of the Penrith Sandstone Formation. Its local derivation means that **Brockram** is highly heterolithic and can vary considerably in its clast assemblage. In the Scottish outcrop, around Annan, two units have been differentiated from small, isolated outcrops. The **Kelhead Breccia** is up to 15 m thick and comprises angular clasts of Visean limestone and sandstone. The **Kettleholm Breccia** is some 20 m thick and is composed mostly of Lower Palaeozoic wacke sandstone and mudstone clasts. Both units rest unconformably on Lower Palaeozoic strata and can be regarded as part of the wider Brockram lithofacies. These locally derived, basal clastic strata accumulated as small alluvial fans shed either from low hills that rose above the general peneplain or from the higher ground of the Southern Uplands massif to the north (Figure 50a).

The red, aeolian sandstone of the **Penrith Sandstone Formation** has only a very restricted presence in the Scottish outcrop of the Carlisle Basin succession. A small outcrop of red sandstone with 'millet seed' grains is seen on the south bank of the River Esk at Canonbie [NY 392 761], and there forms the base of the Permian sequence. This sequence is only about 35 m thick, but to the south the formation has an extensive outcrop in the Vale of Eden, on the south side of the Carlisle Basin, where it attains a maximum thickness of over 500 m. There, much of the formation is aeolian sandstone (with wind direction from the east or south-east) with interbedded fluvial sandstones and alluvial fan breccias (Brockram). There is little or no biostratigraphical evidence for the age of the Penrith Sandstone Formation but it is likely to be mid Permian from its conformable relationship with the overlying, and demonstrably middle to upper Permian, Eden Shales Formation of the Cumbrian Coast Group.

Cumbrian Coast Group

Strata of the Cumbrian Coast Group succeed the Appleby Group in the Annan to Gretna sector of the Carlisle Basin's northern margin (Figures 46 and 47). Only one of the group's divisions, the **Eden Shales Formation**, is present in the Scottish outcrop, where it is about 100 m thick. Exposures in the Kirtle Water [NY 248 745] show the formation to consist principally of siltstone, commonly micaceous and sandy, fine-grained sandstone and silty mudstone, all mostly purplish-red or reddish-brown. Interbeds of more coarsely grained red sandstone, conglomerate and breccia also occur. Grains of gypsum and anhydrite are widely disseminated throughout these rocks and cross-cutting fibrous gypsum veins are common. In the lower part of the formation, beds and nodules of gypsum–anhydrite have been identified in a series of boreholes sunk at the site the former Chapelcross Power Station [around NY 220 700]. These were the first recorded occurrences of Permo-Triassic evaporites in onshore Scotland. The boreholes showed that there is an increase in the proportion of sandstone towards the top of the formation, which is gradational into the overlying St Bees Sandstone Formation.

Throughout much of the formation, interbedded siltstone and fine-grained sandstone are either irregularly or regularly laminated, whereas the less common, finer-grained siltstone and mudstone appear structureless. Similar rock types are typical of this formation in its more extensive outcrop in Cumbria and have also been proved by boreholes in the offshore

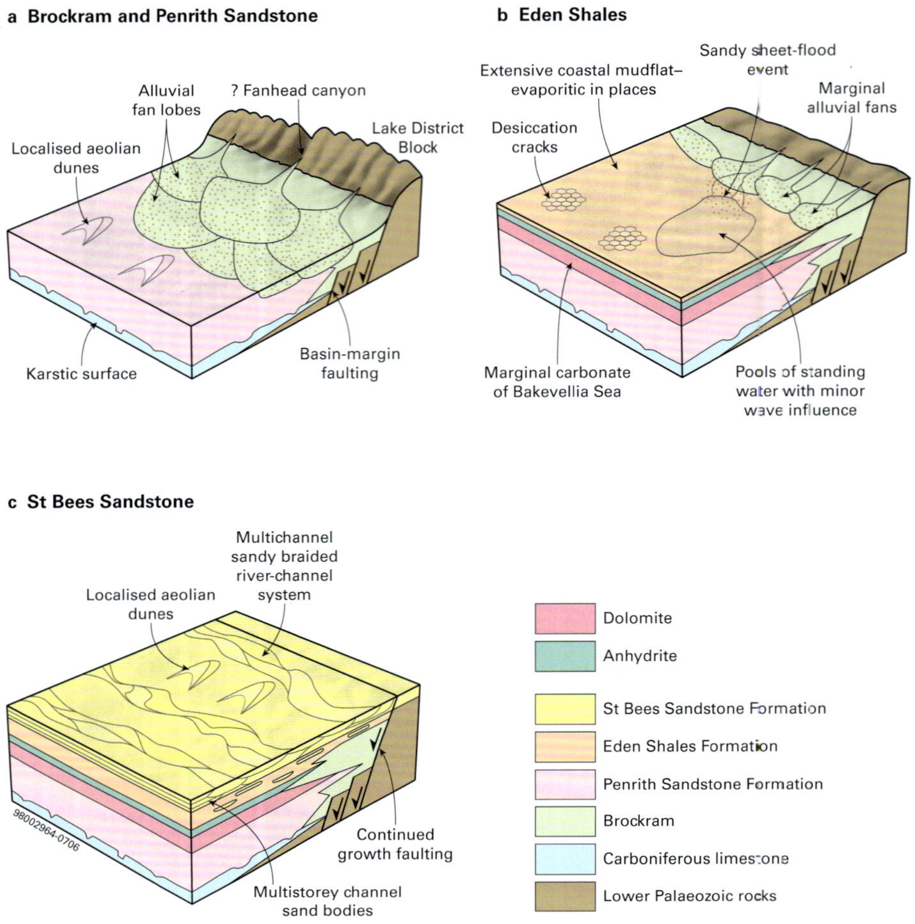

Figure 50 *Cartoons illustrating the changes in depositional regime through Permian and Early Triassic times (adapted from Akhurst et al., 1997. The geology of the west Cumbria district. BGS Memoir).*

Solway Firth Basin (Figure 46) and farther south in the East Irish Sea Basin. The Eden Shales are thought to have been deposited by accretion of fine wind-blown and sheet-flood detritus across mudflats, with the intermittent establishment of evaporitic conditions in sabkhas and shallow lakes (Figure 50b). An arid climate is indicated by the presence of the evaporite beds, by the evidence for periodic desiccation such as mud cracks, and by the local reworking of mud flakes into sandstones and conglomerates.

With the exception of the lower, evaporitic (gypsum–anhydrite) part of the formation, the sequence and lithology of the Eden Shales in the Chaplecross boreholes broadly correspond to what is seen at outcrop, albeit the transition into the St Bees Sandstone Formation is not well exposed. In addition, the boreholes allow, from a combination of geophysical log signatures and core lithology, three informal lithostratigraphical divisions to be recognised in the 'Scottish' Eden Shales. The lower and middle divisions fine upwards overall, but the highest of the divisions is a coarsening-upwards sequence passing into the overlying St Bees Sandstone.

The Chapelcross boreholes have provided the first record of late Permian palynomorphs from Scotland. Miospores are the sole or dominant component of the assemblages and comprise associations dominated by saccate pollen, principally of conifer origin. The miospores originated from the contemporary land flora and allow correlation with rocks in the Kazanian and Tatarian stages of the classical type-Permian succession in Russia. Of particular note is the bisaccate pollen *Lueckisporites virkkiae*, a distinctive species that appears first in the mid Guadalupian in its type region and ranges to the top of the Permian sequence there and elsewhere. Hence a Guadalupian (mid Permian) to Lopingian (late Permian) age range is likely for the Eden Shales Formation.

Sherwood Sandstone Group

The Eden Shales Formation is succeeded by the sandstone-dominated **St Bees Sandstone Formation,** the lowest division of the Sherwood Sandstone Group. The sandstone is typically reddish brown and fine grained, and in the restricted Scottish outcrop it was extensively worked for building stone near Annan, at Corsehill [NY 205 701] and nearby at several quarries close to the Kirtle Water; one of these, Cove Quarry [NY 254 710], was still being worked early in the 21st century. The sandstone is sporadically micaceous and/or silty, some beds grade up to medium-grained, and intraformational mud clasts are common at some levels. Many of the sandstone beds are tabular and composite, with flat or gently undulating erosion surfaces at their base. A wide range of sedimentary structures is displayed, including parallel and low-angle cross-lamination, planar tabular and trough cross-bedding and convolute bedding. There are numerous, generally thin partings of siltstone, silty mudstone and mudstone, particularly towards the basal transition into the underlying Eden Shales Formation. The maximum thickness of the St Bees Sandstone in its Scottish outcrop is about 170 m, proved in a borehole at Staffler [NY 3398 7226], but it is much thicker to the south where its outcrop extends over much of onshore west Cumbria and the Carlisle and Vale of Eden basins.

The lower part of the formation marks a transition from the mudflat deposition of the Eden Shales to the fluvial environment of the St Bees Sandstone. By comparison with the similar strata in west Cumbria, the lower part of the formation is thought to record sedimentation by unconfined flood events fed by a large fluvial system that regional palaeocurrent data suggest probably flowed towards the north. This fluvial system — the Budleighensis River — was the dominant depositional influence for the St Bees Sandstone across much of north-west England. Higher in the formation, but proved only in boreholes at Staffler, Cranberry [NY 3072 6949] and Becklees [NY 3517 7158], the fluvial regime progressively evolved into a channel sandstone facies association. This produced a thick, sand-filled channel complex (Figure 50c), the sandstone bodies still largely derived from the south and interbedded with less common overbank siltstone and mudstone.

Around Gretna, the St Bees Sandstone contains a scattering of wind-rounded quartz grains and relatively little interbedded mudstone. This might indicate a transition into the overlying Ormskirk Sandstone Formation (the Kirklinton Sandstone in some older literature), a higher division of the Sherwood Sandstone Group that has a mixed fluvial and aeolian origin, as recognised in the Carlisle Basin to the south of the Border.

eight

Jurassic to Palaeogene

The youngest strata preserved in the Carlisle Basin belong to the Lower Jurassic Lias Group, but their outcrop is restricted to an area west of Carlisle that does not extend north of the Scottish border. Almost 190 million years then elapsed between deposition of these Lias Group rocks and the onset of the Quaternary glacial episodes, yet the geological record across the south of Scotland provides little evidence of the original distribution and character of any deposits laid down during this time. It is widely agreed that Mesozoic rocks once covered at least some of the region, well beyond their present distributions in the Carlisle Basin and the offshore Solway Firth Basin. There are also extensive Palaeogene to Neogene successions in the offshore basins, but their onshore outcrop is very limited. None are preserved in southern Scotland, though at least 600 m of clay, sand and lignite form the Palaeogene Lough Neagh Clays Group in Northern Ireland.

POST-TRIASSIC SUBSIDENCE

It is likely that the Jurassic sedimentation pattern that had been established in the Carlisle and offshore basins continued into Early Cretaceous times. Comparison with regions to the south and east, where the Jurassic to Early Cretaceous stratigraphical record is more complete, suggests that the Carlisle and Solway Firth extensional basins would have continued to develop through normal faulting well into Early Cretaceous times. In west Cumbria, fault displacement demonstrably postdates the Sherwood Sandstone Group within the hanging-wall block of the Lake District Boundary Fault, whilst several faults in the Carlisle Basin displace Lias Group rocks. Supporting evidence for post-Triassic movements on the northern margin of that basin is provided by a radiometric (U-Pb) date of 185±20 Ma derived from uraninite veins within small north-west orientated tear faults in the footwall of the North Solway Fault at Sandyhills Bay [NX 890 546].

Early in Cretaceous times, areas such as the Lake District and Pennine blocks experienced another interval of erosion as the widespread, late Cimmerian unconformity developed; it seems highly probable that the Southern Uplands massif also experienced erosion at that time. Regional subsidence then dominated across the southern UK during Late Cretaceous times and probably extended to northern Britain, resulting in deposition of a relatively uniform Cretaceous sequence, dominantly of the Chalk Group. The Southern Uplands massif may possibly have formed an emergent landmass throughout this interval, but the presence of a Chalk Group sequence in Northern Ireland (Ulster White Limestone Formation) about 130 m thick, and of chalk clasts in volcanic vents in Arran, suggests that chalk deposition encroached across at least the lower lying parts of southern Scotland. Maximum post-Variscan burial of the region is likely to have been attained towards the end of the Cretaceous Period, prior to renewed uplift commencing in Late Cretaceous to Early Paleocene times.

PALAEOGENE MAGMATISM

The south of Scotland was affected by distant events of epic proportions during the Cenozoic Era which triggered an episode of erosion that has continued, probably with little interruption, until the present day. The cause was thermal uplift along the north-west margin of Europe as a precursor to the formation of new oceanic crust and opening of the North Atlantic Ocean. Uplift was initiated by the the proto-Icelandic mantle plume acting on the base of the lithosphere in a pre-Atlantic region that included the west coast of Scotland, Northern Ireland and eastern Greenland. This area became the focus of intense magmatism from about 60 to 55 Ma, during which time immense volumes of basaltic magma were erupted from fissures and central volcanoes, with the accompanying intrusion of central-complexes, dyke swarms and sills.

Palaeogene mafic dykes cut across the Southern Uplands from north-west to south-east, as part of a regional swarm of high-level minor intrusions that occurs throughout western, central and southern Scotland and extends well into northern England (Figure 51). Abundant coeval dykes are found in Northern Ireland and the Isle of Man, whence they can be traced by their aeromagnetic anomalies into north-west England. The Scottish dykes have been generally thought to emanate from the Mull Centre and include a section of the 250 km long Cleveland Dyke, which has given radiometric ages (K-Ar) in the 56 to 59 Ma range. Other dykes in the central Southern Uplands appear to link the Blyth and Sunderland subswarms of Northern England with their putative origin at the Mull Centre. The more prominent Southern Uplands dykes are usually a few metres wide, with an exceptional 23 m recorded for part of the Cleveland Dyke. In northern England both the Cleveland and the Acklington dykes reach a width of 30 m. In the western part of the Southern Uplands and the Girvan area there are, in addition, many examples of thin Palaeogene dykes (<2 m across) cutting the Lower Palaeozoic rocks (Plate 50). These thin dykes appear to have no great length extent and may have originated as *en echelon* clusters of intrusions. Farther west, in the Firth of Clyde, thin Palaeogene dykes cut the microgranite of Ailsa Craig, itself dated at 61.5 ± 0.5 Ma (Rb-Sr whole-rock). This provides a maximum age for intrusion of the dyke swarm, which can be regarded as a Paleocene event.

The dykes are typically composed of microgabbro or basalt to basaltic andesite, with phenocrysts of plagioclase, clinopyroxene and orthopyroxene contained within an altered glassy groundmass. Margins to the thicker dykes are chilled whereas the central portions are composed of coarser-grained microgabbro, consisting of randomly orientated to weakly aligned plagioclase laths with intergranular ophitic to subophitic clinopyroxene and interstitial quartz. Olivine, where

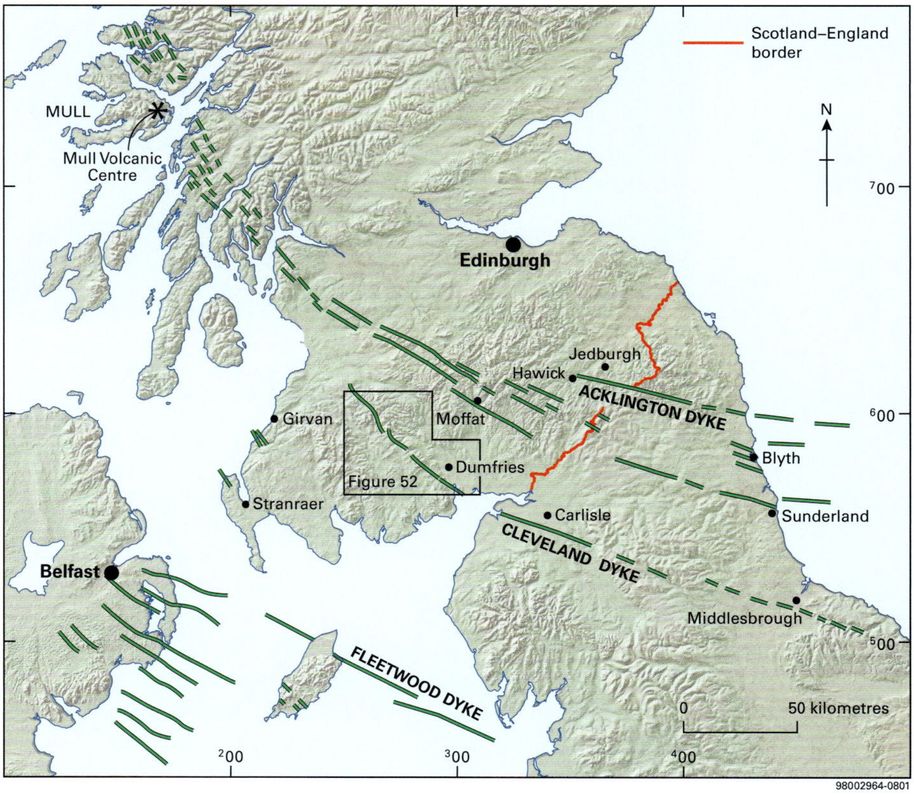

Figure 51 *Distribution of Palaeogene dyke swarms across the south of Scotland and surrounding regions.*

Plate 50 *Palaeogene dolerite dyke cutting altered serpentinite of the Ordovician Ballantrae Complex at Balcreuchan Port [NX 098 877] on the north side of Bennane Head (P005984).*

present, is typically replaced by chlorite and serpentine. Some of the larger intrusions, such as the Cleveland Dyke, show more compositional variation and are andesitic in places. The contact metamorphic aureoles adjacent to the dykes, even the largest of them, are typically very narrow.

The larger dykes, such as Cleveland and Acklington, represent huge volumes of magma. For the Cleveland Dyke alone this has been estimated to be at least 85 km^3. Until recently, the preferred mechanism for their intrusion involved lateral emplacement from one of the major volcanic centres. Modelling of the Cleveland Dyke had suggested that it represents a single pulse of magma that spread laterally from a reservoir beneath Mull at a velocity of up to 18 km per hour, reaching its farthest extent in less than 5 days. Only the clusters of small, *en echelon* dykes were thought to have been intruded vertically, perhaps as offshoots from deeper, laterally emplaced bodies. However, detailed work on the Cleveland Dyke in Scotland is now taken to favour more general vertical intrusion.

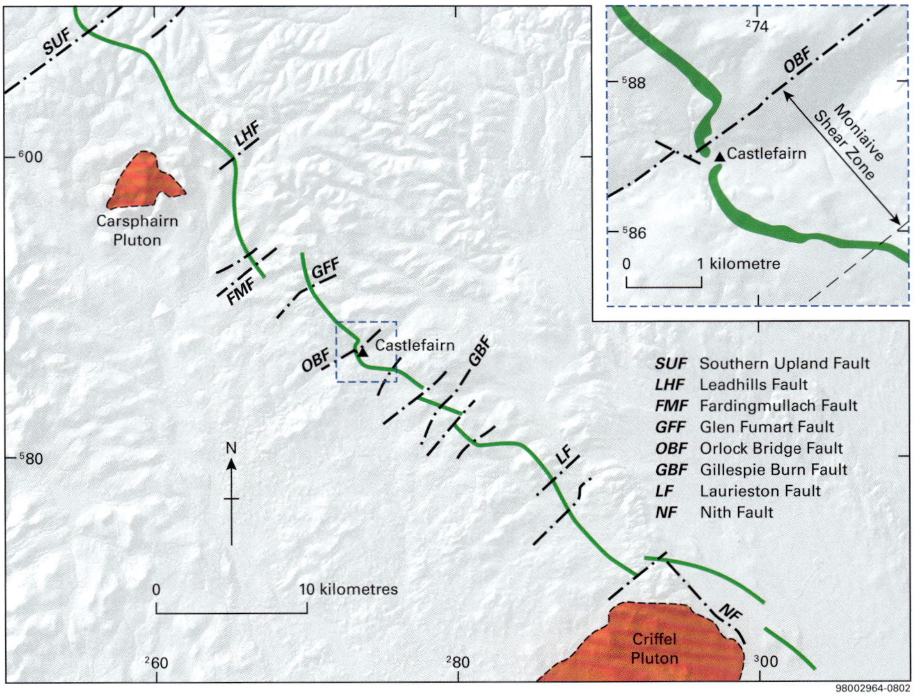

Figure 52 *Outcrop of the Cleveland Dyke across south-west Scotland, as determined from ground magnetic traverses (adapted from Dagley et al., 2008).*

It has now been established that the Cleveland Dyke, at least in its Scottish outcrop, shows unusual and considerable compositional heterogeneity along its length, being andesitic to dacitic in places. It is also a more complex intrusion than previously thought; at some localities it appears as a single feature and at others as two or more subparallel dykes that may differ in composition. Individual dyke segments sometimes overlap to produce an *en echelon* arrangement with, in some cases, offsets of more than 1 km between segments and considerable variation in trend between them. The major dyke offsets generally occur at the intersections with the north-east to south-west, Caledonian tract boundary faults (Figure 52). Curiously, a comparison of the Cleveland Dyke's aeromagnetic signature with that derived from ground traverses shows that the two anomalies are not everywhere symmetrical, or even parallel, and that the subsurface body is broader than the dyke outcrop. The dyke outcrop would seem to be produced by relatively thin 'blades' rising from a larger body at depth. All of this information is more readily reconciled with vertical intrusion of the dykes rather than with lateral emplacement. It now seems most likely that the Palaeogene dykes of the south of Scotland were sourced from small, high-level magma chambers fed by a regional reservoir. The latter most probably arose through magmatic underplating of the lithosphere by the proto-Icelandic mantle plume.

CENOZOIC UPLIFT AND EROSION

During Neogene (probably Miocene) times, a further major episode of uplift affected the Solway Firth and Carlisle sedimentary basins, probably as a distant effect of the Alpine Orogeny. This arose from the collision, away to the south, of the African and European plates. The basins' pre-existing, extensional boundary faults were reactivated with a reverse sense of movement, which also resulted in the folding of adjacent basin strata. As a result of the two major tectonic uplift events during the Cenozoic, and the subsequent erosion that they initiated, it is estimated that between 700 and 2500 m of strata have been stripped from parts of northern England, including the entire cover of Upper Jurassic and Cretaceous rocks.

At least the lower-lying parts of southern Scotland, peripheral to the Southern Uplands massif, would be expected to have experienced a similar history, albeit that the possible Mesozoic cover might have been thinner. However, it is worth noting that the minimum figure of 700 m of erosion derived from northern England came from the Scafell area of the central Lake District, where the current surface altitude (ca 950 m) exceeds that of the Southern Uplands. The Scafell data were obtained by apatite fission track studies, a technique that measures the radiation damage suffered by individual crystals and the degree to which it has been annealed by the rising temperature experienced during burial. The greater thickness, 2500 m, was calculated to have been removed from the Carlisle Basin sequence, with the Lias Group alone accounting for as much as 1500 m of the loss.

No Cenozoic sediments are known to have survived in the south of Scotland, though the products of pervasive weathering during this time may have been preserved locally. Possible examples occur in the Cheviot Hills, where both the Devonian volcanic rocks and the granite are intensely altered and disintegrated in places to depths of between 2 and 50 m; the residual deposit of sand and clay is referred to as saprolite. The preservation in the saprolite of original igneous and volcanic textures shows that transformation occurred in situ. The formation of such deeply weathered rock profiles would have been aided by the warm, humid conditions that prevailed for much of Cenozoic time, most notably during the Eocene Epoch (about 34–56 Ma).

nine

Quaternary

LANDSCAPE EVOLUTION

The present landscape of southern Scotland owes its origins to tectonic and lithological controls on drainage. The Quaternary Period spans the last 2.6 Ma and is subdivided into an earlier epoch, the Pleistocene, during which there were repeated alternations of glaciation and warm phases, and the later, postglacial Holocene Epoch. The long-term evolution of the landscape of the Southern Uplands prior to the last glaciation is not well known. However, the presence at outcrop of igneous intrusions of Paleocene age testifies to a thicker sequence of older rocks at the time of intrusion than is preserved at present. In addition to the remnants of the Carboniferous succession, which originally would have been far more extensive geographically, it is likely that the region was also the site of Cretaceous sedimentation. None of the Upper Cretaceous (Chalk) strata that were probably once present now survive, and it seems likely that from the Neogene onwards the Southern Uplands region has remained an area of relatively high ground undergoing rapid erosion. As the modern drainage system has developed, it has cut down through the postulated Chalk cover and the Upper Palaeozoic strata into the Lower Palaeozoic 'basement', locally exhuming valleys that originally formed during Late Palaeozoic times. Preferential erosion cutting back these pre-existing features appears to have resulted in capture of the headwaters of several of the major rivers, with accompanying adjustment of watersheds.

Many of today's rivers such as the Cree, Dee, Nith and Annan flow nearly normal to the marked north-east to south-west structural 'grain' of the Southern Uplands' Lower Palaeozoic rocks (Figure 1). The river valleys owe their origin, at least in part, to the presence of the north and north-westerly aligned, late Silurian faults that were reactivated to determine the geometry of the small half-grabens infilled with Upper Palaeozoic, and perhaps originally some Mesozoic strata. Of the main rivers in the Southern Uplands, the Nith has a composite history comprising a mature river valley in its lower reaches south of Thornhill through the Permian basins of Thornhill and Dumfries, a rock-cut gorge in its middle section north of Carronbridge, but a relatively wide alluvial valley farther upstream to the north, through the Sanquhar Coalfield and southern Midland Valley. Its tributary, the Carron Water, cuts through the soft Permian sandstones of the northern part of the Thornhill Basin, but owes its origins to capture of headwaters at the head of the spectacular 'V' shaped Dalveen Pass. To the east, the Potrail Water is a small misfit stream that flows through a mature valley of low gradient to join the River Clyde in its upper reaches. The latter rises as one of two streams known as the Clydes Burn at Beattock Summit, just east of the M74, whilst another adjacent Clydes Burn has been captured as a headstream of the southward flowing Evan Water. At Biggar, the River Clyde flows north out of the Southern Uplands but during ice-sheet deglaciation it formed a major tributary, via Broughton, of the easterly flowing River Tweed. The Tweed rises to the north of the Devil's Beef Tub, a deeply eroded south-facing bowl lying 8 km north of Moffat. The Beef Tub forms the source of the River Annan which flows southwards through the partly fault-bounded Permian Moffat Basin, to be joined south of Moffat by the Evan Water and the Moffat Water. The latter drains the steep-sided Moffatdale valley, defined by a major north-east to south-west, Caledonian fault. Modification of the drainage system occurred during deglaciation, with many rivers following routes originally carved out by glacial meltwater. Some rivers were deflected from their preglacial courses, which are now concealed valleys filled with glacigenic material.

Across the summits and higher slopes of the rounded hills of the Southern Uplands, typified by the Moffat Hills, the underlying sandstone and mudstone generally have only a thin cover of superficial deposits. Similarly, the Galloway Mountains, with their granite and granodiorite plutonic centres surrounded by the aureoles of resistant hornfelsed Lower Palaeozoic rocks that form Merrick (Plate 1) and the Rhinns of Kells, have only a veneer of superficial material. Locally the granite surfaces are exposed as smooth, glaciated rock pavements on which are perched glacial 'erratic' granite boulders (Plate 51). In contrast, in the south-eastern part of

Plate 51 *Erratic boulders lying on a glacially polished granite pavement, Loch Doon Pluton, at Craiglee [NX461 802], looking north-west towards Merrick (P774193).*

the region, the Cheviot Hills are mostly covered with a relatively thick blanket of paraglacial and residual deposits. As described in Chapter 8, the andesitic volcanic rocks of the Cheviot Hills have been pervasively and deeply weathered, probably during prolonged subaerial exposure under the warm, humid conditions of the Cenozoic era and/or one or more of the Pleistocene Interglacial stages.

There is little doubt that the repeated glacial and periglacial episodes of the Pleistocene have left a significant imprint on the landscape. Across the south of Scotland there is good evidence for glaciation in the form of small mountain corries, and 'U' shaped valleys (Plate 52), overdeepened rock basins such as those of Loch Trool, Loch Doon and St Mary's Loch, and numerous small lochans in 'knoll and tarn' topography on high ground. Spectacular hanging valleys include that of the Tail Burn which emanates from the moraine-encircled Loch Skene before tumbling down the 100 m waterfall of the Grey Mare's Tail, 14 km north-east of Moffat (Plate 53). Modification of lowland areas occurred mainly through glacigenic deposition, which formed widespread, gently undulating, poorly drained, relatively featureless plains of till interspersed with mounds, ridges and terraces of outwash (glaciofluvial) sand and gravel. Subglacial processes produced swathes of the ice-moulded, drumlin topography and streamlined bedrock landforms that are so characteristic of much of Wigtownshire (Plate 3 and Figure 53) in the western Southern Uplands and Berwickshire in the east.

RECORD OF CLIMATE CHANGE

Evidence of global Quaternary environmental change has been found in deep-sea sediments, in cores of ice taken from the Greenland and Antarctic ice caps and from extended sequences of interbedded loess and organic deposits from continental Europe and Asia. The onset of glaciation in the Northern Hemisphere probably began in the Late Miocene, suggesting that the climate of the British Isles had begun to deteriorate long before the beginning of the Quaternary Period at 2.6 Ma, when ice-rafted debris first appears in the North Atlantic deep ocean record.

The frequency, rapidity and intensity of climatic change are key features of the Quaternary, with climate alternating between **'glacial'** and **'interglacial'** modes; at least 50 significant 'cold–warm' oscillations have been recognised. The driving force of climatic change is the long-term cyclical variation in the incidence of solar energy caused by the Earth's orbital

Plate 52 *The spectacular glaciated valley of Blackhope Glen leading down into Moffatdale, viewed towards the south-east from Cold Grain below the summit of Hart Fell [NT 122 135] (P774194).*

periodicities, which run over roughly 23 ka, 41 ka and 100 ka intervals. These fluctuations have been amplified substantially by additional factors involving physical, biological and chemical interactions together with 'feedback loops' between the atmosphere, oceans and ice-sheets. Of particular importance to the British Isles are changes in the position of the Gulf Stream. This north-eastward-flowing current of warm surface water is compensated by the return southwards of cold, dense water at depth. Sudden changes in this circulation pattern had an immediate and major impact on climate.

An invaluable proxy record of global climate is provided by the relative proportions of the two common isotopes of oxygen contained in the skeletons of calcareous microfossils recovered from deep ocean sediment cores. During glacial periods the oceans' water becomes relatively enriched in the heavy isotope of oxygen (^{18}O), and the marine isotope stages (MIS) thus obtained now provide a universal means of dividing the Quaternary (Figure 54). The oxygen isotope record indicates that during the early Quaternary, when glaciers possibly first developed in the Galloway Mountains, each of the principal cold–warm cycles lasted about 40 ka. Following a major change at about 780 ka BP there have been seven longer, more rigorous glacial–interglacial cycles, although substantial ice-sheets appear to have grown in only three or four of them in the Northern Hemisphere. Each of these glacial episodes lasted between 80 and 120 ka and was followed abruptly by an interglacial lasting 10–15 ka; the rapid deglaciations are described as 'terminations' (Figure 54). The glacial periods included long, cold intervals, termed **'stadials'**, and less cold, or even warm, **'interstadials'** lasting for a few thousand years. Most terrestrial evidence preserved in southern Scotland relates to the last major glacial–interglacial cycle (Devensian and Holocene stages), but older deposits may be preserved locally.

High-resolution evidence from Greenland ice cores, coupled with the MIS record, confirms that dramatic climatic changes have occurred on the millennial (and possibly even decadal) scale. Some 24 interstadial intervals are now identified in the Devensian stage alone, compared with the five or six that were recognised previously from the pollen record and formalised in the traditional British chronostratigraphy (Figure 54). These short interstadials started with abrupt warming and typically cooled over a period of 1 to 3 ka, several being grouped together and superimposed on longer cycles during which temperatures declined gradually (Figure 55). The

Plate 53 Grey Mare's Tail waterfall, Moffatdale [NT 184 148]. Tail Burn drains into Moffatdale from a hanging valley occupied by Loch Skene (P001082).

culmination of each overall cooling phase was a glacial readvance of regional extent coinciding with a massive discharge of icebergs into the North Atlantic (known as a 'Heinrich event'), as indicated by the appearance of abundant ice-rafted debris in deep ocean sediment cores.

LITHOSTRATIGRAPHY

Glacigenic deposits

The British onshore glacigenic deposits of Quaternary age are subdivided into the **Albion Glacigenic Group** (deposits of the Anglian Glaciation) and the **Caledonia Glacigenic Group** (deposits of the Devensian Glaciations). Although the south of Scotland was glaciated during the Anglian Stage there are no known deposits in this district, and any evidence of age from sections exposing two or more tills is equivocal (see below). The **Caledonia Glacigenic Group** embraces all tills, gravels, sands, silts and clays that form surface deposits within the limits of the Devensian ice-sheet. These deposits commonly have distinct morphological expression and are equivalent to the **'Newer Drift'** of previous classifications. Subgroups within the Caledonia Glacigenic Group (Figure 56; Table 8) are based on the provenance of the constituent till, which strongly reflects the composition of both the underlying bedrock, and that of the particular suite of rocks that the ice crossed before the subgroup till was deposited. The composition of morainic glaciofluvial and glaciolacustrine deposits is similarly influenced, though to a lesser extent.

Tills of the **Southern Uplands Glacigenic Subgroup** are typically yellowish or greyish brown diamictons with clasts derived mostly from the Southern Uplands mountains: lithologies present include ubiquitous Lower Palaeozoic clastic lithologies, whilst tills derived partially from the Galloway plutons contain varying proportions of granite and granodiorite.

The **Irish Sea Coast Glacigenic Subgroup** includes tills, glaciofluvial and glaciolacustrine deposits derived in part from the Permo-Triassic basins of southern Scotland and the Vale of Eden together with lithologies from the floor of the Solway Firth and Irish Sea (and locally in south-west Scotland from the Firth of Clyde). These deposits occur typically along the coastal fringes of the Solway Firth and around the Rhins of Galloway. The tills are typically vivid reddish brown or grey, containing clasts of red and yellow Permian sandstone and dark grey Lower Palaeozoic wacke-type sandstone. Along coastal regions of the inner Solway Firth, the

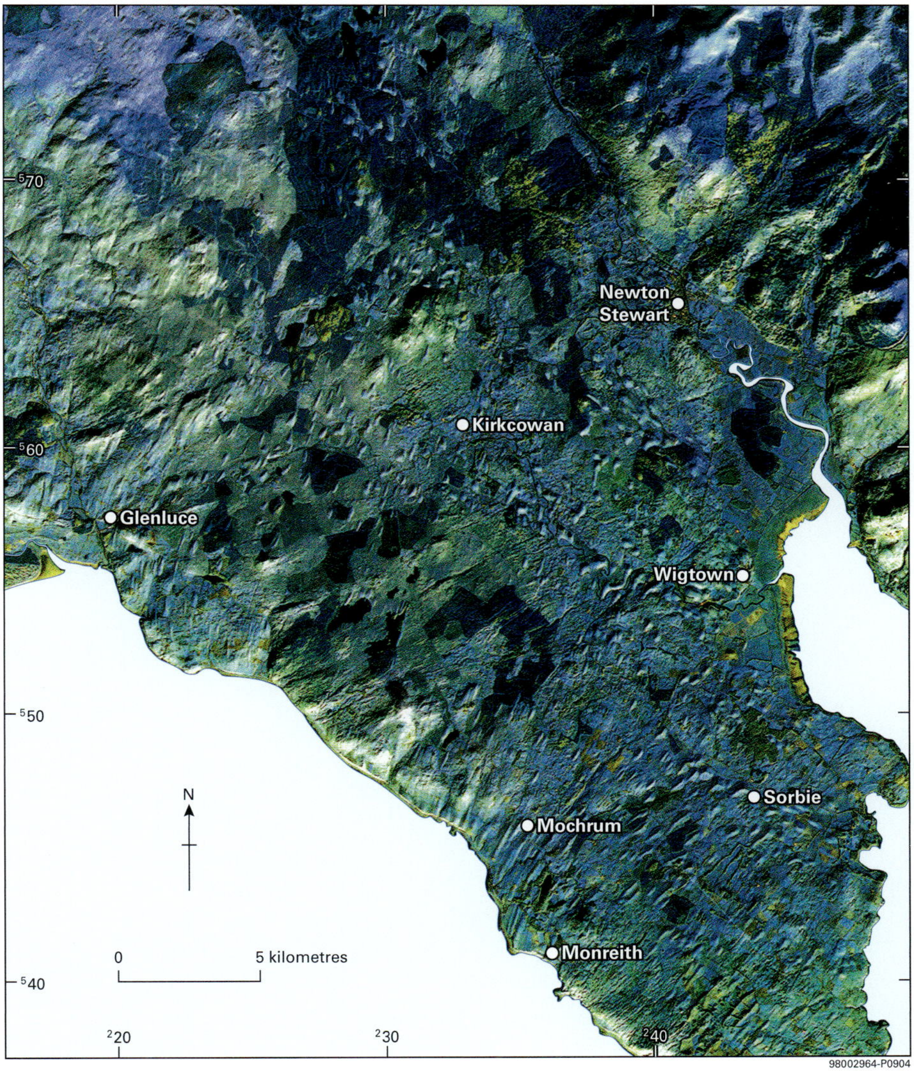

Figure 53 *LANDSAT™ image of the drumlin fields in the area around Wigtown, Galloway.*

clast assemblage also includes varying proportions of Upper Palaeozoic sandstone, siltstone and limestone, and granite and granodiorite from the Galloway plutons. Grey tills containing marine shell fragments are present on the western side of the Rhins of Galloway and around Loch Ryan. This material is associated with distinctive erratics of arfvedsonite-microgranite derived from Ailsa Craig, and so probably originated from the floor of the Firth of Clyde.

The **Borders Glacigenic Subgroup** includes a suite of brownish grey to reddish brown till and glaciofluvial deposits that contain clasts derived predominantly from the Devonian to Carboniferous rocks of Berwickshire (yellow, pink, grey and white sandstones, mudstone and limestone, basalt), and the Lower Palaeozoic and Devonian rocks of southern Scotland and the Cheviot Hills (wacke sandstone and mudstone, granite, andesite, red sandstone) The subgroup was deposited by ice sourced mainly in the Scottish Borders that flowed eastwards through the Merse of Berwickshire.

The **Cheviot Glacigenic Subgroup** comprises sandy diamictons, sand, gravel, silt and clay distributed over the Cheviot massif. The deposits contain clasts derived predominantly from volcanic (basalt, andesite) and intrusive (granite) rocks of the Cheviot Hills. Varying proportions of Lower Palaeozoic wacke sandstone and siltstone are also present.

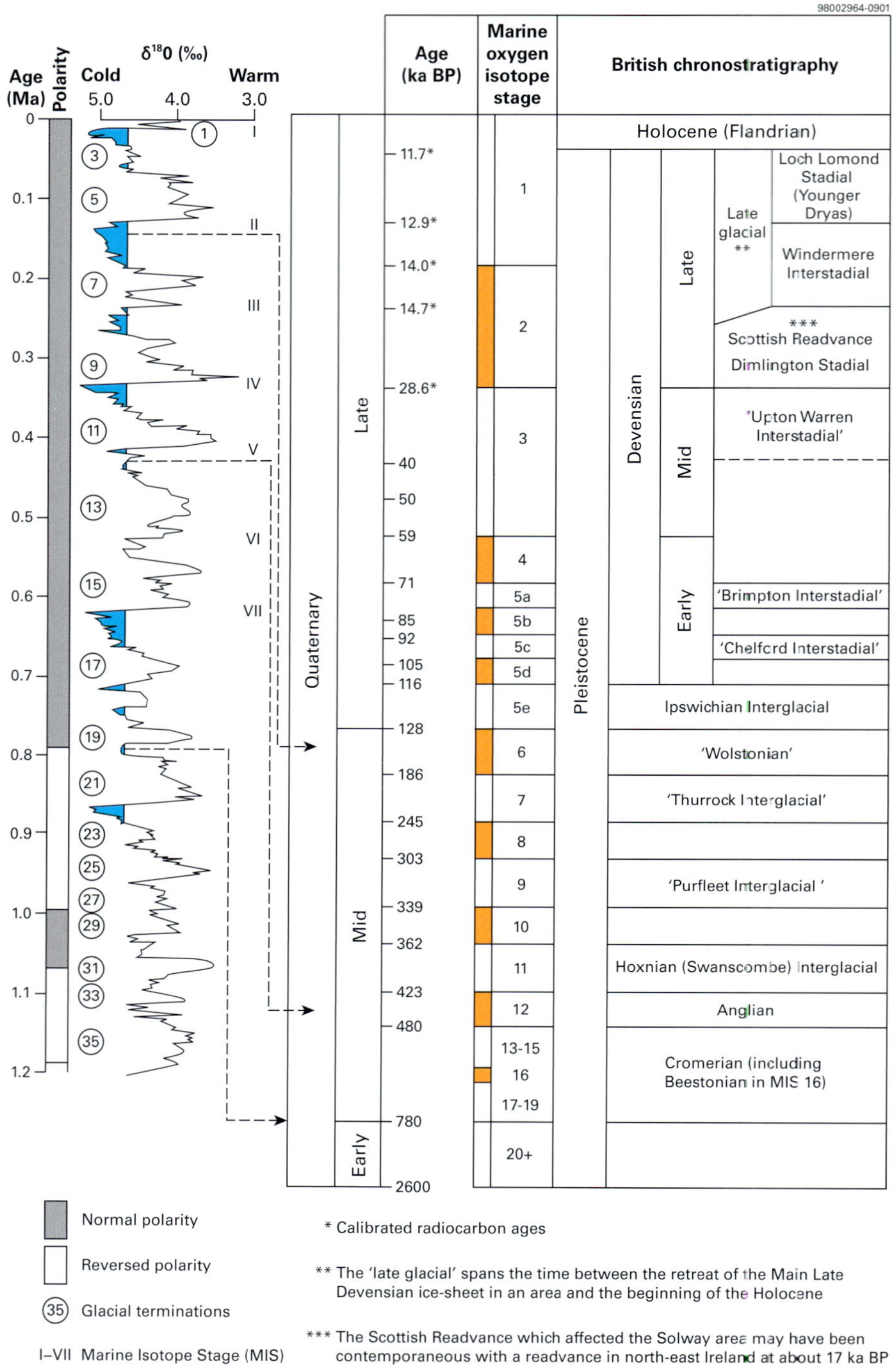

Figure 54 *British chronostratigraphy, geomagnetic polarity and a representative oxygen isotope record (ODP 677) from the North Atlantic tuned to orbital timescales. Calendar ages based on historical records, annual layering in ice cores, tree rings, varve counting etc. are older than conventional radiocarbon ages, which have been calibrated to take this disparity into account using the radiocarbon calibration program of Stuiver et al. (2005) CALIB 5.0 [WWW program and documentation]. Conventional radiocarbon ages on shell material have also been adjusted by subtracting about 405 years to take into account the 'marine reservoir effect' for British waters.*

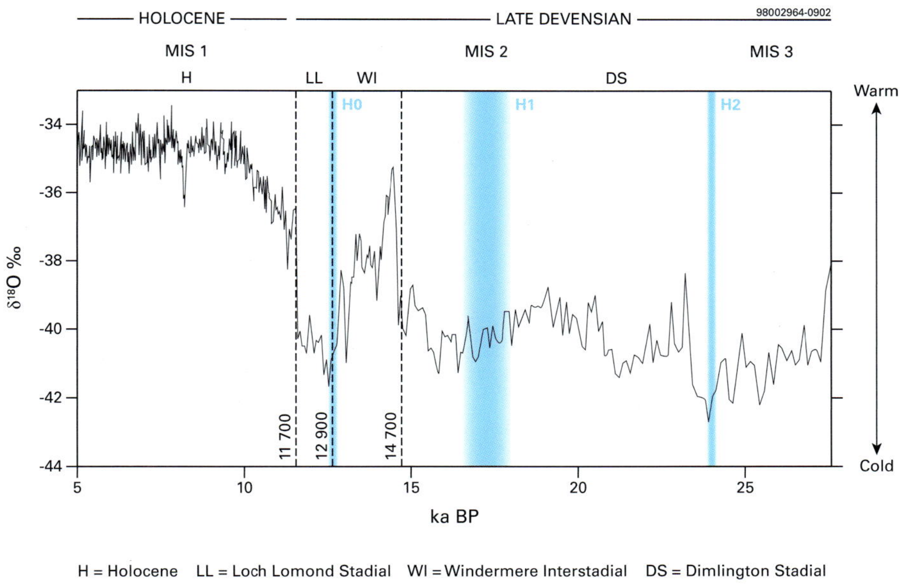

Figure 55 Proxy climate record of the last glacial termination based on an ice core (GISP2) obtained from the Greenland ice-sheet summit. Blue bands denote periods of time within which Heinrich Events (H0 to H2) occurred. Ice core data provided by the National Snow and Ice Data Center, University of Colorado at Boulder, and the World Data Center for Paleoclimatology, National Glaciophysical Data Center, Boulder, Colorado, USA.

Non-glacigenic deposits

Non-glacigenic deposits are classified under two groups (Table 8). The **British Coastal Deposits Group** includes all raised marine, marine and coastal deposits. Most of these are currently classified as morpho-lithogenetic units and are unnamed, but the Loch Lomond Stadial to Holocene, Carse Clay deposits of the Inner Solway Firth area are assigned to the Newbie Silt Member of the **Carse Clay Formation** (defined from the estuaries of the rivers Forth and Tay).

The **Britannia Catchments Group** includes all fluvial and mass movement silt, sand and gravel, together with organic peat. Stratigraphically important late glacial to Holocene organic beds of the Solway district are referred to the **Blelham Peat Formation**, originally defined in west Cumbria. The south of Scotland spans three catchments (Figure 1) and the fluvial deposits (alluvial and river terrace silts, sands and gravels) are accordingly assigned to three catchment subgroups. The **Solway Catchments Subgroup** includes the **Solway Esk Valley Formation** together with fluvial deposits of the valleys of the rivers Cree, Fleet, Dee, Nith and Annan. The **Tweed Valley Formation** of the **Tweed Catchments Subgroup** includes the deposits of the River Tweed and its tributaries. The **Clyde Catchments Subgroup** includes deposits of the Clyde Valley Formation and of other, smaller river valleys draining northward from the Southern Uplands.

IPSWICHIAN INTERGLACIAL

No lower and middle Quaternary deposits have been uneqivocally identified in the south of Scotland, but some idea of the likely conditions at the start of the Late Quaternary, during the interval of climatic amelioration known as the Ipswichian Interglacial, can be extrapolated from records in Northern England. These show that the land was covered by mixed deciduous forest, which developed to include a large proportion of hornbeam and alder in addition to birch, pine, oak and holly.

DEVENSIAN GLACIATIONS

The continental European record indicates that the Ipswichian Interglacial was halted by rapid climatic deterioration at about 116 ka BP. This was followed by a warmer period at about 100

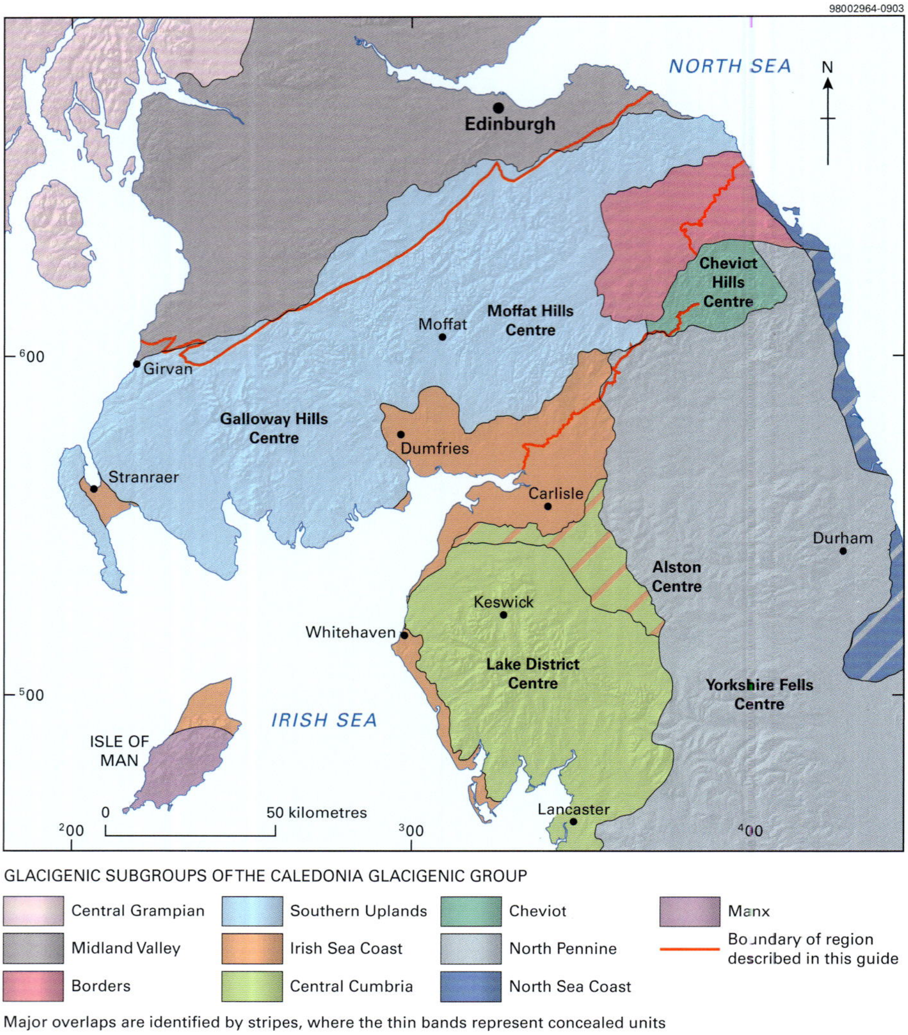

Figure 56 *Distribution of glacigenic groups and subgroups. The geographical boundaries are approximate, but will be refined as knowledge of the distribution of the defining formations of till is improved.*

ka BP (MIS 5c), which probably correlates with the **Chelford Interstadial**, when mixed birch, pine and spruce forest developed over the region. Cooling recommenced at about 90 ka BP, followed by another warmer period at about 80 ka BP (MIS 5a), the so-called **Brimpton Interstadial**. Significant cooling then occurred from about 70 ka BP (MIS 4), when an Early Devensian glaciation may possibly have affected the region. However there is no definitive record and an alternative model indicates that an ice-free, tundra-like environment developed between 90 and 45 ka BP during which there were short-lived, warmer interludes at 76, 57 and 50 ka BP. The main features of the Devensian glacial–interglacial cycles are summarised in Table 9.

Most of the glacigenic deposits preserved in the south of Scotland were laid down during the **Main Late Devensian (MLD) Glaciation**, between about 29 and 14.7 ka BP, when the district was overwhelmed entirely by ice. There is growing evidence from north-west Europe that the **Last Glacial Maximum (LGM)** occurred relatively early in the Late Devensian, from about 27 ka BP to 22 ka BP, but the British ice-sheet appears to have achieved its maximum southerly position on the east coast of Britain shortly after 21.6 ka BP, between Humberside

Table 8 *Lithostratigraphical divisions of Quaternary deposits in the south of Scotland.*

GROUP	SUBGROUP	Defining formations		MIS	Deposits	Distribution of defining formations
BRITISH COASTAL DEPOSITS GROUP	Unassigned	Carse Clay Formation	Newbie Silt Member	2 to 1	Estuarine silt and silty clay	Solway
	Unassigned	Currently classified as morpho-lithogenetic units (unnamed)		2 to 1	Unnamed raised marine, marine and coastal clay, silt, sand and gravel	Coasts of the Firth of Clyde, the Rhins of Galloway, Solway and North Sea
BRITANNIA CATCHMENTS GROUP	Clyde Catchments Subgroup	Clyde Valley Formation and other river valley formations		2 to 1	Alluvium and river terrace silt, sand and gravel	River Clyde and tributaries
	Tweed Catchments Subgroup	Tweed Valley Formation		2 to 1	Alluvium and river terrace silt, sand and gravel	River Tweed and tributaries
	Solway Catchments Subgroup	Solway Esk Valley Formation and other river valley formations		2 to 1	Alluvium and river terrace silt, sand and gravel	River Esk and tributaries
	Unassigned	Blelham Peat Formation		2 to 1	Lacustrine and organic peat, silt, sand and gravel	Solway
	Unassigned	Currently classified as morpho-lithogenetic units (unnamed)		2 to 1	Mass movement deposits: peat, silt, sand and gravel. Lacustrine to organic	South of Scotland
CALEDONIA GLACIGENIC GROUP	Southern Uplands Glacigenic Subgroup	Kirkbean Sand and Gravel Formation		2	Glaciofluvial subglacial, ice-contact and sheet (outwash) sand and gravel	Solway and Moffat district
		Dalswinton Moraine Formation		2	Bouldery and sandy diamicton	Solway
		Mouldy Hills Gravel Formation		2	Boulders and gravel	Solway
		Langholm Till Formation with Hoghill Gravel Bed at base	New Abbey Till Member	2	Matrix-supported, gravelly, sandy, silty clay diamicton, yellowish brown to grey	Dumfries and Galloway
		Currently classified as morpho-lithogenetic units (unnamed)		2	Glacigenic, glaciofluval and glaciolacustrine diamicton, boulders, gravel, sand, silt and clay	South of Scotland
	Irish Sea Coast Glacigenic Subgroup	Kilblane Sand and Gravel Formation		2	Glaciofluvial subglacial, ice-contact and sheet (outwash) sand and gravel	Solway
		Cullivait Silt Formation		2	Interbedded silty clay and sandy silt, reddish brown	Solway
		Kerr Moraine Formation	Marchfield Moraine Member	2	Interbedded diamicton, boulder gravel, sand, silt and clay	Solway
		Gretna Till Formation		2	Matrix-supported, sandy, silty clay diamicton, reddish brown	Solway
		Plumpe Sand and Gravel Formation	Plumpe Farm Sand Member	2	Silty sand with gravel, reddish brown	Solway
			Loganhouse Gravel Member		Sandy gravel, reddish brown	Solway
		Chapelknowe Till Formation		2 or 4	Matrix-supported, gravelly, sandy, silty clay diamicton, reddish brown	Solway
	Borders Glacigenic Subgroup	Greenlaw Gravel Formation, Norham Till Formation		2	Matrix-supported, gravelly, sandy, silty clay diamicton, dark red	Tweed Basin
	Cheviot Glacigenic Subgroup	Kale Water Till Formation		2	Matrix-supported, gravelly, clay diamicton, grey-brown	Cheviot Hills

98002964-T0901

and The Wash. An instructive coastal section is seen at Dimlington, to the east of Hull in Holderness, and as a result this glacial episode has been referred to as the **Dimlington Stadial**. It was followed by a period of glacial retreat before significant readvances, particularly involving coastal ice streams. One such event, the so-called **Scottish Readvance**, affected the Solway area (Figure 57). It followed a dramatic reorganisation of ice-flow patterns within the Vale of Eden and Solway lowlands, and may have been contemporaneous with a readvance in north-east Ireland at about 17 ka BP. The whole of southern Scotland was probably ice free by 14 ka BP (Figure 54).

Directions of ice flow have been obtained mainly from the distribution of erratics and the orientation of striae, drumlins and other ice-moulded landforms. However, the ice-flow indicators clearly relate to more than one phase of active glaciation, and generalised directions of ice flow commonly conflict with local data derived from detailed mapping, till fabric analysis, satellite imagery or digital terrain models.

Table 9 *Representative regional deposits and events of the last two glacial–interglacial cycles to affect northern Britain.*

Approx age (Ka BP)	Chronostratigraphical stage	A	B	Marine isotope stage	Climate	Environment	Human cultures	Typical large herbivore fauna	Representative deposits	Characteristic features	Lithostratigraphy
1	FLANDRIAN/HOLOCENE	Sub-Atlantic	III	1	Temperate	Mixed forest (alder, oak, elm, birch beech)	Historical		Made ground; Peat bog growth	Urbanisation; Peat erosion	Britannia Catchments Group; British Coastal Deposits Group
2							Iron Age	Domestic animals	Silt alluviation on floodplains		
3		Sub-Boreal			Temperate	Mixed forest (elm decline)	Bronze Age		Colluvium	Speleothem growth	
4.3							Neolithic	Red deer	Tidal flat and coastal deposits	Tree clearance; Sub-meandering rivers	
5.6, 6.7		Atlantic	II		Temperate moist, relatively cold	Oak, elm, lime, alder forest		Aurochs	Peat accumulation*		
7.8, 8.7		Boreal			Temperate, dry	Pine-hazel forest	Mesolithic	Red deer	Estuarine silts; Carse Clay Formation	Sea levels restored; Landscape stablisation; Incision of glacigenic deposits	
10		Pre-Boreal	I		Warm continental	Birch-pine forest		Aurochs		Forest expansion; Rapid sea-level rise	
11.7	DEVENSIAN – Late	Loch Lomond Stadial			Cold; Local glaciation	Tundra - glacial	Late Upper Palaeolithic	Reindeer, bison, horse	Landslips, cover sands, head, screes, tufa, till, moraines	Cirque glaciers, screes; Ice-wedge polygons; Moraines	Southern Uplands Glacigenic Subgroup
12.9		Windermere Interstadial			Warm continental	Birch forest		Elk, giant deer, aurochs	Kettlehole deposits, organic silts*	Demise of large herbivores; Initiation of modern drainage	
14.7		Dimlington Stadial		2	Glaciation, dry polar desert	Arctic tundra grasses and sedges - glacial	Man absent	Mammoth, woolly rhino, steppe bison, horse	Extensive tills and other glacigenic deposits; Head, screes	Kettleholes, till plains, meltwater channels, drumlins, eskers, end moraines, valley sandur	Southern Uplands and Irish Sea Coast Glacigenic subgroups
29	DEVENSIAN – Middle	Upton Warren		3	Variable, generally cold	Open, treeless	Early Upper Palaeolithic	Mammoth	No known deposits in the south of Scotland		
40											
59	DEVENSIAN – Early			4	Arctic	Tundra	Middle Palaeolithic				
70		Brimpton		5 a	Continental temperate to polar desert	Pine, birch, spruce forest		Reindeer, wolverine, steppe bison, woolly rhino			
80, 90				b							
100		Chelford		c							
				d	Cold	Open, treeless					
116	IPSWICHIAN			e	Warm temperate	Mixed forest	Man absent	Hippo, straight-tusked elephant, narrow-nosed rhino, aurochs, red deer, fallow deer			
128											

* Blelham Peat Formation (Britannia Catchments Group)

98002964-T0902

The sequence of events that occurred during the MLD glaciation is not fully understood since there is insufficient geochronological control on the formation of landforms and deposits that clearly result from more than one phase of glaciation, problems compounded by a stratigraphical record beset with difficulties of regional correlation. Local centres of ice accumulation over the Galloway Mountains, the Moffat Hills and the Cheviot Hills formed important elements of the MLD ice-sheet. In these areas, the ice largely remained relatively sluggish, depositing little till and causing only limited glacial erosion, but ice emanating from them coalesced with more active ice flowing south from the Grampian Highlands. Overall, the MLD ice-sheet was dynamic with migrating ice divides, corridors of fast-flowing, topographically constrained ice streams and fluctuating margins that locally surged into proglacial lakes and across the adjacent sea bed (e.g. Irish Sea Basin and Tweed Basin extension). Good evidence for fast flowing palaeo-ice streams is provided by features of the Lochmaben and Tweed basins (Figure 58).

MLD glaciation of the Southern Uplands

The evidence in the south of Scotland for glacigenic deposits of possible Early or Mid Devensian age is equivocal. Commonly in the Southern Uplands basal tills rest directly on bedrock. Locally, pockets of frost-shattered rock and rubbly, periglacial deposits occur beneath tills of presumed Main Late Devensian age, and may have formed in the predominantly cold, Mid Devensian substage. Rare examples where intercalated peat is preserved, as at Sanquhar (Plate 54), may enable a maximum age of the overlying till to be estimated, but no radiocarbon age determinations are currently available. Likewise, where sequences of two or more tills are preserved, there is no direct evidence of the age of the lowest units. In general, the earliest tills (of presumed Late Devensian age) were laid down by ice flowing from the Grampian Highlands. At Nith Bridge [NS 594 141], New Cumnock, immediately north of the Southern Upland Fault, three tills are recorded, the lower two of 'Midland Valley', northern provenance and the upper sourced from the Southern Uplands to the south. On the Rhins of Galloway at Clanyard Bay [NX 102 380] and Port Logan Bay [NX 098 405], a blue-grey lodgement till contains bivalves, foraminifera and crustacea, derived from the bed of the Firth of Clyde. It is overlain by a sandy diamicton with a high proportion of locally derived sandstone clasts that possibly formed as ice melted in situ.

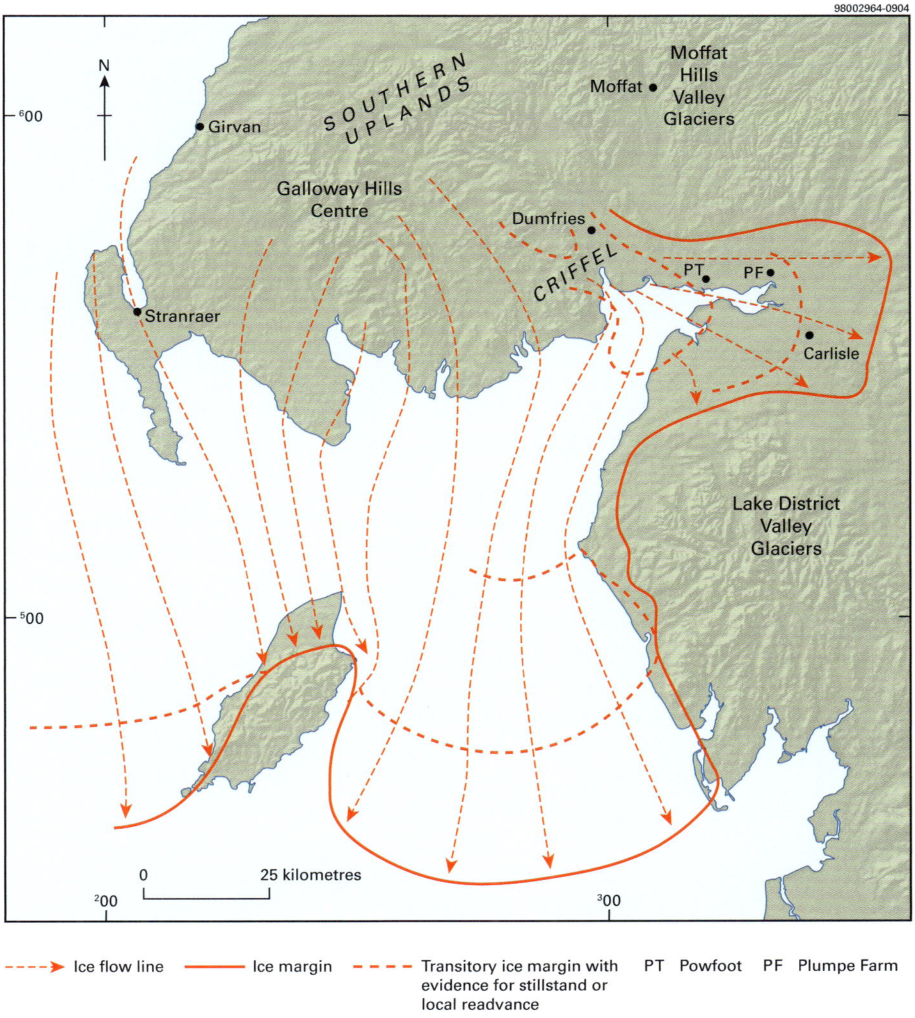

Figure 57 Speculative reconstructions of the last ice-sheet during the Scottish Readvance into the Solway area and the northern Irish Sea Basin.

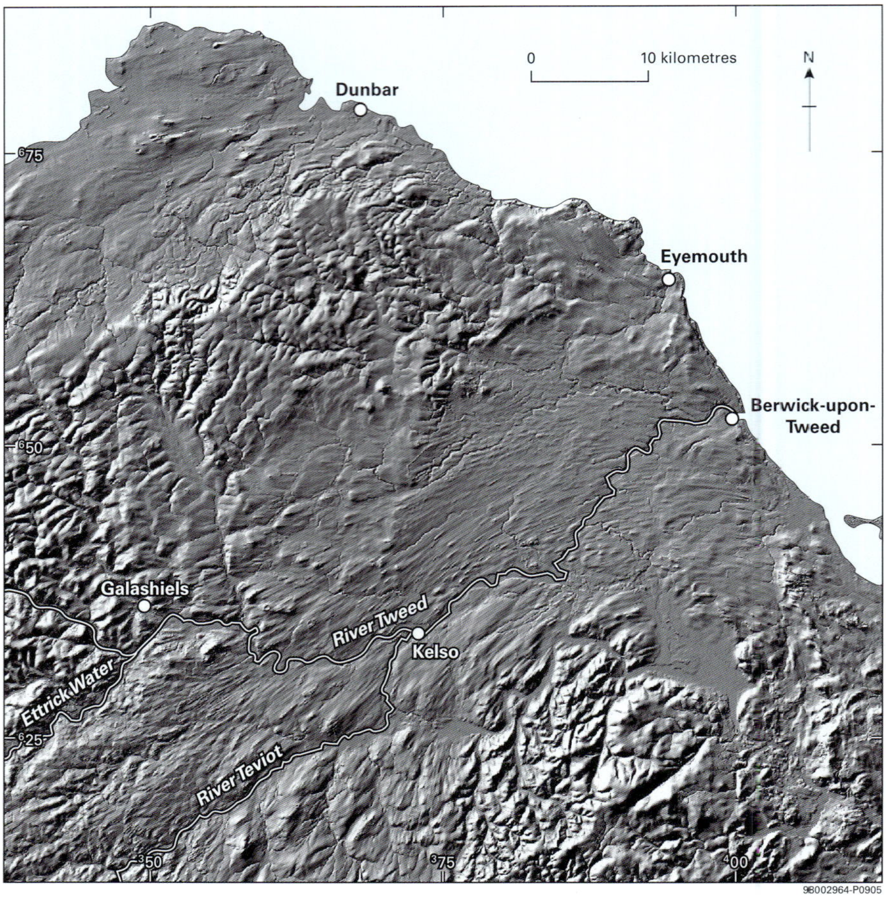

Figure 58 *A fast-flowing, palaeo-ice stream evidenced by streamlined megadrumlins in the Tweed Basin. Here illustrated by a NEXTMap Digital Terrain Model (5 m vertical resolution) (NEXTMap Britain elevation data from Intermap Technologies).*

Basal tills locally present in the Gretna district of the Solway estuary are probably no older than Devensian, and are currently assigned to the **Caledonia Glacigenic Group**. At Chapelknowe [NY 3110 7181], the lower of two tills (the **Chapelknowe Till Formation (Irish Sea Coast Glacigenic Subgroup)**: see below, Figure 59 and Plate 55) could be the product of an early phase of the Main Late Devensian glaciation or of an earlier event during the Early or Mid Devensian. In the higher parts of the Southern Uplands, tills of the **Southern Uplands Glacigenic Subgroup** are thin and patchy, especially across interfluves. Locally thicker accumulations occur in the lee of bedrock highs and within steep-sided valleys. In the Langholm area and parts of the valleys of the Nith and Annan, the till of the MLD glaciation is referred to the **Langholm Till Formation** (Figure 59), a stiff, pale yellowish brown to pale grey, stony, sandy silty clay diamicton containing subangular to subrounded clasts of wacke sandstone and siltstone. Near Hoghill Farm [NY 3820 8905], north of Langholm, the base of the Langholm Till Formation is represented by 2.5 m of dense clast-supported diamicton with angular to subangular wacke and siltstone clasts held in pale, yellowish brown, silty sand. This unit, the **Hoghill Gravel Bed** is interpreted as a gelifractate (scree), and rests on reddish brown diamicton that possibly correlates with the Chapelknowe Till Formation. A granite-rich variant of the Langholm Till Formation, the **New Abbey Till Member**, is defined for tills deposited in the vicinity of the Criffel Pluton. Tills of the western and eastern Southern Uplands remain unnamed, their compositions in large part reflecting local bedrock lithologies.

Plate 54 *Quarry at Sanquhar brickworks, reported as being near the railway station [NS 782 102], pictured in the early 1920s. In the upper part of the working face a dark layer of peat is present within the till deposit (P001449).*

Scottish Readvance in the Solway lowlands

Two major glacial readvances of Scottish ice occurred across the Solway lowlands and the coast of west Cumbria during the later stages of the last glaciation, with the second event referred to as the **Scottish Readvance** (Figures 54 and 57). In the Solway area evidence for this readvance and deglaciation is provided by landforms and by the preservation of 'tripartite sequences' comprising 'lower' and 'upper' tills separated by a 'middle' unit of sands, silts and clays. The sedimentary sequences (shown schematically in Figure 59) are assigned to the **Irish Sea Coast Glacigenic Subgroup** and are best preserved in south-trending valleys, cut in Permian and Triassic strata, which lie at right angles to the general direction of ice flow, as demonstrated at Chapelknowe (Plate 55) and Plumpe Farm, Gretna (Plate 56).

At the Chapelknowe section (Plate 55) the uppermost unit is 2.5 m of very stiff, reddish brown, massive to crudely stratified, matrix-supported, sandy silty clay diamicton with some large boulders and slabs of red sandstone **(Gretna Till Formation)**. The till grades down over 10 cm into very dense, clast-supported gravel with a matrix of reddish brown, clayey fine to coarse-grained sand and secondary infillings of red clay **(Loganhouse Gravel Member of the Plumpe Sand and Gravel Formation)**. The gravel rests unconformably on at least 2 m of extremely stiff, reddish brown and orangey yellow mottled, matrix-supported, sandy silty clay diamicton of the **Chapelknowe Till Formation**. Constituent clasts include wacke sandstone, siltstone, red sandstone, granodiorite and amygdaloidal basalt. Many of the clasts of siltstone and basalt in the lower till are weathered.

The section at Plumpe Farm [NY 3344 6813], Gretna (Plate 56) shows the upper part of the tripartite sequence and also provides evidence for its wet-based, subglacial deformation. The upper unit comprises 1.5 m of very compact, reddish brown, massive, matrix-supported, clayey, fine sandy diamicton **(Gretna Till Formation)**. At its base, there is a 0.5 m thick unit of very compact, interlaminated sand, silt, clay and diamicton. This subglacially sheared unit rests on over 2.5 m of reddish brown, silty fine-grained sand that coarsens downwards **(Plumpe Farm Sand Member** of **Plumpe Sand and Gravel Formation)**.

In broad terms, the 'tripartite sequences' can be explained by a model involving a dynamic ice-sheet with shifting ice divides. Early flow of Scottish ice was up the Vale of Eden and

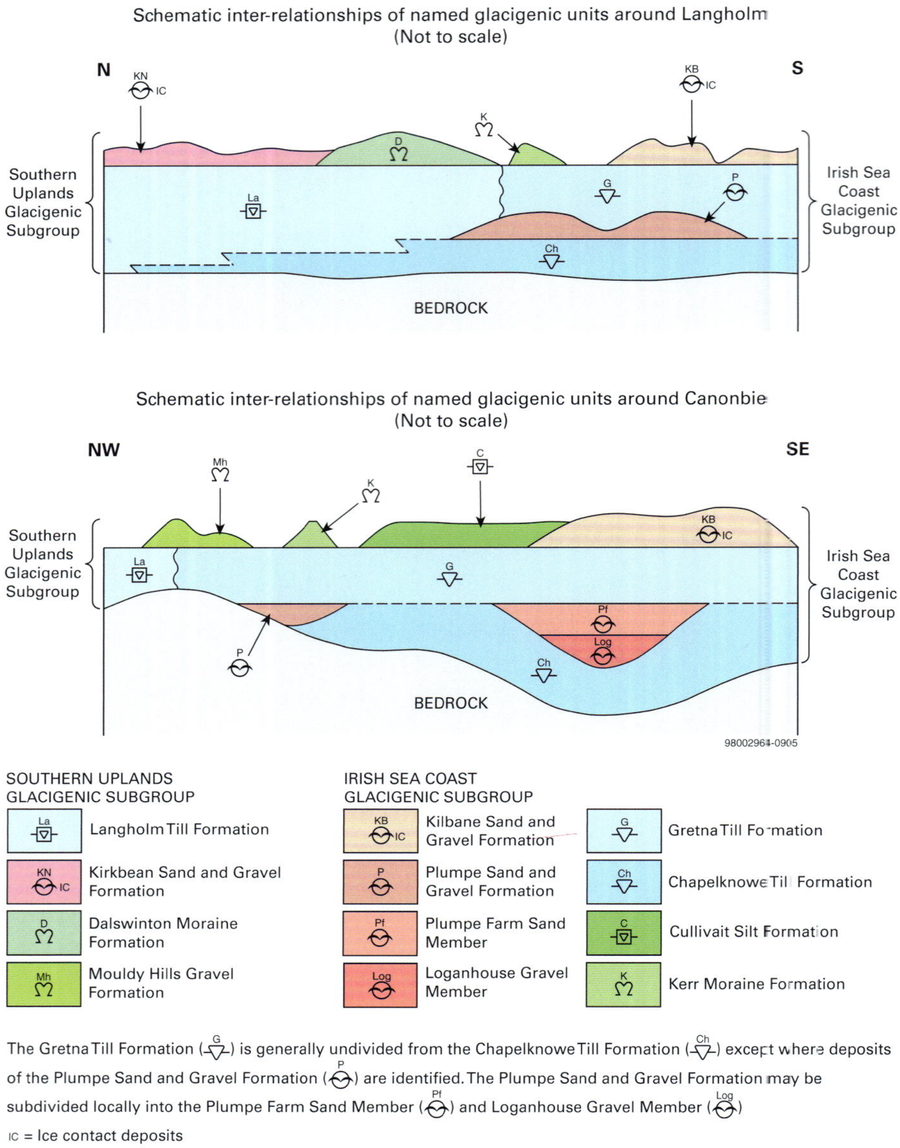

Figure 59 *Schematic transects across 'tripartite' successions in the Langholm and Canonbie areas, showing the lithostratigraphical relationships.*

subsequently eastwards across the Solway lowlands and through the Tyne Gap of Northumberland. This was followed by ice flow westwards from the Vale of Eden towards the Irish Sea, after partial deglaciation and a substantial glacial reorganisation. Next came the Scottish Readvance, with a minor subsequent readvance or stillstand in the vicinity of Powfoot (Figure 57).

Fine-grained glaciolacustrine deposits of the **Cullivait Silt Formation** accumulated in 'Glacial Lake Carlisle', which occupied the Carlisle area during the Scottish Readvance when ice blocked drainage within the Solway lowlands. Glaciofluvial sands and gravels deposited in and around the lake to the north have been assigned to the **Plumpe Sand and Gravel Formation**. Both these and the glaciolacustrine sediments of the Cullivait Silt Formation have been locally disturbed glacitectonically (Plate 56) and are partially capped by the **Gretna Till Formation**, a red diamicton dominated by Scottish clasts (Figure 59).

Plate 55 *River cliff section of the Logan Burn [NY 3110 7181], 2 km south of Chapelknowe, Gretna exposing a 'tripartite sequence': a) The full section, note spade for scale (P543828); b) Diamicton of the Gretna Till Formation forms the top 2.5 m of the section then grades down into the Loganhouse Gravel Member of the Plumpe Sand and Gravel Formation (P543825); c) The Loganhouse Gravel Member overlies diamicton of the Chapelknowe Till Formation which forms the lowest 2 m of the section, the spade rests across the stratigraphical boundary here and in 55a (P543826).*

Deglaciation of south-west Scotland

The products of deglaciation following the LGM comprise predominantly moundy, glaciofluvial sand and gravel, together with glaciolacustrine silts and clays. In the Stranraer district extensive ice-contact kamiform deposits are fringed to the south by outwash sandur deposits which locally were laid down as prograding delta foresets. South of Stranraer, at the former Clashmahew tile-works [NX 062 594] the sands and gravels are overlain by shelly marine laminated clay with dropstones. There is insufficient evidence to confirm whether these and similar clays at the former Terally Brickworks [NX 121 406], north of Drummore, correlate

Plate 56 *Plumpe Farm section, Gretna [NY 3344 6813] showing a deformation till (a glacitectonite) of the Gretna Till Formation overlying glaciofluvial sand of the Plumpe Farm Sand Member, Plumpe Sand and Gravel Formation (P543597).*

with the Errol Clay Formation of eastern Scotland (pre-13 ka BP) or the younger Clyde Clay Formation (about 13–11.5 ka BP) of the Firth of Clyde.

The most extensive morainic, glaciofluvial, or glaciolacustrine deposits occur in the valleys of the Rivers Nith, Annan and Esk (Figure 1). There, the glaciofluvial sands and gravels are referred, on clast content, to the **Kilblane Sand and Gravel Formation** (Irish Sea Coast Glacigenic Subgroup) with clasts predominantly of Permo-Triassic sandstones, and the **Kirkbean Sand and Gravel Formation** (Southern Uplands Glacigenic Subgroup), which contains clasts predominantly of Lower Palaeozoic sandstone. Locally, esker gravels developed in subglacial channels close to an ice margin. Glaciolacustrine, interlaminated red brown silts and clays of the **Cullivait Silt Formation** are present around Dumfries. Many morainic ridges, closely associated with ice-marginal glacial drainage channels, are composed of mixed diamicton, boulder gravel, silt and clay. The deposits are referred to the **Kerr Moraine**, **Mouldy Hills Gravel** and **Dalswinton Moraine** formations (Table 8, Figure 59).

Cheviot Hills and Berwickshire

At the LGM, an independent ice cap was positioned over the Cheviot Hills, where it was almost encircled by a combination of ice flowing from the south-west and a substantial ice stream flowing from the north-west through the Tweed Basin. The Tweed Valley palaeo-ice stream, together with ice from central Scotland, became deflected south-eastwards, parallel to the coast of north-east England. The strongly drumlinised land surface of the lower Tweed basin is interpreted as the product of a fast flowing palaeo-ice stream (Figure 58). It is mainly underlain by the **Norham Till Formation** (Borders Glacigenic Subgroup: Figure 56) which forms a gently undulating, 5 to 10 m thick sheet covering predominantly Upper Palaeozoic strata of the Tweed Basin over much of lowland Berwickshire. Blocky periglacial deposits and decomposed bedrock are particularly common within the Cheviot Hills, where there has been relatively little glacial erosion. Till of the **Kale Water Till Formation** (Cheviot Glacigenic Subgroup: Figure 56), typically preserved in valley floors, is characterised by a high proportion of volcanic clasts derived from the Cheviot massif.

Ice-contact proglacial sands and gravels were laid down in parts of Berwickshire during deglaciation following the LGM. Situated on the southern flanks of the Lammermuir Hills, north of Greenlaw Moor, the remarkable **Bedshiel 'Kaims'** is an upland esker at 210 m above sea level. This 3 km long, isolated ridge rises up to 15 m above the surrounding moorland (Figure 60) and records the early stages of englacial and subglacial meltwater drainage of the Tweed Valley ice stream. Gravels of this esker are referred to the **Greenlaw Gravel Formation** (Borders Glacigenic Subgroup). Later stages of deglaciation are recorded by moundy, glaciofluvial sand and gravel, together with glaciolacustrine silts and clays within river valleys draining to the east. The most extensive of these deposits lie in the wide valley floor of the River Tweed. Gravels of the highest river terraces were probably laid down by meltwater, whereas lower terraces and alluvium are the product of subsequent Holocene fluvial processes.

LATE GLACIAL PERIOD

The period of time between the diachronous retreat of the MLD ice-sheet from any given area, and the beginning of the Holocene at 11.7 ka BP, is referred to as **'late glacial'**. It includes the Windermere Interstadial and the Loch Lomond Stadial, both periods of climatic instability. Deglaciation commenced in a cold, arid environment and high parts of the district probably witnessed several thousand years of ice-free, periglacial conditions before the onset of rapid warming at about 14.7 ka BP, the beginning of the **Windermere Interstadial**. The abrupt amelioration in climate occurred when temperate waters of the Gulf Stream returned to the western coasts of the British Isles. River terraces and alluvial fans formed across the district by paraglacial processes, which swept away loose glacial debris before soils became stabilised by vegetation. Masses of ice buried within glacigenic sediments during all phases of the MLD glaciation melted out relatively slowly to form kettleholes. Organic sequences preserved within kettleholes, loch basins and bog sites may contain pollen, spores and the remains of beetles (coleoptera) and midges (chironomids), and hence provide a valuable record of environmental change through the late glacial and Holocene. At Whitrig Bog [NT 624 349], west of Smailholm in south-east Scotland, the establishment of a late glacial chironomid stratigraphy has enabled mean July air temperatures to be inferred. The thermal maximum occurred early in the interstadial when the temperature reached about 12°C. There was then a downward trend to about 11°C punctuated by four distinct cold oscillations. At the beginning of the **Loch Lomond Stadial** the mean July temperature fell to about 7.5°C, but gradually increased to about 9°C before a rapid rise at the onset of the Holocene.

Good sections in late glacial sediments in the Solway district are known from Bigholms Burn [NY 316 812], 6 km south of Langholm, and Redkirk Point [NY 302 650] on the coast of the Solway Firth, 11 km east of Annan. At Bigholms Burn, grey sandy silt with disseminated organic material from the Windermere Interstadial is overlain by stratified gravel with lenses of peat. These deposits, together with the overlying peat of Holocene age, are referred to the **Blelham Peat Formation**, originally defined in west Cumbria and a division of the **Britannia Catchments Group** (Table 8).

Freshly deglaciated ground was first colonised by a pioneer vegetation of open-habitat, Alpine species, followed by the immigration of crowberry heath, juniper and dwarf varieties of birch and willow. Eventually open birch woodland developed with juniper and isolated stands of Scots pine locally. An oscillatory climatic deterioration occurred throughout the Windermere Interstadial (Figures 54 and 55) and it is possible that glaciers had already started to build up in the Galloway and Moffat hills before more sustained cooling began at about 13 ka BP. Giant deer flourished during the Interstadial, but then became extinct. The Gulf Stream current

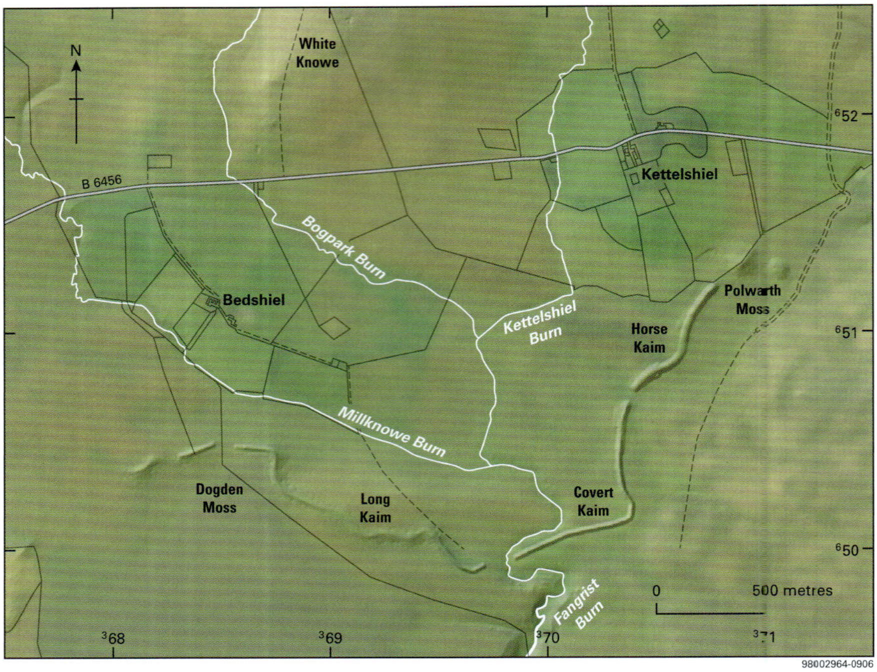

Figure 60 *Shaded relief map based on a NEXTmap Digital Terrain Model illustrating the course of the large esker, near Greenlaw, Berwickshire, known as the Bedshiels Kaimes. Illuminated from the north-west [from NT 679 505 eastward to NT 707 512].*

retreated abruptly to the latitude of northern Portugal at about 12.9 ka BP, heralding the start of the **Loch Lomond Stadial**, known in much of continental Europe as the Younger Dryas stage (named after *Dryas octopetala*, a late glacial tundra flower). On the basis of evidence provided by well-formed, hummocky moraine fields, small corrie glaciers of Loch Lomond Stadial age are thought to have formed in Galloway (Plate 57) and the Moffat hills. Across southern Scotland a tundra environment was ubiquitous and only plant communities tolerant of the Arctic conditions were able to colonise the unstable soils. The stadial witnessed the demise of large herbivores, including the woolly mammoth and woolly rhinoceros, but whether or not hunting by the Upper Palaeolithic human population played any part in the extinctions is still uncertain.

Periglacial processes destroyed the immature soils that had developed during the Windermere Interstadial, creating a range of rubbly deposits collectively known as 'Head' that are especially well developed in the Cheviot Hills. There was extensive development of permafrost and ground ice, the latter melting to create fossil ice wedge casts and pingos (circular depressions) in lowland areas, particularly within spreads of glaciofluvial sand and gravel (e.g. at Lochmaben and Stranraer) (Plate 58). Both fluvial and debris-flow activity were enhanced, especially during springtime snowmelts, and slopes were particularly prone to failure. The sand and gravel that underlies the loamy floodplain alluvium in many of the larger valleys was probably deposited from braided rivers during the stadial.

Fossil periglacial landscape features are widespread, particularly on slopes where intermittent viscous flow of seasonally thawed soils (gelifluction) has occurred. On steep, upland slopes, tear-shaped gelifluction lobes have been washed out to form arcuate lobes of boulders. The lower slopes of valleys and headwater basins commonly take the form of smooth, gently sloping gelifluction terraces that terminate in bluffs along the streams. The periglacial climate was particularly severe in upland areas, causing many rock falls and producing abundant frost-shattered rock debris. Many talus fans formed on hillslopes during the stadial though most have since become inactive and vegetated.

HOLOCENE

The Holocene began abruptly at about 11.7 ka BP when the warm Gulf Stream current became re-established, providing an ameliorating influence on the climate of the British Isles (Figure 55). Average summer temperatures rose by about 8°C within 100 years. The widespread occurrence of bare, unstable soils at the beginning of the Holocene led to intense fluvial erosion and deposition with enhanced debris flow activity on mountain sides and extensive formation of landslips as the ground thawed. The rivers were generally braided with gravelly beds. Soils gradually became more stable following the establishment of vegetation, firstly by pioneering herbs, shrubs and scrub communities similar to the succession of the Windermere Interstadial, and later by woodland. During the Holocene, the river terrace and alluvial deposits developed to shape the present-day valley floors.

Sediment cores from lake basins and bogs throughout the south of Scotland have provided extensive pollen records through the Holocene, which complement coleopteran and pollen records from the Bigholms Burn and Redkirk Point sites. At Loch Dungeon [NX 525 845] on the east side of the Rhinns of Kells, the pollen sequence extends from the late glacial–Holocene boundary to the last 100 years. Based on evidence from this and other sites, trees colonised the south of Scotland from the south, at first mainly via the coastal lowlands, which were much more extensive than today owing to lower sea level in the early Holocene (see below). At the end of the Loch Lomond Stadial, pioneer species-rich, herbaceous assemblages were rapidly replaced by fern-rich juniper and birch scrub. Birch and hazel together with willow, aspen and rowan woodland expanded in the early Holocene by about 9 ka BP. An expansion of elm and oak took place at about 8.5 ka BP, especially in low-lying fertile areas. Pine spread into Galloway from around 7.5 ka BP but was restricted to upland areas free from competition with other species. The arrival of alder at about 7.4 ka BP preceded the transition at about 7.0 ka BP from the drier, 'Boreal' climate period to the wetter, 'Atlantic' one in the mid Holocene (Table 9). All the elements of the mixed mid Holocene deciduous forest were present by 6 ka BP when dense tree cover dominated the landscape of all but the highest parts of southern Scotland. Thereafter, a wetter climate with consequent bog growth resulted in the decline of pine, with both its extinction and a concomitant decline in elm (the 'Elm Decline') at about 5.0 ka BP. Expansion of grasses and sedges accompanied deforestation from 5.0 ka BP and lime arrived shortly after with ash, maple, yew and beech all increasing later in the Holocene as a result of human intervention.

Plate 57 *Morainic mounds composed of granite boulders, gravel and sand near Loch Valley in the Galloway mountains [NX 436 818] (P774195).*

***Plate 58** Ice wedge cast in glaciofluvial gravels of the Kirkbean Sand and Gravel Formation, Halleaths Gravel Pit, Lochmaben [NY 0868 8344] (P774196).*

Throughout the Holocene the burgeoning human population has had a profound influence on the landscape. The **Mesolithic** hunter–gatherers began to burn forests in the earliest Holocene, perhaps to increase the availability of food resources, but substantial clearance did not begin until about 6 ka BP with the arrival of cereal farming in the **Neolithic**. This preceded the 'Elm Decline', an important biostratigraphical marker in pollen records that defines the end of maximum forest extension at about 5 ka BP. Forest clearance increased between 4 and 2.8 ka BP in the **Bronze Age**, with the most extensive deforestation occurring between 2.5 and 2 ka BP in the **Iron Age**. Clear evidence for these changes is preserved in the many wetland sites that have survived throughout much of the Holocene.

Although the imprint of glaciation and periglaciation remains dominant (and though periglacial processes continue to operate on ground above about 450 m OD) postglacial processes have superimposed distinctive modifications on the landscape of southern Scotland. Tidal estuaries and smaller inlets on the Solway Firth are now occupied by muddy estuarine alluvium and salt marsh deposits. Steep hillsides of the Southern Uplands have been modified by gullying, slope failure, soil creep and debris flow, and valley floors have been sculpted by fluvial erosion and deposition. Extensive gullying associated with both climate change and human impact occurred at intervals during the late Holocene (the Sub-Boreal to Sub-Atlantic climate periods) notably at 2500–2200, 1300–800 and after 500 years BP. Locally, deep differential erosion has taken place, for example along the sides of the valley of the Back Water (Fairy Castle Dean), Oldhamstocks (Plate 31), which was most probably initiated as a subglacial drainage channel.

Peat growth accelerated at the beginning of the Atlantic climate period with extensive ombrogenous blanket mires forming over the wet uplands and within poorly drained topographical depressions. The single-thread, submeandering stream patterns of the present day

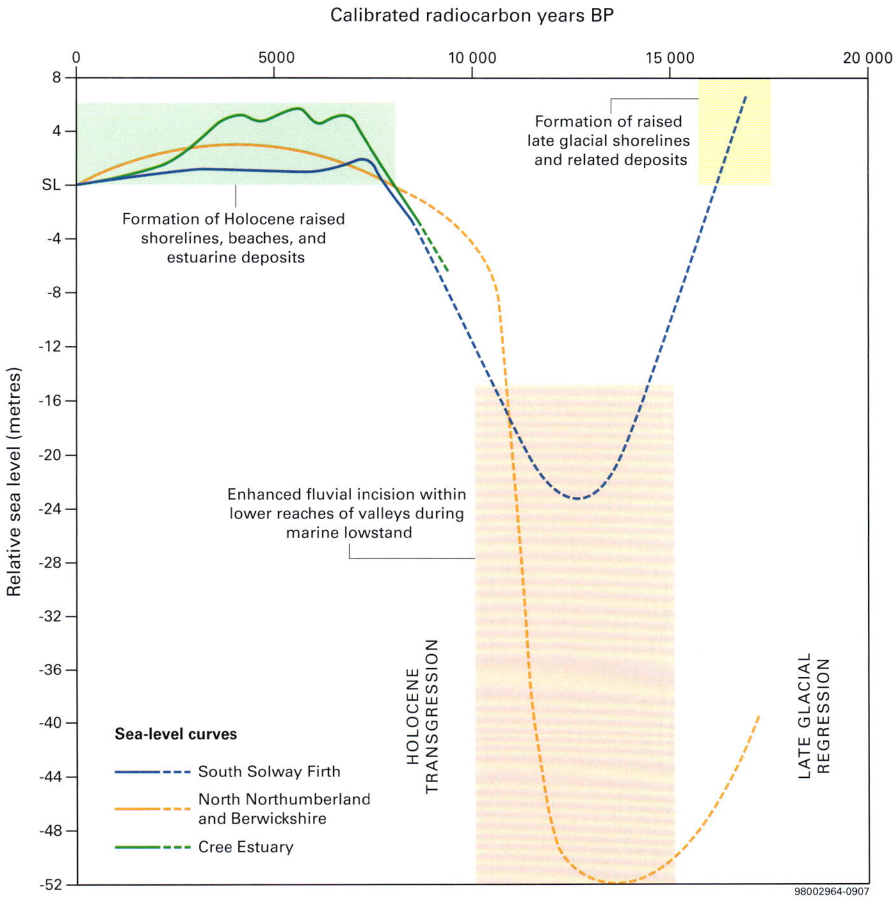

Figure 61 *Representative relative sea-level curves. Solid lines are based on detailed bio-stratigraphical evidence; broken lines are based on predictive glacio-hydro-isostatic modelling.*

became established early in the Holocene once soils had been stabilised by vegetation, but catchments were then profoundly affected by subsequent deforestation, land drainage, cultivation and sheep grazing. New (20th century) forest plantations, mining (metalliferous in the 19th century; coal in the 19th and 20th centuries), gravel extraction and industrial and infrastructure development have all effected further local changes to the landscape.

SEA-LEVEL CHANGE

Present sea level is relatively high compared to its position during most of the Quaternary, so much evidence of past fluctuations has been either destroyed by marine erosion or submerged. Relative sea levels around southern Scotland have been determined by both depression of the land under the ice load during glaciation and its subsequent recovery — isostatic effects — and by changes in global sea levels — eustatic effects. The latter are mainly determined by the amount of water contained within continental ice-sheets and global sea level has risen from a lowstand of about 120 m below current OD some 20 ka BP as those ice-sheets have melted. The interaction of isostatic and eustatic effects means that former sea levels, as portrayed in sea-level curves, vary considerably around the coast of Britain (Figure 61). The rate of isostatic uplift was greatest during and immediately after deglaciation and has since fallen exponentially.

Two distinct sets of raised beach and estuarine deposits occur along the coast of the Solway Firth, where the amount of glacio-isostatic recovery was sufficiently great to bring about an early period of falling sea level. The older, Late Devensian ('late glacial') set was created during and shortly after retreat of MLD ice, whereas the younger, far more extensive set formed during the mid to late Holocene. Although some fragmentary raised beaches, rock platforms and marine planation surfaces in this area have been claimed as Late Devensian, the evidence is equivocal. Equally, claims that deglaciation of the Irish Sea Basin was accompanied by widespread deposition of glaciomarine sediments up to 150 m above present OD cannot be substantiated.

The sea-level curves for southern Scotland reveal lowstands in the early Holocene as glacio-isostatic rebound outstripped global sea-level rise. Sea level may have been as low as 60 m below OD within the northern Irish Sea Basin. The subsequent **Main Postglacial Transgression** followed a reduction in the rate of isostatic rebound and resulted in the formation of the set of Holocene raised beaches and associated estuarine silts, fine-grained sands and clays comprising the **Carse Clay Formation** (Table 8). Peat beds and tree stumps ('submerged forests') interbedded with the estuarine deposits are exposed periodically on the foreshores of the Solway (e.g. at Redkirk Point [NY 302 650]) and near Girvan on the Firth of Clyde, where they provide clear evidence of Holocene sea-level rise.

Along the northern shores of the Solway sea-level rose towards a distinct mean tidal range highstand of about +3 to +4 m OD by 7.7 to 7.4 ka BP in the mid Holocene (Figure 61). There, and in other nearby coastal areas, this rise created the **Main Postglacial Shoreline**, which is represented by fragmentary raised beach deposits of well-rounded shingle backed by a degraded cliff line (Plate 59), with extensive raised tidal flats and saltmarsh deposits up to about 9 m OD. Since their formation, these features have been uplifted by continuing, slow isostatic rebound. Elsewhere around the coast, small accumulations of blown sand occur, the most extensive being the dunes at the head of Luce Bay.

On the rocky eastern seaboard of southern Scotland, south of Cockburnspath, there is limited raised beach evidence for sea-level change. In contrast there is a more or less continuous former coastline northwards towards Dunbar. There, the Holocene coastal sediments rest on a rock platform that may have a polyphase history of Quaternary planation.

Plate 59 *Raised beach and relict sea cliff on the coast of the Firth of Clyde to the south of Bennane Head [NX093 855] looking south towards Ballantrae. The rocky outcrops protruding from the gently sloping raised beach are the remains of sea stacks (P774197).*

ten

Geology and man

After the final deglaciation of southern Scotland, early humans spread into the region. Evidence for Mesolithic occupation is tantalisingly sparse, but by Neolithic times, about 5000 years ago, there was a well-established population. In that period, and the succeeding Bronze Age, the local rock was utilised for chambered cairns, cup and ring carvings, stone circles and individual standing stones, the latter showing a particular concentration in the Wigtown–Whithorn peninsula. The town of Whithorn was a centre of early Christianity, founded by St Ninian in 397 AD. There, the local Silurian wacke sandstone of the Hawick Group was used for carved religious monuments, a practice that spread across Galloway and beyond from the 5th century onwards.

Since that early time, the interaction between man and the geology of southern Scotland has developed and ramified. Much of the region, particularly that underlain by Lower Palaeozoic rocks and granite, is relatively unproductive upland suited mainly for sheep farming and forestry and is therefore sparsely populated. The bulk of the population is located in scattered towns and villages along the main valleys and on lower ground near the coast where arable farming is possible. Some villages such as Leadhills and Wanlockhead owe their existence to the local lead-zinc mineralisation and would not be there but for that reason. Similarly, towns such as Kirkconnel in Nithsdale and Rowanburn near Canonbie grew rapidly during the heyday of the mined coal industry, but then saw their commercial importance reduce as that industry declined.

The traditional transport links across the region skirt the higher ground, following the coastal lowlands, but the main north–south route crosses the Southern Uplands, rising over the watershed between the catchments of the rivers Annan and Clyde. It was radically developed at the end of the 20th century with the construction of the M74 motorway. One of the main challenges faced by this major engineering project involved the excavation and stabilisation of large cuttings through the Lower Palaeozoic sandstone–mudstone succession. Much of the same route is followed by the railway and other infrastructure features such as gas distribution pipelines. In stark contrast, the Southern Upland Way, a 340 km long-distance footpath, runs from west to east linking Portpatrick, on the Rhins of Galloway, with Cove and Cockburnspath on the North Sea coast. For much of its length the Southern Upland Way traverses the rolling hills underpinned by Lower Palaeozoic strata, and in the west runs through the Galloway Forest Park where it skirts the Doon and Fleet granitic plutons. Celebrated mountain-biking routes are also found in the Forest Park, which comprises about 780 km^2 largely devoted to forestry. Its remoteness is well illustrated by its designation in November 2009 as Europe's first 'Dark Sky Park'. The Galloway Forest Park attracts about 850 000 visitors each year and demonstrates the importance to the local economy of 'wilderness tourism'.

FUEL AND ENERGY

Coal

There is a long history of coal mining in the south of Scotland though the two coalfields, Sanquhar and Canonbie, are relatively small by comparison to those to the north (Ayrshire and Douglas) and south (Cumbria, in north-west England) to which, respectively, they are geologically linked. The geology of the coal-bearing successions is described in Chapter 6 of this account.

The earliest workings at Sanquhar were opencast pits or adits driven into the hillsides, but by the end of the 18th century deeper underground mining was in operation. The 'New Statistical Account of Scotland', published in 1845, reported three pits in operation exploiting the Splint, Calmstone and Creepie coals. Deep, underground mining was much extended in the late 19th century, supplying good-quality steam coal, and continued through the early part of the 20th century. Some idea of the intensity of the mining operations is given by

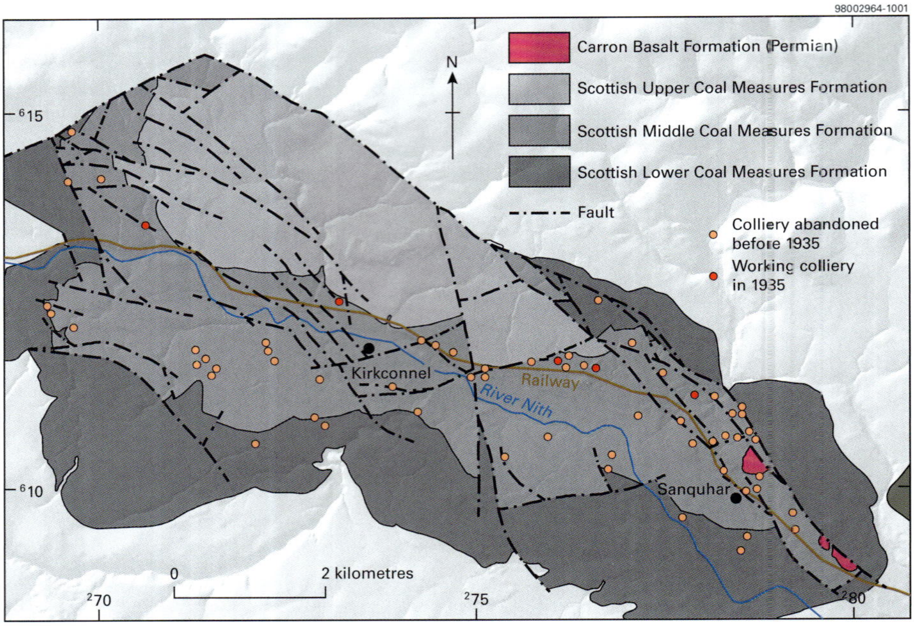

Figure 62 *Distribution of collieries in the Sanquhar coalfield in about 1935.*

Figure 62, which shows the distribution of collieries in about 1935. Deep operations came to an end in the early 1970s, but at the beginning of the 21st century there was still limited opencast production in the Kirkconnel area.

At Canonbie there are records of exploratory boreholes from as far back as the late 18th century, and three seams were being worked underground as early as 1805. However, deep mining there was always much more limited in scale than that at Sanquhar, and effectively ceased in 1922, with only intermittent, small-scale operations continuing for a few years thereafter. The small outcrop of the Canonbie coalfield has a buried extension, concealed beneath Permo-Triassic strata, southward from Rowanburn [NY 410 771] and thence beneath the border and the Solway Syncline until it links with the Cumbria coalfield in north-east England. It is only from the 1970s, and the application of modern seismic methods controlled by deep boreholes, that the significant extent of this concealed coalfield has become apparent. Though there is currently no mining activity at Canonbie a substantial coal resource remains in the concealed coalfield. Also of interest is its potential for coalbed methane generation, since boreholes have established relatively high seam-methane levels, and the contiguous Cumbria coalfield is well known to be gas-rich. An associated possibility arising from methane generation would be the sequestration of CO_2 into the coal seams.

The character and quality of a coal is determined by the conditions in the swamp in which it was formed and by its subsequent burial history. The effects of temperature and pressure over long periods of geological time tend to expel water and volatile constituents from coal. Thus, a coal that has been subjected to elevated temperatures typically exhibits a low volatile content, a high carbon content and high calorific value; it is said to have a high 'rank'. Conversely, coals that remain high in volatiles and have a relatively low carbon content and calorific value are said to be of low 'rank'. The coals in southern Scotland's Canonbie and Sanquhar coalfields are broadly middle ranking.

Peat

Peat has been used for fuel in the past on a minor scale throughout the south of Scotland, basically wherever proximity of habitation and available resources made it economic. The only large area of peat which has been worked on a commercial scale is the Lochar Moss [NY 040 715], south-east of Dumfries, though this is mainly for use in horticulture rather than as a fuel.

Alternative energy sources

At the present time, early in the 21st century, nuclear energy is generated in the south of Scotland region considered here at the Torness power station, on the North Sea coast south-east of Dunbar [NT 745 752]; an older nuclear power station, at Chaplecross to the north-east of Annan [NY 216 697], has ceased operations and is being decommissioned. In the search for replacement 'green' energy, many of the more exposed upland areas have recently seen the development of windfarms to take advantage of the natural meteorological conditions. These hilltop sites commonly have only a thin veneer of superficial deposits, fragile soil and a vegetation cover that regenerates only slowly; all are vulnerable to any disruption of the natural surface drainage. Hence it is important to minimise disturbance during construction phases if serious and long-term problems of upland erosion are to be avoided. As the engineering challenges are met, there is an increasing tendency to site windfarms offshore, and such developments have already begun in the Solway Firth.

Hydroelectricity is generated from the far-sighted Galloway Power Scheme which was initiated in 1929 and completed in the early 1930s. Water flow into the River Ken catchment is regulated by dammed reservoirs in the upper part of the valley and to the west of New Galloway at Clatteringshaws Loch, which was much extended as part of this enterprise; from Clatteringshaws, water is fed into the Ken catchment via a tunnel to Glenlee. Electricity is generated at several stations including the Tongland Power Station [NX 697 538], just to the north of Kirkcudbright, where a visitor centre is located. Over the decades, the formerly stark, concrete dams, powerhouses and associated reservoirs have gradually blended into the landscape so that they now form an attractive visual asset. The scheme is small in comparison with those in the Highlands but has proved a good investment and at the beginning of the 21st century was upgraded with additional turbines to maximise generation capacity.

BULK MINERALS

Sand, gravel and crushed rock are extracted for use as construction materials, as a concrete aggregate, and for roadstone. The working of limestone for the production of agricultural lime has been of historical importance locally.

Sand and gravel

Sand and gravel are principally derived from glaciofluvial outwash deposits in the major valleys, including those of the rivers Nith, Annan and Tweed. The available resources at any one locality are usually restricted, and the small-scale operations that they sustain supply mainly local markets, a distribution pattern encouraged by the high cost of transport in relation to the relatively low value of the materials. The most extensive operations work the deposits around Dumfries, Lochmaben, and to the south of Stranraer. Most of the gravels are compositionally dominated by the local wacke-sandstone, which is a lithology that can give rise to shrinkage in concrete.

Crushed rock aggregate

For crushed rock, both the ubiquitous Lower Palaeozoic wacke-type sandstones as well as a range of igneous rocks have been worked at a number of localities. The igneous rocks range from the upper Silurian to Lower Devonian granites of the south-west to the Carboniferous basaltic lavas and microgabbro intrusions of the Border country in the south and south-east of the region. The more important of the quarries active in the late 20th and early 21st centuries are described briefly below.

The largest wacke sandstone quarry in the region is at Morrinton [NX 870 815], some 12 km north-west of Dumfries. Here, sandstone of the Gala Group is crushed for roadstone and concrete aggregate, with an adjacent block-making facility. Another large quarry in the south west is at Barlockhart [NX 213 565] near Glenluce, where Gala Group sandstone has been baked and hornfelsed in proximity to the Glenluce diorite intrusion. A quarry at Boreland Fell [NX 348 594] near Kirkcowan has produced crushed wacke-sandstone from the Mindork Formation, whilst another at Coatsgate [NT 064 053], near Moffat, works wacke-sandstone of the Queensberry Formation, both units within the Gala Group. The Queensberry Formation sandstone was also worked nearby at Garpol Linn [NT 069 030] where a large temporary quarry was opened during construction of the M74 motorway. To the north-west of Langholm, Hawick Group (Carghidown Formation) wacke sandstone was formerly worked from a quarry at Peden's View [NY 345 862]. In the east of the Southern Uplands, a series of quarries in Gala Group wacke sandstone along the A7 road north of Galashiels, including Hazelbank [NT 425 505], Bower [NT 429 501] and Craigend [NT 444 461], mostly ceased operation during the last half of the 20th century, though some work at Hazelbank continued until the beginning of the 21st century.

Crushed granite is produced at Craignair [NX 819 608], near Dalbeattie, an area which was once renowned for its output of granite blocks used in substantial engineering projects (see below). Tongland Quarry [NX 698 544] near Kirkcudbright produces aggregate from both sheet-like bodies of intrusive diorite and the host, contact-metamorphosed wacke sandstone. Kirkmabreck Quarry [NX 479 565], near Creetown, is an intermittent producer of crushed granite but is used mostly to extract large blocks of granite from time to time, as and when required.

The quarry at Tormitchell [NX 234 944], about 6 km south-east of Girvan, is unusual in the south of Scotland in working a limestone for aggregate (the Stinchar Limestone of the Ordovician Barr Group) as well as the adjacent wacke sandstone units. Some Carboniferous limestone produced from quarries near Canonbie has also been used locally for roadstone, though these operations were primarily for the production of agricultural lime (see below).

Near Duns, in the east of the region, Borthwick Quarry [NT 770 543] works a large, Carboniferous intrusion of basalt on an intermittent basis. The Carboniferous Kershopefoot Basalt has been worked for roadstone to the north-east of Langholm at Kersehope Quarry, adjacent to Kersehope Bridge [NY 500 834].

Limestone

Wherever Carboniferous limestones are exposed there have been numerous small workings and associated kilns for the production of agricultural lime. Many of these were 18th or 19th century operations, and during the early 20th century production became centred on larger quarries. In the Oldhamstocks Basin succession, to the south of Dunbar, many of the Visean limestones within the Strathclyde and Clackmannan groups have been worked locally and on a small scale; well-preserved limekilns survive in places, for example at Catcraig [NT 717 773], adjacent to the old quarries. More recently, large-scale quarrying for cement production has been developed at sites around Oxwell Mains [NT 707 764], collectively known as Dunbar Quarry and situated about 2–3 km south-east of the town. The Hurlet and Blackhall limestones of the Lower Limestone Formation (Clackmannan Group) are utilised there, together with some of the associated mudstones.

In south-west Scotland, a similar 18th and 19th century history of small-scale quarrying pertains in the Canonbie area, with more recent operations most notable at Harelawhill [NY 426 789], where Brigantian limestones of the Alston Formation (Yoredale Group) were extracted by both opencast quarrying and through underground workings (Plate 60) until about 1970. In the Thornhill outlier of Permo-Carboniferous strata, the Visean (upper Asbian to Brigantian) Closeburn Limestone Formation was of sufficient thickness and quality to have been mined at several sites near Closeburn and burned for agricultural lime in the late 18th century; well-preserved limekilns remain near the Croalchapel quarry site [NX 911 915]. Underground, stoop-and-room excavation of the Closeburn Limestone was employed farther west at Barjang [NX 882 902], albeit on a limited scale.

Plate 60 *Old limestone workings at Harelawhill Quarry [NY426 789], near Canonbie, in about 1949 (P217567).*

Brickmaking

There has been much small-scale exploitation of localised lacustrine, alluvial or glacigenic clay deposits for the manufacture of bricks and tiles. Such operations were largely resticted to the 18th and 19th century, but some continued until more recent times. Typical was the Tarrasfoot Tile Works near Langholm which operated from the mid 18th to the mid 20th century, initially producing bricks but in later years turning out drainage pipes for local agricultural use.

In common with the practice in many other coal mining districts, bricks were produced from mudstones in the Coal Measures succession in the Sanquhar Coalfield, especially at Sanquhar and Kirkconnel. Near the railway station in Sanquhar, the Buccleuch Terra Cotta Works produced high quality terracotta bricks and tiles of an attractive red hue, which can still be seen incorporated into many of the local buildings. The mudstone workings are shown in Plate 54 as they appeared in about 1920.

BUILDING STONES

Throughout the south of Scotland, local building stone has been used from early times, for example in the construction of defensive Iron Age earthworks, but a more organised extraction and distribution system was required in medieval times to allow construction of castles and the Border abbeys. Castles and defensive tower houses were usually built mainly from locally available rock rubble, but with dressed features such as corner stones and lintels carved from softer sandstone sourced farther afield if necessary. A good example is provided by Smailholm Tower, a 15th century Borders tower house near Kelso (Plate 61), which is largely built of the local Carboniferous basalt, with the red sandstone dressed features made from Devonian sandstone.

Sandstone quarrying on a substantial scale was clearly in progress during the construction of the Border abbeys, for which a large quantity of dressed building stone was required. Local sources were exploited. For example, work on Sweetheart Abbey, to the south of Dumfries, started in 1273 utilising the local, red Permian sandstone (Plate 62). Farther east, Dryburgh Abbey had a protracted history of construction between the 12th and 16th centuries using the local Devonian sandstone, mostly a cross-bedded and a warm pinkish

Plate 61 *Smailholm Tower [NT 638 346], constructed of local basalt with red, Devonian sandstone door arches, window surrounds and quoins (P596389).*

grey variety that was most probably sourced locally from long-disused quarries at Ploughlands [NT 633 307]. Melrose Abbey had a similarly protracted history of building, alteration and rebuilding, closely bound up with the destruction caused during intermittent warfare and cross-border skirmishing on the frontier with England, only 40 km to the south, during the 14th, 15th and 16th centuries. Much of the earlier construction work in the 12th and 13th centuries utilised a brown to grey-green volcanic agglomerate obtained from the nearby Chiefswood volcanic neck; of the two likely quarry sites, the closest [NT 339 542] is only about 700 m from the abbey. This quarry continued in use until the early 20th century. The later phases of construction at Melrose used a wider range of local rock types, including various Devonian sandstones, at least some of which were probably obtained at Bourjo [NT 543 327] on the north-west flank of the Eildon Hills, and some igneous rocks from the Eildon Hills Carboniferous intrusion.

In the urban landscape, granite buildings are concentrated, not unexpectedly, in the vicinity of the main granite plutons in the south-west, and so are especially numerous in the Galloway towns. Dumfries, however, has a plethora of fine buildings built from the local Permian red sandstone, as presently quarried at Locharbriggs. Carboniferous sandstones were extensively worked in the eastern part of the region, producing a variety of attractive pink and pale grey building stones that are well represented in Borders towns.

Granite

The largest of the Galloway granite quarries, near Dalbeattie, worked the Criffel intrusion, whose coastal location assisted transport of the extracted stone. Large quantities of building and dimension stone were exported and used in cities on the west coast of Britain for major engineering works such as the construction of Liverpool Docks. At the peak of output, around 1900, the quarries employed about 400 men. A granite from Kirkmabreck, Creetown — known commercially as Silver-Grey (a similar granite from Dalbeattie is shown in Plate 63) — also makes an attractively speckled ornamental stone when polished. Many of the Galloway towns — Dalbeattie, Castle Douglas, Creetown, Gatehouse of Fleet, Newton Stewart — contain fine examples of granite buildings. However, despite this wide use of the stone for building purposes, granite setts for road building were traditionally the principal output of the Dalbeattie quarries (Plate 64).

Lower Palaeozoic sandstone

Since most of the area is underlain by Lower Palaeozoic sandstones, it is inevitable that these rocks are used in many buildings. However, it is not an easy stone to work, generally lacking an orthogonal set of joints, so that much of the walling is built using irregular rubble held together by lime mortar. Only in relatively few instances, such as some high-value buildings in places like Moffat, Peebles and Innerleithen, is the wacke sandstone roughly squared into blocks, and even here they are only used for the main wall panels rather than for door and window jambs and quoins.

The Ordovician Shinnel Formation is characterised by the thick sequences of laminated grey siltstone that are interbedded with the more abundant sandstone. However, in the Leadhills area the siltstone is locally the dominant lithology and is there known informally as the

Plate 62 *Ruins of Sweetheart Abbey, 10 km to the south of Dumfries, constructed of Permian red sandstone from the Locharbriggs Sandstone Formation (P596386).*

Plate 63 *Granite from the Criffel Pluton at Craignair Quarry, Dalbeattie (P521374).*

'Lowther Shales'. A bed-parallel cleavage is commonly developed in these siltstones and they have consequently been worked as an inferior roofing 'slate' in numerous small and medium-sized quarries, now all disused; Stobo Quarry [NT 158 365], about 7 km west-south-west from Peebles, is the largest of these.

Devonian and Carboniferous sandstones

The Devonian sandstones of the eastern Borders region are typically red-brown, though there is considerable variation in colour locally. The stone was quarried in a large number of locations around larger towns such as Jedburgh, Melrose, Coldstream, Hawick and Kelso, and used therein for building. The generally grey to buff and pale brown sandstones of Carboniferous age in the eastern Borders have also been extensively quarried for building stone and used widely in towns such as Duns and Kelso. Many of the older buildings in Langholm are constructed from the local Whita Sandstone. Throughout the region, the sandstones have been quarried on a small scale for local use in the villages and farms, and their natural variation enhances the vernacular architecture and its contribution to the landscape.

Now, at the beginning of the 21st century, an attractive honey-coloured stone is being quarried on a small scale near Swinton [NT 854 486] with sawn and other worked blocks produced for both modern building and replacement work. Stone was first produced on a significant scale from the Swinton quarries in the 1790s, and much was used for construction work in Edinburgh in the late 19th and early 20th centuries. After a period of disuse, Swinton Quarry was reopened at the end of the 20th century.

Permian and Triassic sandstones

The attractive, bright red Permian sandstones of the Dumfries and Lochmaben basins have long been worked for building stone and it has been exported widely from quarries such as Locharbriggs [NX 990 810] at Dumfries and Corncockle [NY 085 870] near Lochmaben.

Many of the public buildings in nearby towns such as Dumfries and Lockerbie were inevitably constructed with this stone. In the Dumfries Basin, the Locharbriggs Sandstone Formation was worked, and in the Lochmaben Basin, the Corncockle Sandstone Formation was utilised. A similar lithology in the Thornhill Basin, from the Thornhill Sandstone Formation, was quarried at Gatelawbridge [NX 902 965] and Closeburn [NX 892 910].

Some of the modern quarries working the Permian sandstone were active as early as the 17th century, but saw a rapid expansion during the late 19th century as the railway network developed and provided a ready means of transport. The red sandstone was particularly popular for the expanding urban areas in the central belt of Scotland, and fine examples of its use as a building stone can be seen in both Edinburgh and Glasgow. Much stone was also exported to other parts of the UK and as far afield as North America. During the peak period of production, around 1900, the Locharbriggs quarry employed about 260 men. Thereafter production of sandstone declined in the south of Scotland, and had virtually ceased in the late 20th century, by when most of the quarries had closed. However, Locharbriggs and Corncockle continued to be worked intermittently, so maintaining supply, and at the beginning of the 21st century are satisfying an upsurge in demand from both modern building and restoration work; Newton Quarry at Gatelawbridge is also intermittently active.

The Triassic St Bees Sandstone was worked in quarries at Corsehill [NY 205 701] and Cove [NY 254 710], which supplied much local building stone to Annan and Gretna. Corsehill sandstone was widely used in Scottish cities from the late 19th century onwards, with the Cove quarry still active early in the 21st century.

METALLIFEROUS AND ASSOCIATED MINERALS

Although there is currently no metalliferous mining in the south of Scotland, the lead-zinc veins of the Southern Uplands have provided an important resource that was extensively exploited from the 17th until the early 20th century, with records of mining at Leadhills going back as far as 1239 and the probability that there was some even earlier activity. Of greatest importance were the mines of the Leadhills–Wanlockhead district which, working just over 70 lead-zinc veins, was the largest metal mining operation in Scotland and produced about

Plate 64 *Sett making at Craignair Quarry, Dalbeattie, in about 1939 (P000110).*

400 000 tonnes of metallic lead, 10 000 tonnes of zinc and 25 tonnes of silver between 1700 and 1958. In addition, this same general area has produced gold from small-scale alluvial sources that were worked from the mid 16th century onwards. Farther to the south-west, in Galloway, the Woodhead mines near Carsphairn were an important source of lead in the mid 19th century, whilst base metal veins between Newton Stewart and Gatehouse of Fleet were mined from the late 18th to the early 20th centuries. From the Galloway mines, it is estimated that about 30 000 tonnes of lead were produced, together with smaller amounts of zinc and copper. Away from the principal mining districts are a range of mineralised locations, the more important of which, together with the Leadhills–Wanlockhead, Carsphairn and Galloway sites, are shown on Figure 63. Some of the more isolated sites have been worked on a small scale and others have been trialled, but most are too small to be of economic interest. Several different mineralisation styles are represented.

Lead-Zinc-Copper

Leadhills–Wanlockhead district

The 70 or so veins of the Leadhills–Wanlockhead mining field are spread over about 6 km of the regional strike of the host Lower Palaeozoic strata, which are mainly late Ordovician sandstones of the Portpatrick Formation. The majority of the veins run broadly north–south, though a few, which apparently formed later, have an approximately north-west to south-east trend (Figure 64). The north–south veins dip steeply to the east, the north-west trending veins dip steeply to the south-west. The vein orientations approximate to the late Caledonian fault

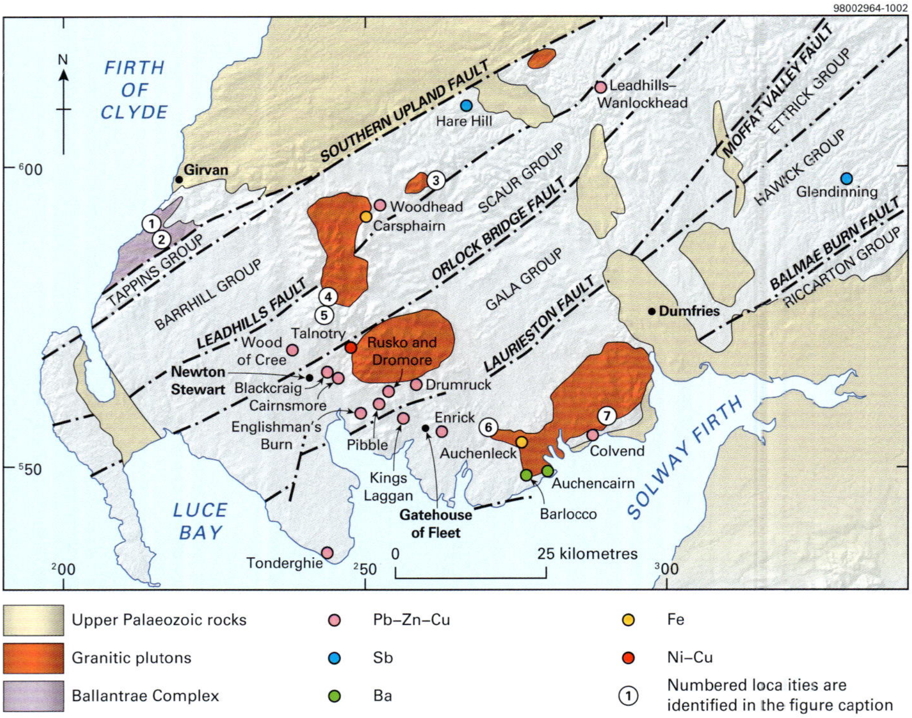

Figure 63 *Distribution of mineralisation in south-west Scotland. The best known mines and trial workings (all long-abandoned) are named on the map. All are vein deposits with the exception of the Ni-rich magmatic segregation at Talnotry. Other notable examples of mineralisation with no history of exploitation, but described in the text, are numbered as follows: 1 Pinbain (Cr); 2 Poundland Burn (Cr); 3 Moorbrock Hill (As-Au-Pb-Zn); 4 Glenhead Burn (As-Au-Pb-Zn); 5 Penkiln Burn (Pb-Zn-As-Cu); 6 Black Stockarton Moor (Cu-Mo-As-Au); 7 Needle's Eye (U-Cu-Bi-Fe-Hydrocarbons).*

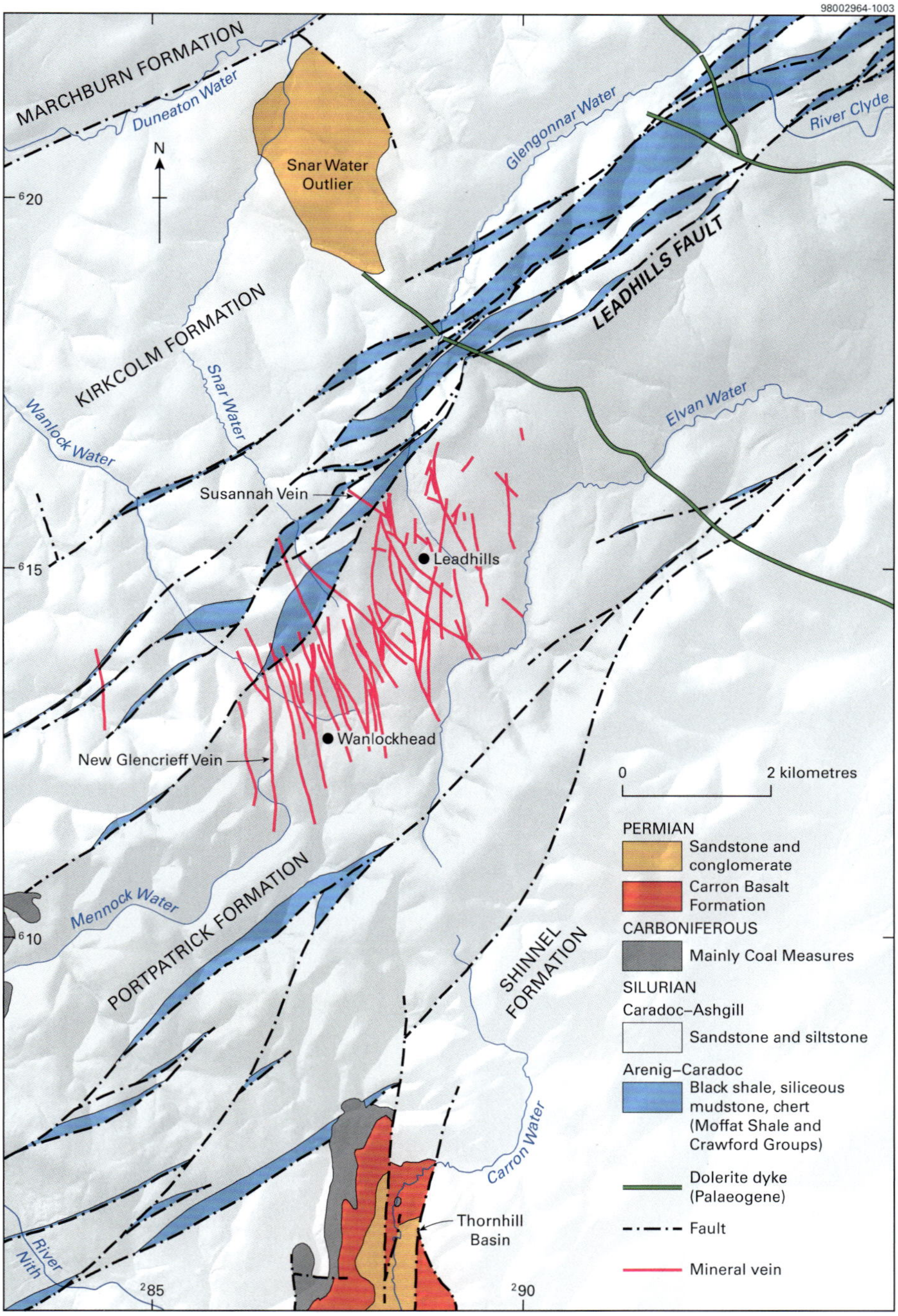

Figure 64 *Principal mineral veins of the Leadhills–Wanlockhead mining field.*

pattern and it is likely that fault reactivation played a role in the localisation of the mineral deposits, most probably during Carboniferous times. Most of the veins terminate in the north against the Leadhills Fault, which is here a reverse structure dipping to the north-west at about 45°. In the hanging wall of the fault, an imbricate zone of crushed chert and mudstone (Crawford and Moffat Shale groups) may have acted as an impervious barrier to the mineralising fluids, whilst providing a fissile transfer plane for the extensional stress. Only a few veins cross the imbricate zone and continue into the overlying, quartzofeldspathic sandstones of the Kirkcolm Formation (Figure 64).

Mineralised fault sections are traceable for up to 3 km, whereas the largest of the workable ore bodies, in the New Glencrieff Vein, extended 360 m along strike and 400 m down dip. More commonly, the strike length of individual ore bodies was about 50 m, extending about 100 m down-dip. Elsewhere, the veins are typically banded fissure fillings, several centimetres to about 5 m in width, consisting of brecciated Ordovician sandstone cemented by quartz, dolomite, calcite and locally baryte, and cut by sulphide stringers containing galena, sphalerite, chalcopyrite and pyrite. The carbonates were the earliest-formed gangue minerals, and their subsequent replacement by quartz was associated with the introduction of a second generation of sulphides. There is some evidence for mineral zonation, with sphalerite increasing with depth relative to galena. Around the mineralised zones, hydrothermal alteration forms a diffuse halo in the Ordovician sandstone country rock.

A great variety of minerals have been identified in the Leadhills–Wanlockhead veins; almost 60 at last count and still rising. A few rare, primary ore minerals accompany the dominant galena-sphalerite association, but most of the unusual minerals reported are secondary. Most were found in an oxidised zone that in places reached a depth of 180 m, suggesting that in the past the water table was much lower than it is at present. Some of the secondary forms present are relatively common, for example the hydrated copper mineral malachite, but rare secondary minerals that have their type locality within the district, and so are named accordingly, include leadhillite, lanarkite, caledonite (Plate 65), and susannite, the latter named from the Susannah Vein. These, and most of the other secondary minerals present, are complex, variably oxidised and hydrated, mixed sulphate-carbonates of lead, zinc and/or copper.

Spoil tips and the remains of headgear and mine buildings still bear testament to the extensive mining activities in the Leadhills–Wanlockhill district and at Leadhills some have now been preserved and incorporated into the Scottish Lead Mining Museum.

Plate 65 *Caledonite, a complex sulphate of lead and copper from the Susannah Vein, Leadhills (P243193).*

Plate 66 *Spoil heaps, surface workings and the ruins of mine buildings at Pibble Mine [NX 5255 6068], near Creetown, Galloway (P001527).*

Galloway district

The most important of the Galloway mineral veins lie in a 7–8 km wide zone extending from the north of Newton Stewart, south-east towards Gatehouse of Fleet (Figure 63). The zone skirts the south-west margin of the Cairnsmore of Fleet pluton, though the mineral veins are hosted by Ordovician, Shinnel Formation sandstones in the north, and Silurian, Gala Group sandstone in the south, all within the pluton's thermal aureole. The veins are of epigenetic origin, with stringers of mostly galena and sphalerite but locally with chalcopyrite dominating, in a dolomite-calcite-baryte-quartz gangue occupying brecciated zones, up to 18 m wide at Blackcraig [NX 445 647], but usually much narrower. Most of the veins trend broadly north-west, though there is considerable local variation ranging from almost north–south at Wood of Cree [NX 386 695], where in addition to the underground mining there was some opencast working of a brecciated zone 4.5m wide, to east-north-east at Rusco [NX 553 615], where the mineralised zone was only 0.3 m wide.

The most important of the Galloway mines were those in the Blackcraig–Cairnsmore complex (Figure 63). These mines most probably all worked the same north-west-trending vein over a strike length of almost 4 km, though individual economic ore runs did not exceed 500 m; to the north-west the vein passes into unmineralised quartz. The Blackcraig–Cairnsmore mines were operated intermittently between 1743 and 1882 with some very limited activity in the early 1920s. The most productive phase was between 1853 and 1881 when the Blackcraig mines are reported to have produced 3820 tons of lead, 1668 tons of zinc-blende (sphalerite) and 28 tons of copper. The Cairnsmore mine had a relatively brief life of only about 15 to 20 years in the mid 19th Century, but in that time a little over 3000 tons of lead ore were produced. Two closely spaced, parallel veins were worked at Cairnsmore, cutting across both sandstone and mudstone, but were apparently only economic where they cut the mudstone. A similar situation was reported farther south-east at the Pibble Mine [NX 527 605] where the most productive sector of the north-north-west-trending vein cuts across mudstone of the Moffat Shale Group, wherein there is some evidence for disseminated mineralisation. Although principally a lead-zinc mine, Pibble was the largest producer of copper in Galloway, albeit this amounted to only a few hundred tons of metal; the remains of the engine house still stand (Plate 66) amidst traces of adits, surface workings and spoil heaps.

Copper, mostly as chalcopyrite, was the main product of several of the smaller and short-lived Galloway mines. For example, at Enrick [NX 619 550] about 2.5 km south-east of Gatehouse of Fleet, a breccia vein in Kirkmaiden Formation sandstone, trending east–west and 1–2 m wide, contained chalcopyrite in a calcite-quartz-dolomite gangue; it was worked at several levels for a few years after its discovery in 1820. Farther east, in a coastal location on the south side of the Criffel pluton, the Colvend Mine [NX 868 528] was active around 1770. It worked a breccia vein about 1 m wide that cut Ross Formation sandstone and intrusive microdiorite sheets; chalcopyrite, malachite and azurite are reported to have been present in a quartz-calcite gangue.

The age and metallogenesis of the Galloway veins is uncertain. The proximity of the majority of veins to the Cairsmore of Fleet granite pluton invites an association with its 'Acadian' intrusion at about 400 Ma and a concentric, compositional zoning of the mineralisation, subparallel to the pluton margin, has been proposed. However, as at Leadhills–Wanlockhead, the orientation of the veins broadly follows the Caledonian, north-west and north-north-east-trending cross faults, and a similar history of mineralisation during Carboniferous fault reactivation could be argued. The Blackcraig vein is apparently closely associated for much of its length with an olivine dolerite dyke, but the relationship between the dyke and the mineralisation is uncertain, as is the age of the dyke, perhaps Permo-Carboniferous, perhaps Palaeogene. The ultimate source of the metals is similarly uncertain, but may be hinted at by the disseminated mineralisation in Moffat Shale Group mudstone described from the Pibble Mine. Disseminated mineralisation has also been reported from Moffat Shale Group strata elsewhere in the region, for example at Penkiln Burn [NX 446 767], 10 km north-north-east from Newton Stewart (Figure 63).

Carsphairn district

About 3 km to the west of Carsphairn village are the remains of the Woodhead lead mine [NX 530 938], which include several shafts in close proximity, spoil heaps and various surface facilities. There are two main veins trending approximately north-west, following a brecciated fault zone in the host sandstone of the Ordovician Kirkcolm Formation. Galena is the principal ore mineral, accompanied by sphalerite and some chalcopyrite, set in a gangue of calcite, dolomite and quartz. Mining (initially opencast) was active from 1838 until late in the 19th century and in the most productive period, 1840 to 1852, annual lead production peaked at 905 tons.

Baryte

Baryte occurs widely as a gangue mineral in the lead-zinc veins of Leadhills–Wanlockhead and Galloway but in these mining areas appears only to have been commercially produced, on a small scale, at the Cairnsmore mine in Galloway. Of greater importance were baryte veins to the east-south-east of Kirkcudbright at Barlocco [NX 785 474] and Auchencairn [NX 816 483] (Figure 63). At Barlocco, two veins trend approximately east–west, are up to 2.4 m wide, and cut Silurian sandstone of the Ross Formation (Hawick Group). About 3000 tonnes of baryte were reported to have been extracted between 1856 and 1920 with some intermittent and small-scale activity continuing until the early 1950s. At Auchencairn, the baryte vein trends east-north-east, is about 0.6 m wide and cuts Carboniferous sandstone of the Rascarrel Member, Fell Sandstone Formation. The Auchencairn mine had a relatively short life, producing about 700 tonnes of baryte in the 1860s.

Gold-arsenic-antimony

At the western margin of the Criffel–Dalbeattie granodioritic pluton, in the Black Stockarton Moor area [NX 725 555], Silurian sandstones of the Hawick Group have been intruded by a complex of intersecting volcanogenic breccia pipes, felsic dykes and small granodioritic bodies (see Chapter 4). The whole complex is associated with weak copper-molybdenum mineralisation of porphyry-copper type with a wide zone of hydrothermal alteration centred on an inner zone with chalcopyrite-bornite veining. This mineralisation is concentrated into brecciated zones which are also enriched in arsenic (arsenopyrite), antimony (stibnite) and gold, though gold assays show only up to 0.06 parts per million (ppm).

Adjacent to the eastern margin of the Carsphairn granitic intrusion, at Moorbrock Hill [NX 620 980], gold values of 1–3 ppm have been reported from quartz veins carrying pyrite, chalcopyrite and arsenopyrite in a north-east-trending zone of intense brecciation and hydrothermal alteration of the host Ordovician (Kirkcolm Formation) sandstone and mudstone. The gold forms small, isolated grains, usually in association with pyrite-chalcopyrite intergrowths. To the south, at the southern margin of the Loch Doon pluton in the Glenhead Burn area [NX 449 780], arsenopyrite, pyrite and native gold occur in thin quartz veins cutting brecciated and hydrothermally altered Ordovician sandstone of the Portpatrick Formation. In the most intensely mineralised material, arsenic levels range up to 3.5 per cent and gold reaches 8.8 ppm. Lead-zinc veins are also present in both the Moorbrock Hill and Glenhead Burn localities and it seems likely that mineralisation was protracted as hydrothermal systems developed at the pluton margins and scavenged metals from the country rock. A similar origin seems likely for the antimony-lead veins at Hare Hill [NS 658 104], which cut a small, hydrothermally altered granodioritic intrusion adjacent to the Southern Upland Fault near New Cumnock. The veins are associated with a zoned development of arsenic-antimony-copper-lead-zinc mineralisation. The Moorbrock Hill and Glenhead Burn mineralisations are relatively recent discoveries with no history of previous working. Trial workings have been driven into the Hare Hill veins, but there is no record of their having been economically mined.

In the south-central part of the Southern Uplands antimony was won from the Louisa Mine [NY 313 966] at Glendinning, about 6 km east of Eskdalemuir. About 200 tonnes of antimony were produced in the late 19th century from a north-east-trending network of thin veins worked by shafts and adits. The antimony is principally contained in stibnite, but other antimony-bearing minerals are present, including semseyite, bournonite and tetrahedrite. Also commonly present are arsenopyrite, pyrite, chalcopyrite, galena and sphalerite, all contained in a quartz-carbonate gangue. The host rocks are Silurian sandstone of the Hawick Group, which modern exploration has shown to contain stratabound disseminations of pyrite and arsenopyrite concentrated into debris flow beds of intraformational conglomerate up to 4 m thick and containing 0.75 per cent arsenic. The arsenopyrite contains up to 0.4 per cent gold and 0.33 per cent antimony, whilst the pyrite contains up to 5 per cent arsenic. These relationships suggest a complex mineralisation history, and a possible source for the vein mineralisation.

Alluvial gold

Gold had been discovered in the stream gravels of the Leadhills–Wanlockhead district by the beginning of the 16th century. One record from 1502 describes a nugget almost 1 kg in weight, and by about 1510 gold extraction was being carried out on an organised basis. Extant documents show that 113 ounces of Leadhills gold were used to augment the Scottish crown jewels between 1537 and 1541, and that over a period of 30 days in 1579, 128 ounces of gold were sent to the Scottish Mint. Thereafter, production declined as the richest deposits were worked-out, but intermittent sluicing operations continued until the mid 20th century, and gold is still panned from the streams by hobbyists.

The two villages of Leadhills and Wanlockhead occupy high ground with radial drainage: the Mennock Water to the south-west, the Wanlock Water, Snar Water and Glengonnar Water to the north, and the Elvan Water to the east (Figure 64). Gold has been recovered from all of these streams and from many of their tributaries, with the early workers reported to favour gravel beds preserved locally between rockhead and the overlying glacial till and exposed in the stream banks. These sources are estimated to have given a yield of between 5 and 10 grams per tonne. Though clearly local, a bedrock source for the alluvial gold has not been established. To compound this difficulty, modern microchemical analyses suggest that the gold might have been derived from at least four different geological environments. The dominant type present is probably of mesothermal shear zone origin and has microscopic arsenopyrite inclusions. Also present are grains thought to have been derived from: gold concentrated along the margins of mafic, Palaeogene dykes; gold associated with ultramafic rock (most probably recycled from the glacial till); and gold concentrated by the downward movement of the oxidising groundwater solutions circulating within the nearby Permian red-bed basins. Some gold grains have characteristics that do not fit readily into any of these categories.

Chromium and nickel

Chrome spinel is a widespread accessory mineral in the serpentinised ultramafic rock of the Lower Ordovician Ballantrae Complex, but is particularly concentrated at two localities, in both cases hosted by serpentinised dunite. Neither of these localities has been deemed of commercial interest in their own right, but their presence has stimulated extensive, though unsuccessful, mineral exploration elsewhere in the Ballantrae Complex. In the northern serpentinite belt, at Pinbain Bridge [NX 138 916], spinel (a true chromite in this case) forms between 30 per cent and 90 per cent of a lensoid body 4–5 m wide and several tens of metres long. In the southern serpentinite belt, at Poundland Burn [NX 170 882], nodular iron-rich chrome spinel (picotite) forms up to 70 per cent of a restricted area of outcrop. Sparse Ni-Cu mineralisation is also widely associated with the ultramafic rock.

Within the Southern Uplands massif, an unusual occurrence of Ni-Cu sulphide minerals forms a small basal lens (c. 1 x 4 x 20 m) in a dioritic, sill-like body intruded into Silurian, Gala Group sandstone at Talnotry [NX 4785 7035]. The principal ore minerals are pyrrhotite, pentlandite, chalcopyrite, nickeline and gersdorffite. The deposit was worked soon after its discovery in about 1885, but though about 100 tons of ore were raised, very little was removed from the site.

Uranium

Pitchblende (uranium oxide) is present in thin polymetallic-carbonate breccia veins adjacent to the North Solway Fault and close to the southern margin of the Criffel–Dalbeattie pluton at Needle's Eye [NX 915 562]. The veins cut thermally metamorphosed, Silurian Riccarton Group sandstone and Carboniferous strata of the Border Group. Associated with the pitchblende are abundant secondary uranium minerals. The pitchblende has been dated at 185 ± 20 Ma, an age which has been taken to show Mesozoic movement on the North Solway Fault (see Chapter 7), but the date has been questioned and a late Caledonian age suggested for the veins based on their close proximity to the granite pluton (c. 397 Ma). The Needle's Eye site has been much studied as a natural analogue for the radionuclide migration that may occur around sites housing radioactive waste.

Haematite

About 3 km to the west of the Woodhead lead mine a substantial haematite breccia vein, up to 3 m wide, transects the margin of the Loch Doon Pluton and its metamorphic aureole in Kirkcolm Formation sandstone. The vein runs north–south for a distance of about 3 km [NX 505 911 to NX 503 938]. It was worked intermittently between 1869 and 1876, but though about 400 tons of ore were raised it proved to be of poor quality and the enterprise was abandoned. Several haematite breccia veins are also known from the western margin of the Criffel–Dalbeattie Pluton, where they cut both the igneous rock and its sedimentary host, sandstone of the Hawick Group. In the largest of these veins, which cuts the granite at Auchenleck [NX 773 525], botryoidal haematite is accompanied by quartz and baryte. The Auchenleck vein trends west-north-west and was reputedly up to 20 m wide. It was extensively worked in the early to mid 19th century, with a report from 1845 that 50–70 tons of ore per week were then being extracted.

The haematite veins may have a similar origin to the much more extensive deposits farther south in west Cumbria, and to the associated veins at Maughold Head in the north-east of the Isle of Man. These deposits are thought to be late Permian or Early Triassic in age and to have been formed from mineralising fluids circulating in the Permo-Triassic sedimentary rocks of the Irish Sea Basin. It is possible that a former covering of permeable Permo-Triassic or Carboniferous rocks acted as a pathway that enabled mineralising fluids to gain access to fractures within the Lower Palaeozoic sandstone and Caledonian granites of south-west Scotland.

GEOLOGICAL HAZARDS

The region's geology exerts a major influence on a wide range of land use and environmental issues, and in some circumstances may introduce natural hazards that constrain or limit development. Some potential hazards are discussed briefly below.

Seismicity

The south of Scotland is an area of very low seismicity, but the region is not entirely seismically quiescent. Historical records describe minor events in the Galashiels area of the Scottish borders in 1650, 1728 and 1844, whilst a series of minor earthquakes have been recorded around Lockerbie and Dumfries from the late 19th century onwards.

From the records of the Galashiels events, the largest seems to have occurred on 1st March 1728, with an epicentre between Galashiels and Selkirk. This earthquake was felt from Cumbria in the south, to Fife in the north, but apparently caused no damage. The magnitude is estimated to have been 4.3 ML and the focus was most probably deep. A series of small events felt in the Leadhills–Wanlockhead area between 1748 and 1828 were most probably initiated by the local mining activity.

Small earthquakes in the Lockerbie–Dumfries district are likely to be related to the regional seismicity of the north of England, which is characterised by a band of relatively intense activity up the spine of the Pennines, with some activity to the west of this but with practically nothing to the east (Figure 65). This pattern terminates in the vicinity of the Anglo–Scottish border, north of which there is only a little activity, for example an earthquake near Dumfries on 26 December 2006 that had an instrumental magnitude of 3.7 ML. In the same general area, the 26 December 1979 Carlisle earthquake, with instrumental magnitude of 4.7 ML, was one of the larger British earthquakes of the 20th century. It was felt over most of southern Scotland, Cumbria and north-east England (Figure 66) and was generated at a depth of about 5 km. There was a marked directionality to energy release, resulting in the earthquake being much more perceptible to the north than to the south. The strongest effects were felt around Carlisle, Longtown (close to the epicentre and about 5 km south of the border) and Canonbie (about 2 km north of the border), with damage caused to roofs and chimney stacks, debris falling and cracks appearing in walls. A series of significant aftershocks continued until 1981. The fault plane solution of the aftershock sequence, and the lineation of aftershock epicentres, shows lateral movement on a near-vertical plane with a strike of 123°.

It seems likely that the pattern of seismicity is related to the geometrical reaction of structural components, most likely the major geological blocks, to the overall pattern of crustal stress. There is currently a maximum compressive stress from the north-west or north-north-west arising from the widening of the Atlantic Ocean. Such a regime could induce generally rotational movements of the Pennine and Lake District blocks, with shearing movements on their flanks and between them and the adjacent Southern Uplands block.

Factors affecting ground stability

Across the south of Scotland, bedrock or glacial till usually provides adequate foundation conditions for conventional structures, though bedrock may be significantly weakened locally by weathering, faulting or mineralisation. Stability problems are most likely to occur either on steep slopes, or within areas covered by alluvium or peat where soft or unpredictable ground conditions may be associated with a high groundwater table. Particular problems of subsidence

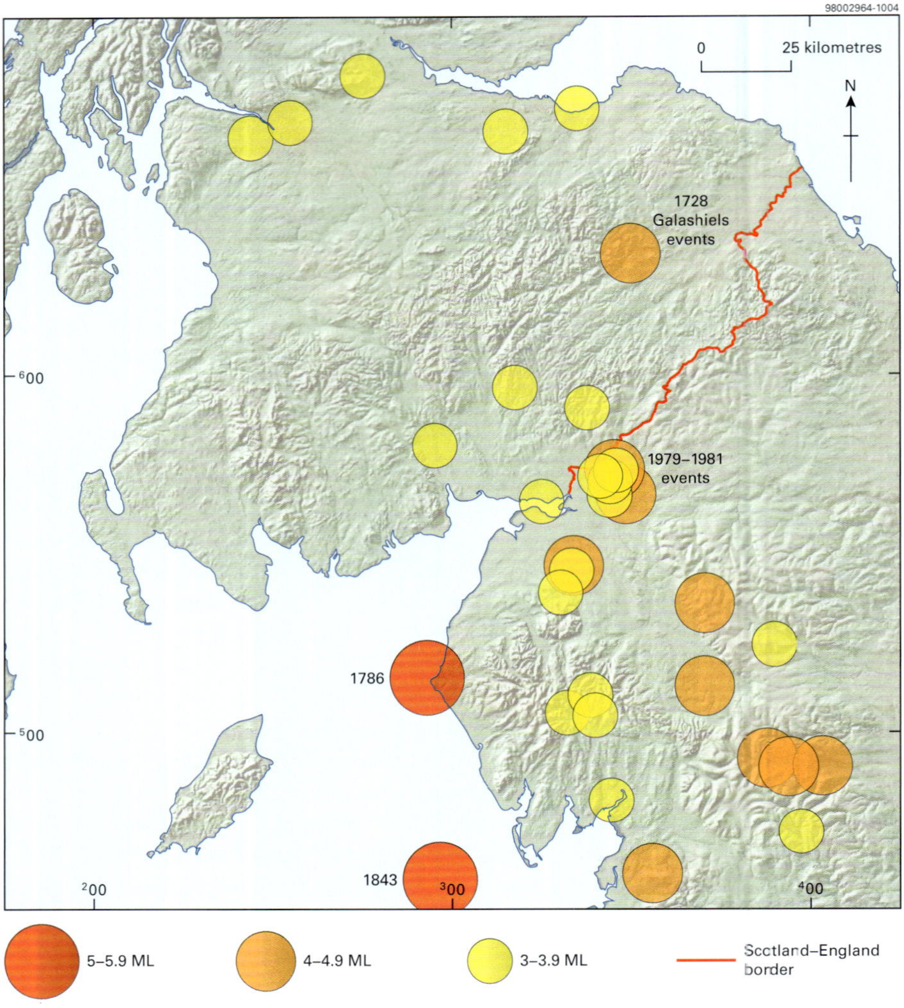

Figure 65 *Distribution of earthquakes above 3.0 ML in southern Scotland and northern England, 1650–2010.*

arise in the coalfields and metal-mining areas, and where limestone has been extracted from underground workings. This issue is also taken up in the section below on 'mining legacy'.

Ancient, inactive landslide deposits are fairly widespread across the region and are generally under-represented on existing geological maps. They were mostly initiated during the late glacial period, when the melting ice-sheets left glacially oversteepened slopes unsupported and unstable. Thawing of permafrost led to water oversaturation and high pore-water pressures that reduced rock shear strength and facilitated failures. The ancient landslides range from deep-seated rotational slides to relatively shallow translational slides and include progressive, multiple slope failures. These landslides may be stable under present-day conditions, though human intervention by excavation or loading, or any alteration of the local groundwater regime, could cause renewed instability.

Today, infrastructure following steep-sided valleys through the Southern Uplands is particularly vulnerable to disruption by landslips. These are usually initiated by periods of unusually heavy rain, and may range from debris flows and surface outwash, commonly coupled with gulley erosion by swollen streams, to larger-scale mass movement. The following two examples of the latter are illustrative.

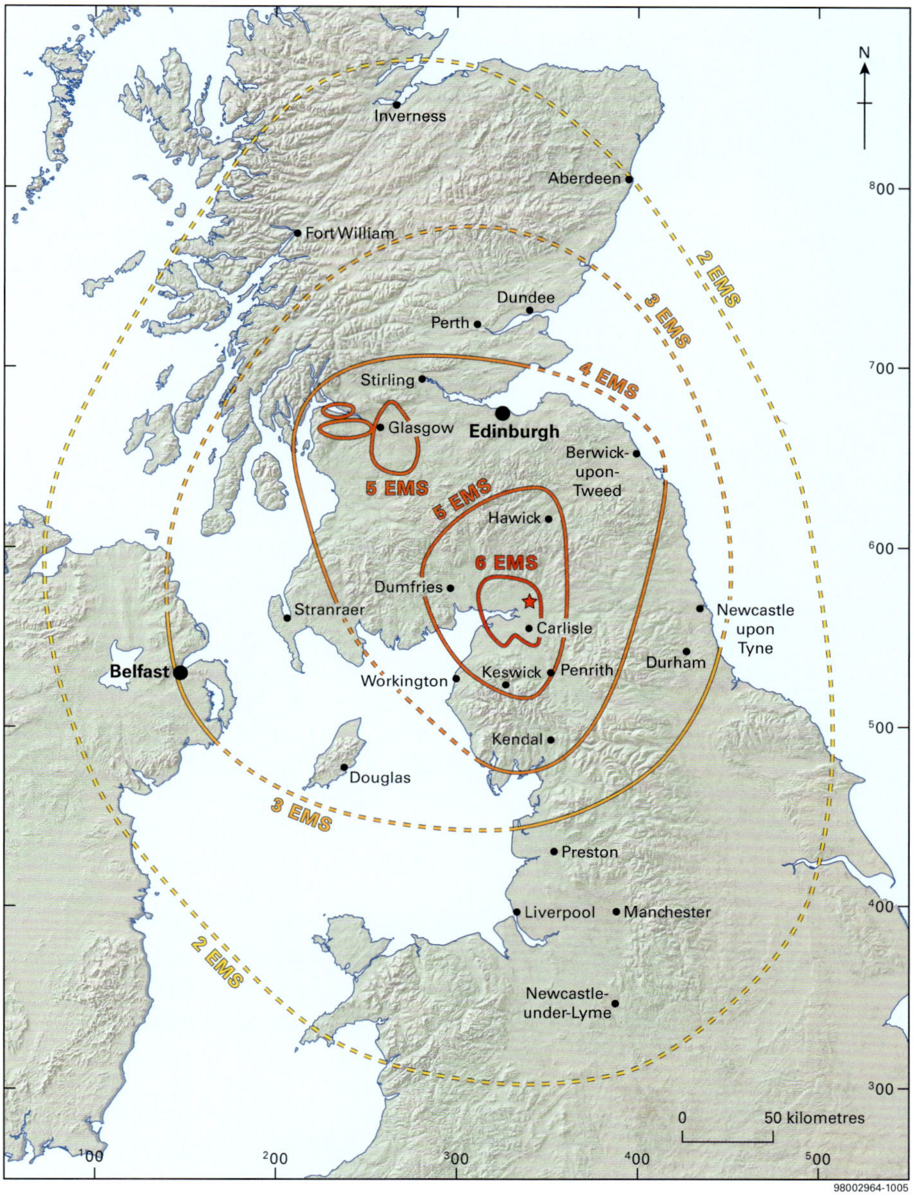

Figure 66 *Isoseismal map of the 26 December 1979, Carlisle earthquake, showing the extent of the felt area. The isoseismal lines enclose areas of equal levels of EMS (European Macroseismic Scale) intensity. The star shows the instrumental epicentre.*

In December 2006, following a period of prolonged, heavy rainfall, the B7068 road linking Langholm and Lockerbie was undermined by a landslide at Scroggs Cleuch [NY 1621 8135]. The relatively small, rotational failure of unconsolidated, superficial deposits was instigated by a combination of factors: elevated pore water pressure through the substrate, a raised water table, and erosional undercutting of the bank beneath the road by the swollen burn. Part of the road was carried away and the rest affected by tensional fissuring. Despite the small absolute size of the slip, its location was damaging and disruptive. Such local events are not unusual.

In July 2008, again following a period of torrential rainfall, the main A7 road through Langholm was blocked by large landslides both to the north and to the south of the town. In each case the landslip originated on the slopes above the road in water-saturated, superficial deposits. It is perhaps a telling feature of the southern landslip, at Auchenrivock [NY 373 805], that it was initiated in an area that had been recently disturbed by engineering works for a new road, though destabilisation of the slope by human intervention was not unequivocally established. The Langholm landslips resulted in the closure of the main road through the Borders for several weeks, with a significant impact on the local economy.

The coastal routes along both the eastern and western sides of the region pass through cliffs and cuttings in places and have a history of disruption by rockfall. In the east, the main East Coast railway linking Scotland and England passes through steep sided cuttings in bedrock, close to the edge of sea cliffs, to the north of Berwick. The difficulties are obvious to any rail passenger. On the west coast, the main A77 road hugs the coastline between Girvan and Ballantrae, utilising the raised beach. In several stretches, the road skirts the relic sea cliffs, which have required much stabilisation over the years and have in places encouraged the re-routing of the road inland.

With all of the issues described above in mind, appropriate engineering solutions were required for the region's largest civil engineering project in recent years, the construction of the M74 motorway linking Glasgow to north-west England across the Southern Uplands via Abington and Beattock. Large-scale, gently inclined cutting faces, and extensive drainage works are much in evidence.

Flooding

Quite apart from its influence on ground stability, periods of unusually heavy rainfall lead to flooding. There is a growing consensus that climate change is likely to cause an increase in the frequency of these extreme events and so flood prevention and mitigation is likely to become a more pressing issue. Alluvial flood plains are, by definition and of necessity, prone to flood (Plate 67). Nonessential development of these areas should be resisted and suitable precautions taken to protect vulnerable infrastructure where it has to cross them. Dumfries has a particular and long-standing problem of flooding in parts of the town centre, especially when high discharge levels in the River Nith coincide with high spring tides in the Solway Firth and an onshore wind. Elsewhere, in the Scottish Borders, some towns and villages are at risk of flash flooding after periods of heavy rain, surrounded as they are by high ground. Selkirk, for example, has suffered regularly, with a road bridge being swept away in 1977, and a major flood defence scheme has been initiated there.

Radon

Most people receive almost 90 per cent of their average annual radiation dose from natural sources. Of those sources linked to geology, the most important are terrestrial gamma radiation from the ground and buildings, and the short-lived decay products of radon gas released from rock and soil.

About 15 per cent of the average annual radiation dose is due to terrestrial gamma-rays which originate chiefly from the radioactive decay of natural potassium, uranium and thorium, all of which are widely distributed in rocks, soils and excavated building materials. Relatively high levels of background gamma radiation are associated with the granite outcrops in Galloway and the Cheviot Hills, and particularly with the uranium mineralisation associated with the Criffel pluton. However, even there, the overall hazard to human health from background gamma radiation is considered to be negligible in comparison to the risk that may be posed, in some circumstances, by radon gas.

Radon is a radioactive gas derived from the decay of the naturally occurring uranium that is found in small quantities in all rocks and soils. It may be responsible for as much as 50 per cent of the average annual radiation dose. The rate of release of radon is largely controlled by the uranium concentration in the source material and the type of minerals in which it resides. Once radon is released it is quickly diluted in the atmosphere and does not normally present a hazard. However, radon that enters poorly ventilated or enclosed spaces

Plate 67 *Flooding between Lochmaben and Lockerbie, 20 November 2009 (P711492). Under normal conditions the bridge in the foreground carries the A709 road across the River Annan. The islands are formed by drumlins that rise above the flood plain.*

can reach high concentrations. The health risk arises from decay of the gas to form solid radioactive particles that may be inhaled, in which case they may lodge in and irradiate the lung.

Regional variations in radon levels are related principally to geology. Across the outcrop of the Lower Palaeozoic turbidite successions the highest radon potential, though not at a level sufficient to cause concern, arises from the Ordovician, Leadhills Supergroup. Curiously, it appears to decrease progressively from north-east to south-west, perhaps reflecting some cryptic change in the bulk composition of the rocks. The Silurian Gala, Ettrick and Hawick groups display low radon potential; but the Riccarton Group outcrop shows higher values similar to those associated with the Ordovician turbidites. Some granites release radon and the Criffel and Cheviot plutons present a limited hazard. The highest levels would be expected inside granite-built properties sited on granite bedrock, such as might be encounterd in the Dalbeattie area of the Criffel pluton. However, of likely greater importance is the well-established link between limestone and high levels of radon emission, with the Carboniferous limestones presenting the greatest hazard in parts of southern Scotland. Here, a moderate risk has been identified in some parts of the Berwickshire Merse, the western Borders between Langholm and Annan, and the southern margin of the Thornhill outlier. Radon monitoring is only considered advisable in these limestone areas.

The nature of the superficial sediment cover on bedrock can influence radon potential. Apart from the obvious compositional influences, the permeability of the unconsolidated deposits is important. Impermeable till and clay will restrict radon flow, whereas permeable sand and gravel will facilitate its release. In general, a good proxy for radon potential is provided by stream sediment geochemistry (e.g. Figure 16). As might be expected, the strongest correlation is with uranium, followed by rubidium, potassium, yttrium, lanthanum and zirconium.

Mining legacy

In common with all coalfield areas with a long history of underground working, parts of the Sanquhar and Canonbie coalfields are at risk of surface subsidence and instability.

Subsidence due to modern mining is usually predictable and manageable but may become less so if the modern, deep workings underlie older, shallow ones. Whilst reliable plans exist for most modern workings, the extent and detail of the earliest mining activity, which was generally by the pillar and stall method, is commonly known only in general terms, and may be completely unrecorded. Voids created during the early work may remain open long after abandonment, with collapse taking place unpredictably over many years. Should such old workings be present above modern longwall workings, the subsidence potential is compounded.

Groundwater levels typically rise following the reduction or cessation of pumping after mine abandonment. In addition to the risk of surface discharge of contaminated water, such groundwater rebound may affect ground stability, causing renewed subsidence in long-abandoned workings. In north-east England, groundwater rebound has been linked to fault reactivation with associated surface instability, and this may prove to be a wider problem. Any mine waters discharged are likely to be acidic and enriched in iron and manganese, plus a range of potentially toxic chemical elements such as lead, arsenic and cadmium. Apart from surface discharge, there is also the possibility that contaminated mine water could access permeable rock formations and so degrade the quality of aquifers.

Most coals and coal-bearing successions generate methane, the cause of many catastrophic underground explosions. Methane continues to be released into old mine workings whence it may reach the surface via abandoned shafts and adits, or via faults or permeable rock formations, driven in part by rising groundwater. Low barometric pressure may also lead to a short-term increase in methane release.

Bings of colliery waste have in the past presented problems of stability and spontaneous combustion. Weathering of colliery waste may also produce high sulphate and acid groundwater. However, as this material is progressively removed, and the areas it once occupied are landscaped, these problems are much reduced.

Some of the subsidence and mine water hazards described above from the coalfields also have limited application in other areas with a history of metal mining. Those mining sites, mostly in the Leadhills–Wanlockhead area and in parts of Galloway, are much more restricted in extent than the coal workings, tend to be relatively shallow, and are in parts of the region where there is only a sparse local population. Apart from the obvious danger of open shafts and accessible adits, the principal hazard probably results from the waste tips and smelting facilities with their elevated levels of toxic base metals. The paucity of vegetation at these sites is testament to the contamination (Plate 68). The underground limestone workings around Harelawhill and Closeburn may also be a potential cause of localised subsidence.

HYDROGEOLOGY AND WATER SUPPLY

High rainfall, leading to an abundant supply of surface water, coupled with a relatively sparse population over much of the region, allows most of the public water supply to be drawn from reservoirs, springs and shallow wells or boreholes. Deeper aquifers are only exploited on a substantial scale in the Dumfries area, which is one of the few areas where high productivity aquifers are found (see below). Elsewhere, the deep valleys and low population density have enabled the construction of several very large, dammed water supply reservoirs, mainly to supply towns and cities in adjacent areas of the Midland Valley. The most extensive of these is the combined Talla-Fruid and Megget schemes, some 20 km north-east of Moffat, which are the principal contributors to Edinburgh's water supply. The Loch Bradan scheme, in the western part of the region, supplies water to domestic and industrial consumers in central Ayrshire, while similar schemes at Daer and Camps, near Crawford, send water to the large towns of South Lanarkshire.

The chemistry of the surface waters is closely related to the regional geology as exemplified by the variation in pH (Figure 67) of stream water, which has a negative correlation with the available Ca, Mg and total carbonate content of bedrock. However, the very low pH (< 6) over the high ground in the west of the Southern Uplands has also been influenced by other factors such as impeded surface drainage and the presence of a thick peat cover. Afforestation has been extensive since the 1960s and has also contributed to surface water

Plate 68 *Remains of smelter flues at Wanlockhead (P266683).*

acidity. Moreover, the pH of surface waters in this part of the region has been significantly modified by 'acid rain', the deposition of atmospheric acids derived from the combustion of fossil fuels and introduced by the prevailing westerly winds. Away from these influences, there is acidic drainage (pH ~ 6) over the Criffel and Cheviot granite outcrops, broadly neutral conditions (pH ~7) over that part of the Southern Uplands underlain by Ordovician wacke-type sandstones or by those of the Silurian Gala Group, and mildly alkaline surface waters (pH ~8) over the carbonate-rich, Silurian sandstones of the Hawick and Riccarton groups. Similarly high pH levels prevail across the Carboniferous, limestone-bearing succession of the Berwickshire Merse, whilst the low pH in the south and across the border into the Kielder Forest area of northern England can be ascribed to the broadly acidic lithofacies within the Border Group, enhanced by the effects of afforestation.

Lower Palaeozoic rocks

Ordovician and Silurian rocks, comprising mainly wacke-type sandstone and mudstone, have by far the largest outcrop area across the south of Scotland. Together with the granitic plutons intruded into them, they form low productivity aquifers with very low intergranular permeability, though that may be enhanced in any well-developed weathered zone present at rockhead. Groundwater storage and flow is almost entirely via fractures, and groundwater flow paths are likely to be relatively shallow, short and localised. Although individual borehole yields are low, there are at least 900 abstractions from the Lower Palaeozoic aquifer across the region, primarily for domestic and farm use. The quality of the water is usually good and potable, although the chemistry can be variable with, for example, higher calcium and bicarbonate in areas where the sandstones have a calcite cement. In a few places, highly mineralised groundwater can be related to mudstone beds within the Lower Palaeozoic succession.

During the 19th century, groundwater from the Lower Palaeozoic aquifer, including some mineralised groundwaters, provided the basis for thriving spa industries at Innerleithen and Moffat. St Ronan's Well at Innerleithen reputedly provided sulphurous, saline and freshwater springs, though the saline springs cannot be located today. Near Moffat, Hartfell Spa is a chalybeate spring whilst the Moffat Well is sulphurous.

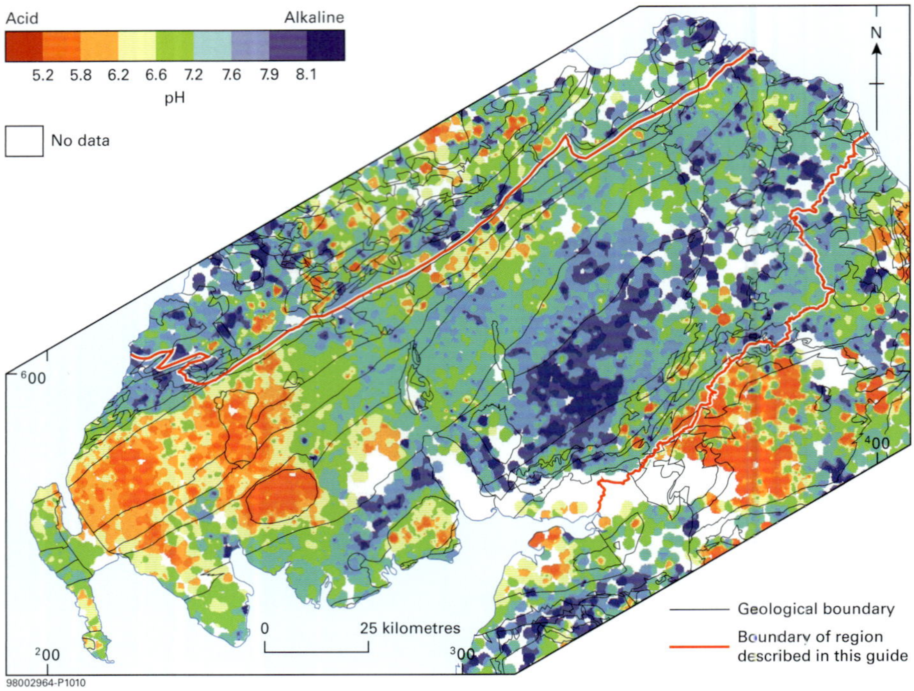

Figure 67 *The pH of surface stream waters across southern Scotland and north-west England.*

Devonian and Carboniferous rocks

The Devonian sandstones are well cemented and have generally low intergranular permeability, but relatively high productivity may be provided locally by fracture flow. Groundwater flow paths are likely to be longer and deeper than those found in the Lower Palaeozoic aquifer. There are approximately 100 small abstractions from the Devonian aquifer in southern Scotland, mostly from boreholes for domestic and farm use in Berwickshire and the southern Borders; there are likely to be many more unrecorded shallow wells and springs.

The lithologically variably Carboniferous successions may form moderately productive aquifers but with a relatively complex, multilayered hydrogeology that is at present only poorly understood; water quality and physical aquifer properties are likely to be highly variable. In general, the sandstone units are likely to act as effectively discrete aquifers, in which fracture flow dominates, separated by finer-grained units of lower permeability. There are at least 50 abstractions from the Carboniferous aquifer in southern Scotland, mostly farm boreholes in the Tweed Basin.

Permian and Triassic rocks

Within the Permian to Triassic basins of Annandale, Lochmaben, Dumfries, Stranraer and Carlisle–Solway, the red sandstones are a well-known aquifer and have been used extensively in the past for industrial enterprises with large water demand such as creameries. These rocks have good potential for public water supply, especially in isolated communities where demand is relatively low and the cost of reservoirs high. At the beginning of the 21st century, the Dumfries Basin, Thornhill and Annandale Permo-Triassic aquifers all support groundwater abstraction for public supply, agriculture and, in some cases, industrial use. Only one borehole is used for public supply in the Thornhill Basin, and three boreholes are currently used for public supply on a standby basis in the Annandale Basin. The Dumfries Basin is the most heavily exploited for public water supply and industrial use, with groundwater abstraction concentrated in the western part of the basin where aquifer productivity is highest (see below). There is some evidence that locally falling groundwater levels here reflect the effects of intense pumping. Also of concern is the rise in nitrate concentration caused by leaching of agricultural fertilisers.

Two main aquifers can be identified in the Dumfries Basin. In the west of the basin lies the breccio-conglomerate and interbedded sandstone of the Doweel Breccia Formation, which interfingers eastwards with the finer-grained Locharbriggs Sandstone Formation. The Doweel Formation breccias have a matrix of lithified clay minerals and have very low intergranular permeability. However, the formation is cut by subhorizontal fissures, commonly along the breaks between sandstone and breccia beds, which significantly increase the rock permeability. Consequently, the Doweel Breccia Formation is one of the most highly transmissive bedrock aquifers in Scotland. Fractures dominate groundwater flow, providing more than 90 per cent of the influx into boreholes. The fracture network also acts as a rapid conduit for groundwater transport towards the River Nith. As a result, the Doweel Breccia Formation aquifer has generally younger water, more active recharge, and higher nitrate concentrations than the Locharbriggs Sandstone Formation aquifer in the eastern sector of the Dumfries Basin. This sandstone has significant intergranular porosity and permeability but few fractures or joints of any hydraulic significance. Consequently, borehole yields are generally lower here than in the west of the basin; storage capacity is higher and groundwater residence times are longer.

The Dumfries Basin is bounded by weakly permeable Silurian sandstones, and is largely concealed by variable superficial deposits. Recharge is either by inflow of surface water from the surrounding catchment via the River Nith and the Lochar Water, or directly from rainfall percolating through overlying superficial sand and gravel. Discharge is predominantly to the rivers in the central part of the basin, rather than directly southward to the sea. Across the basin as a whole, there is a seasonal variation in groundwater levels of about 2 m, with the highest levels in late winter followed by a steady natural decline through spring. The lowest lying parts of the basin, south of Dumfries, are less than 10 m above sea level, and boreholes located within these low-lying areas can be artesian where silty clay deposits overlie the bedrock aquifer. Elsewhere, individual fractures and sandstone horizons may have different groundwater levels.

The natural groundwater chemistry in the Dumfries Basin is a product of maritime rainfall modified by the dissolution of carbonate material in the breccia, sandstone, and superficial deposits. The groundwater is typically hard, moderately mineralised and of calcium-magnesium-bicarbonate type, with bicarbonate concentrations greater than 100 mg/l. Distributions of major ions throughout the basin are related to rock and superficial deposit geochemistry, such as the lower concentrations of calcium, magnesium and bicarbonate in groundwater within the Locharbriggs Sandstone Formation, reflecting the aeolian origin of these sediments and the presence of silica cement. Concentrations of sulphate are highest in the south of the basin beneath raised beach and marine superficial deposits, but may also be related to the presence of thin gypsum lenses or sulphide oxidation in sporadic mudstone horizons.

Superficial deposits

Across the south of Scotland, the hydrogeological properties of superficial deposits are highly variable. Some of the more extensive spreads of glaciofluvial and alluvial deposits form highly productive, local aquifers which are exploited by public supply boreholes in a number of places in the Borders, including Innerleithen and Ettrickbridge. There has also been some commercial exploitation of superficial aquifers as a source of bottled mineral water.

Aquifers within glaciofluvial sand and gravel units may be confined by impermeable, interbedded silts, clays or glacial till, and supplies are likely to fluctuate rapidly in response to precipitation. The water is generally relatively hard compared to surface waters, due to bicarbonate or sulphate concentration, and may contain significant concentrations of iron and manganese locally. Some of the alluvial gravels along the major valleys have also proved highly productive, with high flows of water from boreholes, though for these aquifers it is more difficult to predict the quality of the water and the quantities available. Aquifers within the permeable superficial deposits are particularly vulnerable to pollution, for example from agricultural and industrial discharges. As well as forming productive local aquifers in their own right, these deposits also play a significant role regionally in controlling the volume, timing and chemistry of recharge to underlying bedrock aquifers.

Selected bibliography

Chapter 1 Introduction

Armstrong, H A, and Owen, A W. 2001. Terrane evolution of the paratectonic Caledonides of northern Britain. *Journal of the Geological Society of London*, Vol. 158, 475–486.

Clarkson, E, and Upton, B. 2009. *Death of an Ocean — a Geological Borders Ballad.* (Edinburgh: Dunedin Academic Press.)

Colman-Sadd, S P, Stone, P, Swinden, H S, and Barnes, R P. 1992. Parallel geological development in the Dunnage Zone of Newfoundland and the Lower Palaeozoic terranes of southern Scotland: an assessment. *Transactions of the Royal Society of Edinburgh: Earth Sciences*, Vol. 83, 571–594.

Cope, J C W, Ingham, J K, and Rawson, P F (editors). 1992. Atlas of palaeogeography and lithofacies. *Geological Society of London Memoir*, No. 13.

Kelling, G. 2001. Southern Uplands geology: an historical perspective. *Transactions of the Royal Society of Edinburgh: Earth Sciences*, Vol. 91, 323–339.

Leggett, J K, McKerrow, W S, and Eales, M H. 1979. The Southern Uplands of Scotland: a Lower Palaeozoic accretionary prism. *Journal of the Geological Society of London*, Vol. 136, 755–770.

Stone, P, and Merriman, R J. 2004. Basin thermal history favours an accretionary origin for the Southern Uplands terrane, Scottish Caledonides. *Journal of the Geological Society of London*, Vol. 161, 829–836.

Stone, P, Floyd, J D, Barnes, R P, and Lintern, B C. 1987. A sequential backarc and foreland basin thrust duplex model for the Southern Uplands of Scotland. *Journal of the Geological Society of London*, Vol. 144, 753–764.

Stone, P, Plant, J A, Mendum, J R, and Green, P. 1999. A regional geochemical assessment of some terrane relationships in the British Caledonides. *Scottish Journal of Geology*, Vol. 35, 145–156.

Trewin, N H (editor). 2002. *The Geology of Scotland.* (London: The Geological Society.)

Woodcock, N H, and Strachan, R A. 2000. *Geological history of Britain and Ireland.* (Oxford and Edinburgh: Blackwell Science Publishing.)

Zalasiewicz, J A, Taylor, L, Rushton, A W A, Loydell, D K, Rickards, R B, and Williams, M. 2009. Graptolites in British stratigraphy. *Geological Magazine*, Vol. 146, 785–850.

Chapter 2 Ordovician and Silurian of the Girvan–Ballantrae district

Bluck, B J. 1978. Geology of a continental margin 1: the Ballantrae Complex. 151–162 *in* Crustal evolution in north-western Britain and adjacent regions. Bowes, D R, and Leake, B E (editors). *Geological Journal Special Issue*, No. 10.

BLUCK, B J, and INGHAM, J K. 1992. The Girvan–Ballantrae District. 301–439 *in Geological excursions around Glasgow and Girvan.* LAWSON, J D, and WEEDON, D S (editors). (Glasgow: Geological Society of Glasgow.)

BLUCK, B J, HALLIDAY, A N, AFTALION, M, and MACINTYRE, R M. 1980. Age and origin of Ballantrae ophiolite and its significance to the Caledonian Orogeny and Ordovician timescale. *Geology*, Vol. 8, 492–495.

COCKS, L R M, and TOGHILL, P. 1973. The biostratigraphy of the Silurian rocks of the Girvan district, Scotland. *Journal of the Geological Society of London*, Vol. 129, 209–243.

FLOYD, J D, and WILLIAMS, M. 2003. A revised correlation of Silurian rocks in the Girvan district, SW Scotland. *Transactions of the Royal Society of Edinburgh: Earth Sciences*, Vol. 93, 383–392.

HAMILTON, P J, BLUCK, B J, and HALLIDAY, A N. 1984. Sm-Nd ages from the Ballantrae complex, SW Scotland. *Transactions of the Royal Society of Edinburgh: Earth Sciences*, Vol. 75, 183–187.

INCE, D. 1984. Sedimentation and tectonism in the Middle Ordovician of the Girvan district, SW Scotland. *Transactions of the Royal Society of Edinburgh: Earth Sciences*, Vol. 75, 225–237.

INGHAM, J K. 2000. Scotland: the Midland Valley terrane — Girvan. 41–47 *in* A revised correlation of Ordovician rocks in the British Isles. FORTEY, R A, HARPER, D A T, INGHAM, J K, OWEN, A W, PARKES, M A, RUSHTON, A W A, and WOODCOCK, N H (editors). *Geological Society of London Special Report*, No. 24.

PHILLIPS, E R, SMITH, R A, STONE, P, PASHLEY, V, and HORSTWOOD, M. 2009. Zircon age constraints on the provenance of Llandovery to Wenlock sandstones from the Midland Valley terrane of the Scottish Caledonides. *Scottish Journal of Geology*, Vol. 45, 131–146.

SMELLIE, J L, and STONE, P. 2001. Geochemical characteristics and geotectonic setting of Early Ordovician basalt lavas in the Ballantrae Complex ophiolite, SW Scotland. *Transactions of the Royal Society of Edinburgh: Earth Sciences*, Vol. 91 (for 2000), 539–555.

SMELLIE, J L, STONE, P, and EVANS, J A. 1995. Petrogenesis of boninites in the Ordovician Ballantrae Complex ophiolite, south-west Scotland. *Journal of Volcanology and Geothermal Research*, Vol. 69, 323–342.

STONE, P. 1984. Constraints on genetic models for the Ballantrae Complex, SW Scotland. *Transactions of the Royal Society of Edinburgh: Earth Sciences*, Vol. 75, 189–191.

STONE P. 1999. The Ballantrae Complex. 69–100 *in* Caledonian Igneous Rocks of Great Britain. STEPHENSON, D, BEVINS, R E, MILLWARD, D, HIGHTON, A J, PARSONS, I, STONE, P, and WADSWORTH, W J. *Geological Conservation Review Series*, No. 17. (Peterborough: Joint Nature Conservation Committee).

STONE, P, and RUSHTON, A W A. 1983. Graptolite faunas from the Ballantrae ophiolite complex and their structural implications. *Scottish Journal of Geology*, Vol. 19, 297–310.

STONE, P, and RUSHTON, A W A. 2003. A late Arenig (early Yapeenian) graptolite fauna and the coeval tectonic development of the Ballantrae Complex, SW Scotland. *Scottish Journal of Geology*, Vol. 39, 29–40.

THIRLWALL, M F, and BLUCK, B J. 1984. Sr-Nd isotope and geological evidence that the Ballantrae 'ophiolite', SW Scotland, is polygenetic. 215–230 *in* Ophiolites and oceanic lithosphere. GASS, I G, LIPPARD, S J, and SHELTON, A W (editors). *Geological Society of London Special Publication*, No. 13.

WILLIAMS, A. 1962. The Barr and Lower Ardmillan Series (Caradoc) of the Girvan district of south-west Ayrshire, with descriptions of the brachiopoda. *Geological Society of London Memoir*, No. 3.

Chapter 3 Ordovician and Silurian of the Southern Uplands

CLARKSON, E N K, HARPER, D A T, OWEN, A W, and TAYLOR, C M. 1992. Ordovician faunas in mass-flow deposits, Southern Scotland. *Terra Nova*, Vol. 4, 245–253.

DULLER, P R, and FLOYD, J D. 1995. Turbidite geochemistry and provenance studies in the Southern Uplands of Scotland. *Geological Magazine*, Vol. 132, 557–569.

ELDERS, C F. 1987. The provenance of granite boulders in conglomerates of the Northern and Central Belts of the Southern Uplands of Scotland. *Journal of the Geological Society of London*, Vol. 144, 853–863.

FLOYD, J D. 2001. The Southern Uplands Terrane: a stratigraphical review. *Transactions of the Royal Society of Edinburgh: Earth Sciences*, Vol. 91, 349–362.

FLOYD, J D, and KIMBELL, G S. 1995. Magnetic and tectonostratigraphic correlation at a terrane boundary: the Tappins Group of the Southern Uplands. *Geological Magazine*, Vol. 132, 515–521.

FLOYD, J D, and RUSHTON, A W A. 1993. Ashgill greywackes in the Southern Uplands of Scotland: an extension of the Ordovician succession in the Northern Belt. *Transactions of the Royal Society of Edinburgh: Earth Sciences*, Vol. 84, 79–85.

FLOYD, J D, and TRENCH, A. 1989. Magnetic susceptibility contrasts in Ordovician greywackes of the Southern Uplands of Scotland. *Journal of the Geological Society of London*, Vol. 146, 77–83.

KEMP, A E S. 1986. Tectonostratigraphy of the Southern Belt of the Southern Uplands. *Scottish Journal of Geology*, Vol. 22, 241–256.

LEGGETT, J K. 1987. The Southern Uplands as an accretionary prism: the importance of analogues in reconstructing palaeogeography. *Journal of the Geological Society of London*, Vol. 144, 737–752.

MERRIMAN, R J, and ROBERTS, B. 1990. Metabentonites in the Moffat Shale Group, Southern Uplands of Scotland: Geochemical evidence of ensialic marginal basin volcanism. *Geological Magazine*, Vol. 127, 259–271.

MORRIS, J H. 1987. The Northern Belt of the Longford–Down Inlier, Ireland and Southern Uplands, Scotland: an Ordovician backarc basin. *Journal of the Geological Society of London*, Vol. 144, 773–786.

PHILLIPS, E R, SMITH, R A, and FLOYD, J D. 1999. The Bail Hill Volcanic Group: alkaline within-plate volcanism during Ordovician sedimentation in the Southern Uplands, Scotland. *Transactions of the Royal Society of Edinburgh: Earth Sciences*, Vol. 89, 233–247.

PHILLIPS, E R, EVANS, J A, STONE, P, HORSTWOOD, M S A, FLOYD, J D, SMITH, R A, AKHURST, M C, and BARRON, H F. 2003. Detrital Avalonian zircons in the Laurentian Southern Uplands terrane, Scotland. *Geology*, Vol. 31, 625–628.

STONE, P, and EVANS, J A. 2001. Silurian provenance variation in the Southern Uplands terrane, Scotland, assessed using neodymium isotopes and linked with regional tectonic development. *Transactions of the Royal Society of Edinburgh: Earth Sciences*, Vol. 91, 447–455.

STONE, P, BREWARD, N, MERRIMAN, R J, and PLANT, J A. 2004. Regional geochemistry of cryptic geology: variations in trace element distribution across the Southern Uplands terrane, Scotland. *Applied Earth Science* (Transactions of the Institution of Mining and Metallurgy B), Vol. 113, B43–B57.

STONE, P, BREWARD, N, MERRIMAN, R J, and BARNES, R P. 2006. The interpretation and application of regional geochemistry: lessons from the Paratectonic Caledonides. *Scottish Journal of Geology*, Vol. 42, 65–76.

TUCKER, R D, KROGH, T E, ROSS, R J, and WILLIAMS, S H. 1990. Time-scale calibration by high-precision U-Pb zircon dating of interstratified volcanic ashes in the Ordovician and Lower Silurian stratotypes of Britain. *Earth and Planetary Science Letters*, Vol. 100, 51–58.

WALDRON, J W F, FLOYD, J D, SIMONETTI, A, and HEAMAN, L M. 2008. Ancient Laurentian detrital zircon in the closing Iapetus Ocean, Southern Uplands Terrane, Scotland. *Geology*, Vol. 36, 527–530.

WARREN, P T. 1964. The stratigraphy and structure of the Silurian rocks south-east of Hawick, Roxburghshire. *Quarterly Journal of the Geological Society of London*, Vol. 120, 193–218.

Chapter 4 Caledonian structure and magmatism

AKHURST, M C, MCMILLAN, A A, KIMBELL, G S, STONE, P, and MERRIMAN, R J. 2001. Silurian subduction-related assembly of fault-defined tracts at the Laurieston Fault, Southern Uplands accretionary terrane, Scotland, UK. *Transactions of the Royal Society of Edinburgh: Earth Sciences*, Vol. 91, 435–446.

ANDERSON, T B. 2001. Structural interpretations of the Southern Uplands Terrane. *Transactions of the Royal Society of Edinburgh: Earth Sciences*, Vol. 91, 363–373.

ANDERSON, T B, and OLIVER, G J H. 1986. The Orlock Bridge Fault: a major late Caledonian sinistral fault in the Southern Uplands terrane, British Isles. *Transactions of the Royal Society of Edinburgh: Earth Sciences*, Vol. 77, 203–222.

ARMSTRONG, H A, OWEN, A W, SCRUTTON, C T, CLARKSON, E N K, and TAYLOR, C M. 1996. Evolution of the Northern Belt, Southern Uplands: implications for the Southern Uplands controversy. *Journal of the Geological Society of London*, Vol. 153, 197–205.

BARNES, R P, and STONE, P. 1999. Trans-Iapetus contrasts in the geological development of southern Scotland (Laurentia) and the Lakesman Terrane (Avalonia). 307–323 *in* In sight of the Suture: the Palaeozoic geology of the Isle of Man in its Iapetus Ocean context. WOODCOCK, N H, QUIRK, D G, FITCHES, W R, and BARNES, R P (editors). *Geological Society of London Special Publication*, No. 160.

BARNES, R P, LINTERN, B C, and STONE, P. 1989. Timing and regional implications of deformation in the Southern Uplands of Scotland. *Journal of the Geological Society of London*, Vol. 146, 905–908.

KIMBELL, G S, and STONE, P. 1995. Crustal magnetisation variations across the Iapetus Suture Zone. *Geological Magazine*, Vol. 132, 599–609.

LEAKE, R C, and COOPER, C. 1983. The Black Stockarton Moor subvolcanic complex, Galloway. *Journal of the Geological Society of London*, Vol. 140, 665–676.

MCCURRY, J A, and ANDERSON, T B. 1989. Landward vergence in the Lower Palaeozoic, Southern Uplands–Down–Longford terrane. *Geology*, Vol. 17, 630–633.

MERRIMAN, R J, and ROBERTS, B. 2001. Low-grade metamorphism in the Scottish Southern Uplands terrane: deciphering the patterns of accretionary burial, shearing and cryptic aureoles. *Transactions of the Royal Society of Edinburgh: Earth Sciences*, Vol. 91, 521–537.

NEEDHAM, D T. 2004. Deformation in Moffat Shale detachment zones in the western part of the Scottish Southern Uplands. *Geological Magazine*, Vol.141, 441–453.

NEEDHAM, D T, and KNIPE, R J. 1986. Accretion- and collision-related deformation in the Southern Uplands accretionary wedge. *Geology*, Vol. 14, 303–306.

PHILLIPS, E R, BARNES, R P, BOLAND, M P, FORTEY, N J, and MCMILLAN, A A. 1995. The Moniaive Shear Zone: a major zone of sinistral strike-slip deformation in the Southern Uplands of Scotland. *Scottish Journal of Geology*, Vol. 31, 139–149.

ROCK, N M S, GASKARTH, J W, and RUNDLE, C C. 1986. Late Caledonian dyke-swarms in southern Scotland: a regional zone of primitive K-rich lamprophyres and associated veins. *Journal of Geology*, Vol. 94, 505–522.

RUSHTON, A W A, STONE, P, and HUGHES, R A. 1996. Biostratigraphical control of thrust models for the Southern Uplands of Scotland. *Transactions of the Royal Society of Edinburgh: Earth Sciences*, Vol. 86, 137–152.

THIRLWALL, M F. 1988. Geochronology of late Caledonian magmatism in Northern Britain. *Journal of the Geological Society of London*, Vol. 145, 951–967.

Chapter 5 Devonian: the 'Old Red Sandstone' and associated volcanic rocks

BROWNE, M A E, and BARCLAY, W J. 2005. Siccar Point to Hawk's Heugh, Scottish Borders. 181–187 *in* The Old Red Sandstone of Great Britain. BARCLAY, W J, BROWNE, M A E,

McMillan, A A, Pickett, A E, Stone, P, and Wilby, P R. *Geological Conservation Review Series*, No. 31. (Peterborough: Joint Nature Conservation Committee.)

Browne, M A E , Smith, R A, and Aitken, A M. 2002. Stratigraphical framework for the Devonian (Old Red Sandstone) rocks of Scotland south of a line from Fort William to Aberdeen. *British Geological Survey Research Report*, RR/01/04.

Leeder, M G. 1973. Sedimentology and palaeogeography of the Upper Old Red Sandstone in the Scottish Border Basin. *Scottish Journal of Geology*, Vol. 9, 117–145.

Leeder, M G, and Bridges, P H. 1978. Upper Old Red Sandstone near Kirkbean, Dumfries and Galloway. *Scottish Journal of Geology*, Vol. 14, 267–272.

McMillan, A A. 2005. Southern Scotland and the Lake District. 197–208 *in* The Old Red Sandstone of Great Britain. Barclay, W J, Browne, M A E, McMillan, A A, Pickett, A E, Stone, P, and Wilby, P R. *Geological Conservation Review Series*, No. 31. (Peterborough: Joint Nature Conservation Committee.)

Rock, N M S, and Rundle, C C. 1986. Lower Devonian age for the 'Great (basal) Conglomerate', Scottish Borders. *Scottish Journal of Geology*, Vol. 22, 285–288.

Soper, N J, and Woodcock, N H. 2003. The lost Lower Old Red Sandstone of England and Wales: a record of post-Iapetan flexure or Early Devonian transtension? *Geological Magazine*, Vol. 140, 627–647.

Chapter 6 Carboniferous

Andrews, J E, and Nabi, I. 1994. Lithostratigraphy of the Dinantian Inverclyde and Strathclyde groups, Cockburnspath Outlier, East Lothian — North Berwickshire. *Scottish Journal of Geology*, Vol. 30, 115–119.

Barrett, P A. 1988. Early Carboniferous of the Solway Basin: a tectonostratigraphic model and its bearing on hydrocarbon potential. *Marine and Petroleum Geology*, Vol. 5, 271–281.

Chadwick, R A, Holliday, D W, Holloway, S, and Hulbert, A G. 1993. The evolution and hydrocarbon potential of the Northumberland/Solway Basin. 717–726 in *Petroleum Geology of North-west Europe: Proceedings of the 4th Conference*. Parker, J R (editor). (London: The Geological Society.)

Craig, G Y. 1956. The Lower Carboniferous Outlier of Kirkbean, Kirkcudbrightshire. *Transactions of the Geological Society of Glasgow*, Vol. 22, 113–132.

Craig, G Y, and Nairn, A E M. 1956. The Lower Carboniferous outliers of the Colvend and Rerrick shores, Kirkcudbrightshire. *Geological Magazine*, Vol. 93, 249–256.

Davies, A. 1970. Carboniferous rocks of the Sanquhar outlier. *Bulletin of the Geological Survey of Great Britain*, No. 31, 37–87.

Dean, M T, Browne, M A E, Waters, C N, and Powell, J H. 2011. A lithostratigraphical framework for the Carboniferous successions of northern Great Britain (Onshore). *British Geological Survey Research Report*, RR/10/07.

Deegan, C E. 1973. Tectonic control of sedimentation at the margin of a Carboniferous depositional basin in Kirkcudbrightshire. *Scottish Journal of Geology*, Vol. 9, 1–28.

Guion, P D, Fulton, I M, and Jones, N S. 1995. Sedimentary facies of the coal-bearing Westphalian A and B north of the Wales–Brabant High. 45–78 *in* European Coal Geology. Whateley, M K G, and Spears, D A (editors). *Geological Society of London Special Publication*, No. 82.

Jones, N S, Holliday, D W, and McKervey, J A. 2011. Warwickshire Group (Pennsylvanian) red-beds of the Canonbie Coalfield, England–Scotland border, and their regional palaeogeographical implications. *Geological Magazine*, Vol. 148, 50–77.

Leeder, M R. 1974. Origin of the Northumberland Basin. *Scottish Journal of Geology*, Vol. 10, 283–296.

Leeder, M R. 1975. Lower Border Group (Tournaisian) stromatolites from the Northumberland basin. *Scottish Journal of Geology*, Vol. 11, 207–226.

Leeder, M R. 1976. Palaeogeographical significance of pedogenic carbonates in the topmost Upper Old Red Sandstone of the Scottish Border Basin. *Geological Journal*, Vol. 11, 21–28.

Leeder, M R. 1982. Upper Palaeozoic basins of the British Isles: Caledonide inheritance versus Hercynian plate margin processes. *Journal of the Geological Society of London*, Vol. 139, 479–491.

Leeder, M R, and McMahon, A H. 1988. Upper Carboniferous (Silesian) basin subsidence in northern Britain. 43–52 *in* Sedimentation in a synorogenic basin complex; the Upper Carboniferous of North-west Europe. Besly, B M, and Kelling, G (editors). (London: Blackie.)

MacDonald, R. 1975. Petrochemistry of the early Carboniferous (Dinantian) Lavas of Scotland. *Scottish Journal of Geology*, Vol. 11, 269–314.

Maguire, K, Thompson, J, and Gowland, S. 1996. Dinantian depositional environments along the northern margin of the Solway Basin. 163–182 *in* Recent advances in Lower Carboniferous Geology. Strogen, P, Somerville, I D, and Jones, G L (editors). *Geological Society of London Special Publication*, No. 107.

McMillan, A A, and Brand, P J. 1995. Depositional setting of Permian and Upper Carboniferous strata of the Thornhill Basin, Dumfriesshire. *Scottish Journal of Geology*, Vol. 31, 43–52.

Morton, A, Fanning, M, and Jones, N S. 2010. Variscan sourcing of Westphalian (Pennsylvanian) sandstones in the Canonbie Coalfield, UK. *Geological Magazine*, Vol. 147, 718–727.

Ord, D M, Clemmey, H, and Leeder, M R. 1988. Interaction between faulting and sedimentation during Dinantian extension of the Solway Basin, SW Scotland. *Journal of the Geological Society of London*, Vol. 145, 249–259.

Picken, G S. 1988. The concealed coalfield at Canonbie: an interpretation based on boreholes and seismic surveys. *Scottish Journal of Geology*, Vol. 24, 67–71.

SCHRAM, F R. 1983. Lower Carboniferous biota of Glencartholm, Eskdale, Dumfriesshire. *Scottish Journal of Geology*, Vol. 19, 1–15.

TUCKER, M E, GALLAGHER, J, LEMON, K, and LENG, M. 2003. The Yoredale cycles of Northumbria: high-frequency clastic-carbonate sequences of the mid Carboniferous icehouse world. *Open University Geological Society Journal*, Vol. 24, 5–10.

Chapter 7 Permian and Triassic

ANDERSON, T B, PARNELL, J, and RUFFELL, A H. 1995. Influence of basement on the geometry of Permo-Triassic basins in the north-west British Isles. 103–122 *in* Permian and Triassic Rifting in North-west Europe. BOLDY, S A R (editor). *Geological Society of London Special Publication*, No. 91.

BROOKFIELD, M E. 1978. Revision of the stratigraphy of Permian and supposed Permian rocks of southern Scotland. *Geologischen Rundschau*, Vol. 67, 110–149.

BROOKFIELD, M E. 1979. Anatomy of a lower Permian aeolian sandstone complex, southern Scotland. *Scottish Journal of Geology*, Vol.15, 81–96.

BROOKFIELD, M E. 1980. Permian intermontane basin sedimentation in southern Scotland. *Sedimentary Geology*, Vol. 27, 167–194.

BROOKFIELD, M E. 2004. The enigma of fine-grained alluvial basin fills: the Permo-Triassic (Cumbrian Coastal and Sherwood Sandstone groups) of the Solway Basin, NW England and SW Scotland). *International Journal of Earth Science (Geologischen Rundschau)*, Vol. 93, 282–296.

HOLLIDAY, D W, WARRINGTON, G, BROOKFIELD, M E, MCMILLAN, A A, and HOLLOWAY, S. 2001. Permo-Triassic rocks in boreholes in the Annan–Canonbie area, Dumfries and Galloway, southern Scotland. *Scottish Journal of Geology*, Vol. 37, 97–113.

HOLLIDAY, D W, HOLLOWAY, S, MCMILLAN, A A, JONES, N S, WARRINGTON, G, and AKHURST, M C. 2004. The evolution of the Carlisle Basin, NW England and SW Scotland. *Proceedings of the Yorkshire Geological Society*, Vol. 55, 1–19.

HOLLIDAY, D W, JONES, N S, and MCMILLAN, A A. 2008 Lithostratigraphical subdivision of the Sherwood Sandstone Group (Triassic) of the north-eastern part of the Carlisle Basin, Cumbria, and adjacent parts of Dumfries & Galloway, UK. *Scottish Journal of Geology*, Vol. 44, 97–110.

HOLLOWAY, S. 1985. The Permian. 26–30 *in* Atlas of onshore sedimentary basins in England and Wales: post-Carboniferous tectonics and stratigraphy. WHITTAKER, A (editor). (Glasgow: Blackie and Son.)

MCKEEVER, P J A. 1994. A new fossil vertebrate trackway from the Permian of Dumfries and Galloway. *Scottish Journal of Geology*, Vol. 30, 11–14.

STONE, P. 1988. The Permian successions at Ballantrae and Loch Ryan, south-west Scotland. *Report of the British Geological Survey*, Vol. 9, No 2, 13–18.

Chapter 8 Jurassic to Palaeogene

CHADWICK R A, KIRBY, G A, and BAILY, H E. 1994. The post-Triassic structural evolution of north-west England and adjacent parts of the Irish Sea. *Proceedings of the Yorkshire Geological Society*, Vol. 50, 91–102.

DAGLEY, P, SKELHORN, R R, MUSSETT, A E, JAMES, S, and WALSH, J N. 2008. The Cleveland Dyke in southern Scotland. *Scottish Journal of Geology*, Vol. 44, 123–138.

GREEN, P F. 2002. Early Tertiary palaeothermal effects in northern England: reconciling results from apatite fission track analysis with geological evidence. *Tectonophysics*, Vol. 349, 131–144.

HOLLIDAY, D W. 1999. Palaeotemperatures, thermal modelling and depth of burial studies in northern and eastern England. *Proceedings of the Yorkshire Geological Society*, Vol. 52, 337–352.

JOLLY, R J H, and SANDERSON, D J. 1995. Variations in the form and distribution of dykes of the Mull swarm, Scotland. *Journal of Structural Geology*, Vol. 17, 1543–1557.

MACDONALD, R, WILSON, L, THORPE, R S, and MARTIN, A. 1988. Emplacement of the Cleveland Dyke: evidence from geochemistry, mineralogy and physical modelling. *Journal of Petrology*, Vol. 29, 559–583.

Chapter 9 Quaternary

BOWEN, D Q (editor). 1999. A revised correlation of the Quaternary deposits in the British Isles. *Geological Society of London Special Report*, No. 23.

BROOKS, S J, and BIRKS, H J B. 2000. Chironomid-inferred late glacial air temperatures at Whitrig Bog, south-east Scotland. *Journal of Quaternary Science*, Vol. 15, 759–764.

CHIVERRELL, R C, HARVEY, A M, and FOSTER, G C. 2006. Hillslope gullying in the Solway Firth – Morecambe Bay region, Great Britain: responses to human impact and/or climatic deterioration? *Geomorphology*, Vol. 84, 317–343.

EHLERS, J, GIBBARD, P L, and ROSE, J (editors). 1991. Glacial deposits in Great Britain and Ireland. (Rotterdam: Balkema.)

EVEREST, J, BRADWELL, T, and GOLLEDGE, N. 2005. Subglacial Landforms of the Tweed Palaeo-Ice Stream. Scottish Landform Example, No. 35. *Scottish Geographical Magazine*, Vol. 121, 163–173.

GORDON, J E, and SUTHERLAND, D G (editors). 1993. The Quaternary of Scotland. *Geological Conservation Review Series*, No. 6. (London: Chapman and Hall).

LAMBECK, K, and PURCELL, A P. 2001. Sea-level change in the Irish Sea since the Last Glacial Maximum: constraints from isostatic modelling. *Journal of Quaternary Science*, Vol. 16, 497–506.

LIVINGSTONE, S J, EVANS, D J A, and Ó COFAIGH, C (editors). 2010. The Quaternary of the Solway Lowlands and Pennine escarpment. (London: Quaternary Research Association.)

MCMILLAN, A A, HAMBLIN, R J O, and MERRITT, J W. 2011. A lithostratigraphical framework for onshore Quaternary and Neogene (Tertiary) superficial deposits of Great Britain and the Isle of Man. *British Geological Survey Research Report*, RR/10/03.

MCMILLAN, A A, MERRITT, J W, AUTON, C A, and GOLLEDGE, N R. 2011. The Quaternary geology of the Solway area. *British Geological Survey Research Report*, RR/11/04.

SMITH, D E, CULLINGFORD, R A, HAGGART, B A, TIPPING, R, WELLS, J M, MIGHALL, T M, and DAWSON, S. 2003. Holocene relative sea-level changes in the lower Nith valley and estuary. *Scottish Journal of Geology*, Vol. 39, 97–120.

TIPPING, R. 1999. The Quaternary of Dumfries and Galloway: Field Guide. (London: Quaternary Research Association.)

ZONG, Y, and TOOLEY, M J. 1996. Holocene sea-level changes and crustal movements in Morecambe Bay, north-west England. *Journal of Quaternary Science*, Vol. 11, 43–58.

Chapter 10 Geology and man

GILLANDERS, R J. 1981. The Leadhills–Wanlockhead mining district, Scotland. *The Mineralogical Record*, Vol. 12, 235–250.

HYSLOP, E, MCMILLAN, A, and MAXWELL, I. 2006. *Stone in Scotland*. Earth Science Series: (Paris: UNESCO, IAEG, Queen's Printer for Scotland and British Geological Survey.)

LEAKE, R C, CHAPMAN, R J, BLAND, D J, STONE, P, CAMERON, D G, and STYLES, M T. 1998. The origin of alluvial gold in the Leadhills area of Scotland: evidence from interpretation of internal chemical characteristics. *Journal of Geochemical Exploration*, Vol. 63, 7–36.

MACDONALD, A M, Ó DOCHARTAIGH, B É, KINNIBURGH, D G, and DARLING, W G. 2008. Baseline Scotland: groundwater chemistry of southern Scotland. *British Geological Survey Open Report*, OR/08/62.

MUSSON, R M W. 2004. Early seismicity of the Scottish Borders Region. *Annals of Geophysics*, Vol. 47, 1827–1847.

MUSSON, R M W. 2007. British Earthquakes. *Proceedings of the Geologists' Association*, Vol. 118, 305–337.

MUSSON, R M W, and HENNI, P H O. 2002. The felt effects of the Carlisle earthquake of 26 December 1979. *Scottish Journal of Geology*, Vol. 38, 113–125.

ROBBINS, N S, and BUCKLEY, D K. 1988. Characteristics of the Permian and Triassic aquifers of south-west Scotland. *Quarterly Journal of Engineering Geology*, Vol. 21, 329–335.

SCHEIB, C, APPLETON, J D, MILES, J C H, GREEN, B M R, BARLOW, T S, and JONES, D G. 2009. Geological control on radon potential in Scotland. *Scottish Journal of Geology*, Vol. 45, 147–160.

STONE, P, COOK, J M, MCDERMOTT, C, ROBINSON, J J, and SIMPSON, P R. 1995. Lithostratigraphical and structural controls on distribution of As and Au in south-west Southern Uplands, Scotland. *Transactions of the Institution of Mining and Metallurgy, Section B*, Vol. 106, 79–84.

STONE, P, BREWARD, N, and MERRIMAN, R J. 2003. Mineralogical controls on metal distribution in stream sediment derived from the Caledonides of the Scottish Southern Uplands and English Lake District. *Mineralogical Magazine*, Vol. 67, 325–338.

TEMPLE, A K. 1956. The Leadhills–Wanlockhead lead and zinc deposits. *Transactions of the Royal Society of Edinburgh: Earth Sciences*, Vol. 63, 85–113.

British Geological Survey publications for the South of Scotland

A geological classic and still of great scientific importance

PEACH, B N, and HORNE, J. 1899. The Silurian Rocks of Britain. Vol 1. Scotland. *Memoirs of the Geological Survey of the United Kingdom*. HMSO.

BGS 1:625 000 scale map series

STONE, P. 2008. *Bedrock Geology UK North. An explanation of the bedrock geology map of Scotland, northern England, Isle of Man and Northern Ireland — 1:625 000 fifth edition.* (Keyworth, Nottingham: British Geological Survey).

BGS 1:50 000 scale map series: Memoirs (M) and Sheet Explanations (SE) for the South of Scotland

Sheet No.	Date	Name and authors
1 and 3	1995	Rhins of Galloway (M) Stone, P
2, 4W and 4E	2008	Whithorn, Kirkcowan and Wigtown (M) Barnes, R P
5W, 5E and 6W	2000	Kirkcudbright–Dalbeattie (M) Lintern, B C, and Floyd, J D
7	1986	Girvan (SE) Cameron, I B, Stone, P, and Smellie, J L
8W and 8E	1999	Carrick–Loch Doon (M) Floyd, J D
9W and 9E	2002	New Galloway and Thornhill (M) McMillan, A A
11	1967	Langholm (M) Lumsden, G I, Tulloch, W, Howells, M F, and Davies, A
15W	1999	New Cumnock (SE) Smith, R A
15E	2003	Leadhills (SE) Floyd J D
33W	1985	Haddington (M) McAdam, A D, and Tulloch, W
33E	1986	Dunbar (M) Davies, A, McAdam, A D, and Cameron, I B
34	1988	Eyemouth (M) Greig, D C

BGS economic memoirs

SIMPSON, J B, and RICHEY, J E. 1936. The geology of the Sanquhar Coalfield and adjacent basin of Thornhill. *Memoirs of the Geological Survey, Scotland.* (Edinburgh: HMSO.)

WILSON, G V. 1921. The lead, zinc, copper and nickel ores of Scotland. Memoirs of the Geological Survey, Scotland. *Special Reports on the Mineral Resources of Great Britain.* Vol. 17. (Edinburgh: HMSO.)

BGS subsurface memoirs

CHADWICK, R A, HOLLIDAY, D W, HOLLOWAY, S, and HULBERT, A G. 1995. Structure and evolution of the Northumberland–Solway Basin and adjacent areas. *Subsurface Memoir of the British Geological Survey.*

BGS geochemical atlas

BRITISH GEOLOGICAL SURVEY. 1993. *Regional geochemistry of southern Scotland and part of northern England.* (Keyworth, Nottingham: British Geological Survey).

BGS British Regional Geology Series

CAMERON, I B, and STEPHENSON, D. 1985. *British Regional Geology: Midland Valley of Scotland* (Third edition). (London: HMSO for British Geological Survey).

MITCHELL, W I (editor). 2004. *The Geology of Northern Ireland.* (Belfast: Geological Survey of Northern Ireland).

STONE, P and 6 others. 2010. *British Regional Geology: Northern England* (Fifth edition). (Keyworth, Nottingham: British Geological Survey).

BGS Classical Areas of British Geology Series

STONE, P, and SMELLIE, J L. 1988. *Classical areas of British geology: The Ballantrae area.* (London: HMSO for British Geological Survey).

WEBB, B C, RUSHTON, A W A, and WHITE, D E. 1993. *Classical areas of British geology: Moffatdale and the Upper Ettrick valley.* (London: HMSO for British Geological Survey).

Field excursion guide

(published with the support of the Edinburgh Geological Society).

STONE, P (editor). 1996. *Geology in south-west Scotland: an excursion guide.* (Keyworth, Nottingham: British Geological Survey).

Scottish landscape and geology guides

(published by Scottish Natural Heritage in association with BGS).

MCADAM, A D, and STONE P. 1997. *East Lothian and the Borders — a landscape fashioned by geology.* Scottish Natural Heritage.

MCMILLAN, A A, and STONE, P. 2008. *South-west Scotland — a landscape fashioned by geology.* Scottish Natural Heritage.

Interactive CD of gravity and aeromagnetic data

BRITISH GEOLOGICAL SURVEY. 2006. Regional Geophysics of Southern Scotland and Northern England. (Keyworth, Nottingham: British Geological Survey).

BGS offshore geology reports

CHADWICK, R A, and 16 others. 2001. Geology of the Isle of Man and its offshore area. *British Geological Survey Research Report*, RR/01/06.

GATLIFF, R W, and 11 others. 1994. United Kingdom offshore regional report: the Geology of the Central North Sea. (London: HMSO for the BGS.)

JACKSON, D I, JACKSON, A A, EVANS, D, WINGFIELD, R T R, BARNES, R P, and ARTHUR, M J. 1995. *United Kingdom offshore regional report: the Geology of the Irish Sea*. (London: HMSO for the British Geological Survey.)

Earthquake records

MUSSON, R M W. 1994. A catalogue of British earthquakes. *British Geological Survey Technical Report*, WL/94/04.

Index

Figures are indicated in **bold**, tables and plates in *italic* font

British Geological Survey
NATURAL ENVIRONMENT RESEARCH COUNCIL

Principal Offices

Environmental Science Centre,
Keyworth, Nottingham NG12 5GG
☎ 0115 936 3100

Murchison House,
West Mains Road,
Edinburgh EH9 3LA
☎ 0131 667 1000

Maclean Building,
Crowmarsh Gifford, Wallingford
Oxfordshire OX10 8BB
☎ 01491 838800

BGS London, Natural History Museum,
Cromwell Road, London SW7 5BD
☎ 020 7589 4090

Geological Survey of Northern Ireland,
Colby House, Stranmillis Court,
Belfast BT9 5BF
☎ 028 9038 8462

Columbus House, Greenmeadow Springs,
Tongwynlais, Cardiff CF15 7NE
☎ 029 2052 1962

Parent Body

Natural Environment Research Council,
Polaris House, North Star Avenue,
Swindon SN2 1EU
☎ 01793 411500

Website www.bgs.ac.uk

FSC
www.fsc.org
MIX
Paper from responsible sources
FSC® C023561